DETECTING ECOLOGICAL IMPACTS

DETECTING ECOLOGICAL IMPACTS

Concepts and Applications in Coastal Habitats

Edited by

Russell J. Schmitt
Department of Biological Sciences
University of California
Santa Barbara, California

Craig W. Osenberg
Department of Zoology
University of Florida
Gainesville, Florida

Academic Press
San Diego   New York   Boston
London   Sydney   Tokyo   Toronto

Copyright © 1996 by ACADEMIC PRESS, INC.

Academic Press, Inc.
A Division of Harcourt Brace & Company
525 B Street, Suite 1900, San Diego, California 92101-4495

United Kingdom Edition published by
Academic Press Limited
24-28 Oval Road, London NW1 7DX

Library of Congress Cataloging-in-Publication Data

Detecting ecological impacts: concepts and applications in coastal
  habitats/edited by Russell J. Schmitt, Craig W. Osenberg.
        p.   cm.
  Includes bibliographical references and indexes.
  ISBN 0-12-627255-7 (alk. paper)
  1. Ecological risk assessment.   2. Environmental monitoring.
  3. Coastal ecology.   I. Schmitt, Russell J. II. Osenberg, Craig W.
  QH541.15.R57048   1995
  574.5'2638--dc20   95-18309
                                                        CIP

PRINTED IN THE UNITED STATES OF AMERICA
95   96   97   98   99   00   BC   9   8   7   6   5   4   3   2   1

Contents

SECTION I
AN INTRODUCTION TO
ECOLOGICAL IMPACT ASSESSMENT
Principles and Goals

SECTION II

IMPROVING FIELD ASSESSMENTS OF LOCAL IMPACTS

Before–After–Control–Impact Designs

———

SECTION III
EXTENSION OF LOCAL IMPACTS
TO LARGER SCALE CONSEQUENCES

Contributors[1]

Numbers in parentheses indicate the pages on which the authors' contributions begin.

Khalil E. Abu-Saba (83)
Department of Chemistry
and Biochemistry
University of California, Santa Cruz
Santa Cruz, California 95064

Richard F. Ambrose (345)
Environmental Science and
Engineering Program
School of Public Health
University of California, Los Angeles
Los Angeles, California 90024

James R. Bence (133)
Department of Fisheries and Wildlife
Michigan State University
East Lansing, Michigan 48824

Kerry P. Black (199)
National Institute of Water
and Atmospheric Research, and
Earth Sciences Department
University of Waikato
Hamilton
New Zealand

Robert S. Carney (295)
Coastal Ecology Institute
Louisiana State University
Baton Rouge, Louisiana 70803

Jean Chesson (281)
Resource Assessment Commission
Locked Bag No.1
Queen Victoria Terrace
Canberra, Australian Capital
Territory 2600
Australia

William J. Douros (281)
Energy Division
Department of Planning
and Development
County of Santa Barbara
Santa Barbara, California 93101

A. Russell Flegal (83)
Institute of Marine Sciences
University of California, Santa Cruz
Santa Cruz, California 95064

Charles A. Gray (235)
Fisheries Research Institute
P.O. Box 21
Cronulla, New South Wales 2230
Australia

Judi E. Hewitt (49)
National Institute of Water and
Atmospheric Research
P.O. Box 11–115
Hamilton
New Zealand

Sally J. Holbrook (83)
Department of Biological Sciences
University of California
Santa Barbara, California 93106

Geoffrey P. Jones (29)
Department of Marine Biology
James Cook University
Townsville, Queensland 4811
Australia

Ursula L. Kaly (29)
Department of Marine Biology
James Cook University
Townsville, Queensland 4811
Australia

Michael J. Keough (199)
Department of Zoology
University of Melbourne
Parkville, Victoria 3052
Australia

Michael J. Kingsford (235)
School of Biological Sciences
University of Sydney
Sydney, New South Wales 2006
Australia

Charles Lester (329)
Department of Political Science
University of Colorado
106 Ketchum Hall
Box 333
Boulder, Colorado 80309

Bruce D. Mapstone (67)
CRC: Reef Research Centre
James Cook University
of North Queensland
Townsville, Queensland 4811
Australia

William W. Murdoch (257)
Department of Biological Sciences
University of California, Santa Barbara
Santa Barbara, California 93106

Roger M. Nisbet (257)
Department of Biological Sciences
University of California, Santa Barbara
Santa Barbara, California 93106

Craig W. Osenberg (3, 83, 281, 345)
Department of Zoology
University of Florida
Gainesville, Florida 32611

Frederick M. Piltz (317)
Environmental Studies Program
Pacific OCS Region
Minerals Management Service
770 Paseo Camarillo
Camarillo, California 93010

Richard D. Pridmore (49)
National Institute of Water and
Atmospheric Research
P.O. Box 11–115
Hamilton
New Zealand

Peter T. Raimondi (179)
Coastal Research Center
Marine Science Institute
University of California, Santa Barbara
Santa Barbara, California 93106

Daniel C. Reed (179)
Coastal Research Center
Marine Science Institute
University of California, Santa Barbara
Santa Barbara, California 93106

Russell J. Schmitt (3, 83, 281, 345)
Department of Biological Sciences, and
Coastal Research Center
Marine Science Institute
University of California, Santa Barbara
Santa Barbara, California 93106

Stephen C. Schroeter (133)
Department of Biology
San Diego State University
San Diego, California 92115

Allan Stewart-Oaten (17, 109, 133, 257)
Department of Biological Sciences
University of California, Santa Barbara
Santa Barbara, California 93106

Simon F. Thrush (49)
Ecosystems Division
National Institute of Water and
Atmospheric Research
P.O. Box 11–115
Hamilton
New Zealand

A. J. Underwood (151)
Institute of Marine Ecology
University of Sydney
Zoology Building A 08
Sydney, New South Wales 2006
Australia

Acknowledgments

We gratefully acknowledge Sally J. Holbrook, Helen Murdoch, Colette St. Mary, and Bonnie M. Williamson for their considerable assistance in preparing the manuscript; K. Erlich, H. Lin, and A. Sberze for their clerical assistance; and the anonymous reviewers who provided valuable constructive criticism of individual chapters. We also thank Chuck Crumly, our sponsoring editor, and the entire staff at Academic Press (especially Dana Polachowski and Candie Jamerson) for their considerable assistance in producing this book. It is a pleasure to acknowledge the unfailing encouragement and support of Fred Piltz, Chief of the Pacific OCS Region's Environmental Studies Program of the Minerals Management Service, U.S. Department of the Interior.

Chapters 4 (authored by Thrush *et al.*), 5 (B.D. Mapstone), 6 (Osenberg *et al.*), and 9 (A.J. Underwood) were first published in *Ecological Applications*. We thank the Ecological Society of America for permission to reprint these articles.

The genesis of this book was a special symposium held at the Second International Temperate Reef Conference in Auckland, New Zealand in 1992, although this volume is not the proceedings of that symposium (for the proceedings of the conference, see C. N. Battershill, D. R. Schiel, G. P. Jones, R. G. Creese, and A. B. MacDiarmid (eds.), 1993, *Proceedings of the Second International Temperate Reef Symposium,* NIWA Marine, Wellington, New Zealand). We thank the organizers of the conference (the editors of the proceedings listed above), and especially our New Zealand liaison for the special symposium on environmental impact assessment (R. G. Creese). We also thank the U.S. Minerals Management Service, Department of the Interior, for providing transportation funds for the U.S. participants to the symposium (provided through the MMS–University of California cooperative research program, the Southern California Educational Initiation).

Support to develop this book was provided by the U.S. Minerals Management Service through the MMS–UC Southern California Educational Initiative (MMS contract no. 14-35-001-3071), and by the State of California through the Coastal Toxicology component of the UC Toxic Substances Research and Teaching Program. We are deeply grateful for their sponsorship without which this volume could not have been produced. Views, opinions, and conclusions in this book are solely those of the contributing authors and should not be interpreted as necessarily representing the official policies, either expressed or implied, of the Federal, State, or local governments of the United States, Australia, or New Zealand.

Preface

The increasing public awareness of broad degradation of our natural world has been accompanied by a substantial number of policies aimed at protecting the environment from further harm and at repairing the damage already inflicted. At the core of this effort are field monitoring programs, mandated by an abundance of statutes in virtually all industrialized countries, whose purpose is to provide information on the environmental impacts of various activities to aid in management decisions to safeguard our natural world. In the United States alone, hundreds of millions of dollars are spent annually on these monitoring efforts for just the marine environment. Yet despite this enormous fiscal expenditure, the participation of numerous agencies and deeply committed individuals, and the universal desire for effective management, considerable uncertainty remains about the nature and magnitude of environmental impacts of our activities. This uncertainty has countless adverse repercussions for developers, regulators, the general public, and the natural systems that we attempt to manage. How is this so? In a recent evaluation of current environmental monitoring practices for the marine environment, the U.S. National Research Council concluded that monitoring programs need to be better designed and applied more appropriately if these efforts are to yield information useful for effective management (NRC 1990, *Managing Troubled Waters: The Role of Marine Environmental Monitoring.*, National Academy Press, Washington DC). Thus, issues that constrain the value of current efforts to estimate ecological impacts of localized perturbations are both scientific and institutional in nature.

This book is motivated by the need to redress issues that impede effective appraisal of environmental impacts, particularly those involving field assessments of localized perturbations. Numerous volumes have appeared in the past decade dealing with other aspects of the rather broad domain of environmental assessment such as the emerging science of Ecological Risk Assessment or general policy issues pertaining to the administrative process of Environmental Impact Assessment (EIA). None of these books—even those focused largely on monitoring—deals with scientific or institutional aspects of field assessments of ecological effects in more than a superficial way. Yet the intellectual core of the problem rests with the application of appropriate monitoring designs that can adequately distinguish the effects of anthropogenic activities from natural processes. This issue is substantial and unresolved, and has not been adequately addressed in other books. This book is designed to fill that void by concentrating on the conceptual and technical underpinnings of scientific methodology to reliably estimate local and regional ecological effects of localized disturbances.

This volume is aimed at three audiences. The first group includes regulators, resource managers, and professional scientists who are actively involved in assessment and management of environmental effects arising from our activities. Our intent is not to lay out a "prescription" for the application of science to estimate impacts, but rather to provide an understanding of the "state-of-the-science" of field assessment programs to estimate impacts and to facilitate discussion of the strengths and limitations of alternative approaches. We also identify certain institutional aspects that currently limit the potential contribution of science to the EIA process. Increasing understanding by practitioners of these issues is needed if more appropriate monitoring/assessment programs are to be designed and applied.

The second audience consists of academic and research scientists who focus on developing the scientific applications and approaches to the resolution of environmental problems. For this group we hope this book serves two purposes. Despite marked advances in the science of ecological impact assessment, many critical scientific issues remain unresolved. One intent, then, is to highlight some of these technical and conceptual issues that require further research. The other aim is to apprise the scientific community of some of the institutional or structural aspects that constrain our ability to apply scientifically rigorous methods. Indeed, the greatest impediments at present probably are institutional rather than technical in nature.

The third and perhaps most important audience we wish to reach are graduate students and upper-division undergraduates in the environmental sciences. It is crucially important that the next generation of environmental scientists and policymakers understand well the most frequently used "scientific" tool used in environmental management—field monitoring programs—including some of the consequences of our failure to reduce uncertainty to a level acceptable to the general public. Producing cohorts of scientists, managers, and policymakers who are better prepared to deal with current and future problems associated with our natural world is the single best strategy for improving the effectiveness of environmental management. We deeply hope that this book contributes to that effort.

Russell J. Schmitt
Craig W. Osenberg

AN INTRODUCTION TO ECOLOGICAL IMPACT ASSESSMENT
Principles and Goals

CHAPTER 1

DETECTING ECOLOGICAL IMPACTS CAUSED BY HUMAN ACTIVITIES

Craig W. Osenberg and Russell J. Schmitt

Ecologists and environmental scientists have long sought to provide accurate scientific assessments of the environmental ramifications of human activities. Despite this effort, there remains considerable uncertainty about the environmental consequences of many human-induced impacts, particularly in marine habitats (e.g., NRC 1990, 1992). This is especially surprising when one considers the vast amounts of capital and human resources that have been expended by industry, government, and academia in reviewing, debating, and complying with, a plethora of environmental regulations, which often require extensive study and documentation of environmental impacts. As we face an ever increasing number of environmental problems stemming from human population growth, it is critical that we achieve better understanding of the effects of humans. Due to the variety of human activities that potentially affect ecological systems, it also is imperative that we discriminate among effects of specific types of disturbances (rather than focus on an overall effect without regard to the particular sources), so that we can identify and give adequate attention to those that are most "harmful" (e.g., having the biggest effects, or which affect the most "valuable" resources). This requires approaches that can isolate effects of particular activities from nonhuman sources of natural variation as well as background variation caused by other anthropogenic events. Such approaches should reduce the uncertainty that underlies the documentation of effects of anthropogenic impacts and thus facilitate solutions to many of these problems.

Uncertainty surrounding the effects of anthropogenic activities arises from limitations imposed during the two scientific processes that comprise environmental impact assessment: (i) the predictive process, aimed at detailing the likely impacts that would arise from a proposed activity (most recently termed "Risk Assessment"; Suter 1993), and (ii) the postdictive process, aimed at quantifying the actual impacts of an activity (sometimes called "retrospective risk assessment," and which we will refer to as "Field Assessment"). Instead of standing alone, these two processes should proceed in tandem and build upon each other; resolution of many environmental issues requires both quantification of impacts, as well as accurate prediction of future impacts. Neither process substitutes for the other. For example, a prediction reveals little unless the prediction is an accurate indicator of actual effects; the accuracy of predictions can only be assessed after repeated tests (i.e., comparison with the actual outcomes). Similarly, documenting an impact that has already occurred yields only a limited ability to improve environmental planning (i.e., by avoiding environmental problems, or facilitating environmentally safe activities) unless we use this information to construct or refine frameworks that enable us to accurately anticipate future environmental impacts (e.g., predict their magnitude, and know the likelihood of such impacts based on the type of activity, its location, or the system being affected). The development of such frameworks is crucial to sound decisions being made before an activity occurs.

To date, we have only a limited ability to accurately predict the ecological consequences of many anthropogenic impacts (e.g., Culhane 1987, Tomlinson and Atkinson 1987, Buckley 1991a, 1991b, Ambrose et al., Chapter 18). For example, audits of environmental impact assessments (i.e., comparisons of predicted impacts with those actually observed) often have found relatively good agreement between predictions and reality when focused on physical or engineering considerations (e.g., the amount of copper discharged from a wastewater facility), but poor or limited agreement when focused on biological considerations (e.g., impacts on population density). As Ambrose et al. (Chapter 18) point out, the poor agreement stems both from lack of quantitative (or often qualitative) predictions of ecological change in the predictive phase of assessments, as well as lack of knowledge of the actual impacts (due to the absence or poor design of Field Assessments). Advances in Risk Assessment are likely to promote more precise predictions and thus reduce the first hurdle; however, the second will continue to plague us until Field Assessment studies are designed that better isolate effects of human activities from other sources of variation. Furthering this latter goal is a major theme of this book.

The field of environmental impact assessment is quite broad, requiring expertise from a diversity of fields, including physics, chemistry, engineering, toxicology, ecology, sociology, economics, and political science. In assembling the components of this book, we did not seek a comprehensive treatise on impact assessment. Instead, we focused on a narrower but central topic: the Field Assessment of localized impacts that potentially affect ecological systems (we further use marine habitats to provide the context for the discussions). We chose this conceptual focus because a wealth of books have appeared in the past 10 years that deal well with other aspects of environmental assessment (e.g., Risk Assessment: Bartell et al. 1992, and Suter 1993; general overviews of EIA: Westman 1985, Wathern 1988, Erickson 1994, Gilpin 1994; general introduction to monitoring: Spellerberg 1991; see also Petts and Eduljee 1994). None of these books, however, deals with the issue of Field Assessments in more than a cursory way (but see NRC 1990 for a good introduction to the role of environmental monitoring); there are no detailed discussions of sampling designs that can most reliably estimate the magnitude of the impacts and quantify the power of the designs to detect impacts, nor are there evaluations of institutional and scientific constraints that limit the application of such designs. This book is designed to fill that gap, and in so doing, provides grist for future discussion and advances that are critically needed to better understand effects of anthropogenic activities.

Two major tenets, which we elaborate below, underlie this book: (i) that Field Assessments are absolutely essential to understanding human impacts, in part, because they complement, and provide field tests of, predictions provided by Risk Assessment; and (ii) that improved sampling designs are critical to improving the quality and utility of results obtained from Field Assessments.

The Need for Field Assessments

The emerging field of Ecological Risk Assessment (Bartell et al. 1992, Suter 1993) has led to a tremendous increase in the precision and explicitness of predictions of anthropogenic impacts on ecological systems. These predictions are often based on models derived from laboratory studies of toxicological effects, transport models that describe the movement of contaminants, and population models that attempt to couple physiological and demographic changes with shifts in population dynamics and abundances. However, no degree of sophistication of such models can guarantee the accuracy of the predictions. The quality and applicability of Risk Assessment can only be judged by the degree to which its predictions match the impacts that actually occur. This requires estimation of the magnitude of the impact, not just its detection, and thus requires a Field Assessment that is able to separate natural spatial and temporal variability from variation imposed by the activity of interest. This is not a trivial problem, and many previous assessments have failed in this regard. The small number of successes that exist are too few to permit any sort of rigorous evaluation of Risk Assessment models.

As Risk Assessment models become more complex and sophisticated, it is possible that they will be championed as the final step in environmental impact assessment; follow-up Field Assessments might be deemed a waste of effort (redundant with the effort devoted to obtaining the predictions). While this is a worthy (but elusive) goal, no Risk Assessment model, no matter how sophisticated, is currently capable of accurately predicting ecological change in response to an anthropogenic activity. As mentioned above, this, in part, is due to the lack of knowledge about the actual response of many systems to anthropogenic disturbances, and therefore the general inability to compare predicted and observed change.

Uncertainty in predictions from Risk Assessment models often is acknowledged but typically is limited to two sources: (i) uncertainty about the actual value of parameters that are estimated from the studies that underlie the Risk Assessment model; and (ii) uncertainty about the environmental inputs to the model (e.g., how much freshwater runoff will enter an estuary during an upcoming year). Estimates of these sources of error often are incorporated into a Risk Assessment analysis to estimate the uncertainty associated with the prediction(s) of the model. Another, perhaps more important source of uncertainty rarely is examined: the uncertainty that the model chosen exhibits dynamics that are quantitatively (or even qualitatively) similar to the dynamics exhibited by the actual system. For example, improved laboratory techniques might provide improved quantification of the effect of a toxicant on the fecundity of a focal organism. This information might then be used in a model that links toxicant exposure with fecundity, and fecundity with population growth. However, even if the effect of the toxicant can be accurately extrapolated to field conditions, there is little guarantee that the connection between fecundity and population dynamics has been modeled correctly. More generally, the predicted dynamics may bear little resemblance to the observed dynamics, not because of uncertainty in the laboratory measurements, but due to uncertainty in the structure of the model into which the laboratory data are embedded. Addressing this uncertainty requires extensive field data, including information on the link between physiological changes and behavior (e.g., habitat selection, mate selection, reproductive condition), demographic consequences (e.g., changes in survival, birth rates, migration), population-level responses (e.g., shifts in age-structure, temporal dynamics), community responses (e.g., due to shifts in the strengths of species interactions), and ecosystem properties (e.g., feedbacks between biotic shifts and the physio-chemical aspects of the environment). Ultimately, these field data, together with laboratory data and the Risk Assessment model(s), must be integrated and then tested via comparison with actual responses to specific human activities. This last step requires properly crafted Field Assessment designs of sufficient power to distinguish the effects of the activity from a diverse set of other processes that drive variation in ecological systems.

The (In-)Adequacy of Exisitng Field Assessment Designs

The Goal of Field Assessments

A basic goal of a Field Assessment study is to compare the state of a natural system in the presence of the activity with the state it would have assumed had that activity never occurred. Obviously, we can never know, or directly observe, the characteristics of a particular system (occupying a specific locale at a specific time) in both the presence and absence of an activity. Thus, fundamental goals of the assessment study are to estimate the state of the system that would have existed had the activity not occurred, estimate the state of the system that exists with the activity, and estimate the uncertainty associated with the difference between these estimates (Stewart-Oaten, Chapters 2 and 7). The inability of most studies to accomplish these goals has, in large part, led to tremendous uncertainty regarding the environmental consequences of anthropogenic activities. We briefly review some of these design considerations, beginning with an often misunderstood approach borrowed from modern field ecology—the manipulative field experiment.

The Role of Field Experiments

Manipulative field experiments (with spatial replication of independent subjects, and randomized assignment of subjects to treatment groups) is a common and powerful tool of field ecologists. However, field experiments can do very little to resolve the specific goal of Field Assessments. This issue (i.e., the application of experimental design to assessment) has clouded much of the debate about the design of Field Assessment studies (e.g., Hurlbert 1984, Stewart-Oaten et al. 1986). While field experiments may provide crucial insight into the functioning of systems and the role of particular processes (typically acting over limited spatial and temporal scales), a field experiment cannot reveal the effects of a specific activity on the system at a specific locale at a specific time, which is often the focus of a Field Assessment. A field experiment could provide a powerful way to determine the average effect of a process (e.g., an anthropogenic activity) defined over replicates drawn at random from a larger population of study (assuming that we could conduct such a replicated experiment on the appropriate spatial and temporal scale). However, this field experiment could not tell us about the effect of the treatment on any one of the replicates. Yet, this is analogous to the problem faced in Field Assessments.

To illustrate, consider the possible environmental impacts related to offshore gas and oil exploration, specifically those associated with the discharge of drilling muds. We could conduct an experiment to address whether oil exploration has localized effects on benthic infauna inhabiting a particular region, say the Southern California Bight, by (i) randomly selecting a subset of sites within the Bight and allocating these between "Control" and "Treatment" groups; (ii) drilling exploratory wells and releasing muds in our Treatment sites; (iii) quantifying the abundances of infauna in the Control and Treatment sites after a specified amount of time (e.g., 1 year); and (iv) determining if there is sufficient evidence to reject the null hypothesis of "no effect" (e.g., are the means of the two groups sufficiently different to be unlikely to have arisen by chance?) using standard statistical procedures (e.g., a t-test).

Clearly, this is an unlikely scenario (few oil companies would be willing to have a group of ecologists dictate where they will conduct their exploration), but in some situations, such an opportunity might exist. If so, then we will be in a tremendous (and enviable) position to estimate the average local effect of oil exploration on benthic fauna inhabiting the Southern California Bight. While such a study would provide invaluable information, the results would say nothing about the effect of a single oil platform drilling exploratory wells at a specific site within the Bight. Indeed, it is possible that a significant (and biologically important) treatment effect could be found in our experiment even if there were no effects at a majority of the Treatment sites (so long as the remaining Treatment sites were sufficiently affected). There are, of course, situations where knowledge of the average affect of an activity would be quite useful (e.g., the administrative process of Environmental Impact Assessment). However, in most Field Assessments we are less concerned with the average effect and more concerned with the specific effects of a particular project at a specific locale. This is analogous to the experimentalist pondering the effect of the treatment on a single replicate (and not a collection of replicates).

Furthermore, an oil company certainly does not randomly select sites for exploration. It always could be argued that part of the selection criteria included the need to find sites that not only yield oil or gas, but also are sites in which oil and gas could be found and extracted without any environmental damage; resolution of the issue thus requires specific information about specific locales. Therefore, we require a tool more powerful, or at least more specific, than the replicated field experiment with randomized assignment.

Instead of manipulative field experiments, the basic tools used in Field Assessments involve monitoring of environmental conditions. Many such monitoring designs bear superficial resemblance to one another, but differ in some fundamental aspects. In the next section, we draw on discussions from Underwood (1991) and Osenberg et al. (1992) to clarify some of these distinctions. We illustrate the basic elements of the most commonly used assessment designs, and summarize results from studies with which we have been associated to illustrate where these studies can go wrong.

The Control-Impact Design

Perhaps the most common Field Assessment design involves the comparison of a Control site (a place far enough from the activity to be relatively unaffected by it) and an Impact site (i.e., near the activity and thus expected to show signs of an effect if one exists); a common variant involves a series of Impact sites that vary in their proximity to the activity. This sort of design often is part of the monitoring program required by regulatory agencies for various coastal activities. Environmental parameters typically are sampled at the two sites (with multiple samples taken from each site), and an "impact" is assessed by statistically comparing the parameters at the Impact and Control sites.

We illustrate this approach in Figure 1.1a, which shows that the density of a large gastropod (*Kelletia kelletii*) was significantly greater at a Control site (1.6 km from a wastewater diffuser) than at either of two Impact sites (located 50 and 250 m from the diffuser). This difference might be taken as evidence that the discharge of wastewater had a negative effect on the density of the gastropod.

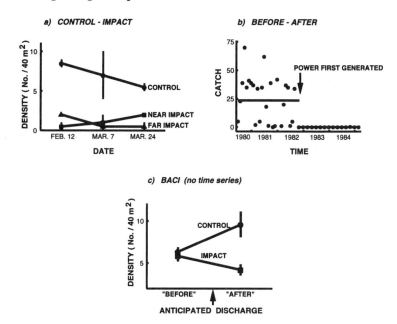

Figure 1.1. Three commonly used assessment designs that confound natural variability with effects of the anthropogenic activity. (a) The Control-After design showing the density of the snail *Kelletia kelletii* at three sites over time. The Near (square) and Far (triangle) Impact sites are located 50 and 250 m downcurrent of a wastewater outfall; the Control site (circle) is 1500 m upcurrent. These data were collected prior to discharge of wastewater. Shown for each date are the mean and range of gastropod density (*n* = 2 band transects per site). (b) The Before-After design showing density (catch per otter trawl) of pink surfperch *Zalembius rosaceus* over time at a location 18 km from the San Onofre Nuclear Generating Station (SONGS). The arrow indicates the first date on which power was generated by two new units of SONGS. Mean densities during the Before and After periods are indicated by the solid lines. (c) The Before-After-Control-Impact (BACI) design of Green (1979) showing the density of seapens *Acanthoptilum* sp. at two sites. The Control site is located 1500 m upcurrent, and the Impact site 50 m downcurrent, of a wastewater outfall. Because of permitting and production delays, discharge of wastewater did not begin when expected; all data were collected prior to discharge. Shown are means (± SE) using all observations within a period as replicates. The figure is adapted from Osenberg et al. (1992).

However, these data were collected prior to the discharge of wastewater. Thus, these differences observed during the Before period simply indicate spatial variation arising from factors independent of the effects of wastewater. To be applied with confidence, the Control-Impact design requires the stringent and unrealistic assumption that the two sites be identical in the absence of the activity. However, ecological systems exhibit considerable spatial variability, and it is extremely unlikely that any two sites would yield exactly the same result if sampled sufficiently. This design fails to separate natural *spatial* variability from effects of the activity.

The Before-After Design

An alternative design requires sampling of an Impact site both Before and After the activity; this avoids problems caused by natural spatial variation. Here, a significant change in an environmental parameter (e.g., assessed either by comparison of one time Before and one time After using within site sampling error as a measure of variability, or sampled several times Before and After and using the variation in parameter values through time as the error term) is taken as evidence of an "impact". Figure 1.1b provides an example for a fish, pink surfperch (*Zalembius rosaceus*), sampled Before and After the generation of power from new, seawater-cooled units of a large nuclear power plant (DeMartini 1987). The precipitous decline in abundance of pink surfperch is suggestive of a dramatic and detrimental impact from the power plant. However, these data are from a Control site 18 km from the power plant (a similar pattern also was seen at an Impact site: DeMartini 1987). Instead of indicating an impact, these data simply reflect the effect of other processes that produce temporal variability (in this case, it was an El Niño Southern Oscillation event that began at the same time as initiation of power generation: Kastendiek and Parker 1988). Applied in this way, the Before-After design fails to separate natural sources of *temporal* variability from effects of the activity.

A more sophisticated Before-After design is possible, and a classic example of intervention analysis (Box and Tiao 1975) provides both an illustration of its successful application and helps identify why the approach is limited for most ecological studies. Box and Tiao estimated the influence of two interventions (a traffic diversion and new legislation) on the concentration of ozone in downtown Los Angeles. Their procedure required that they (i) frame a model for the expected change; (ii) determine the appropriate data analysis based on this model; (iii) diagnose the adequacy of the model and modify the model until deficiencies were resolved; (iv) make appropriate inferences. Their analysis provided estimates of the effect of each intervention on ozone concentration.

There are several features of their system/problem that facilitated their successful analysis: (i) there was a long and intensive time series of ozone samples (hourly readings were available over a 17-year period, which included several years during the pre- and postintervention periods); (ii) the dynamics of ozone concentration were fairly well behaved, with repeatable seasonal and annual patterns; (iii) the number of pathways for the production and destruction of ozone were relatively few. These features contrast markedly with many ecological systems, where (i) we often have little expectation of how the system is likely to respond; (ii) data are sparse (time series are short, and intervals between sampling are long); and (iii) population density (for example) can be influenced by a multitude of processes (including a variety of mechanisms driven by abiotic factors and a wealth of mechanisms involving interactions with other species, each of which is also influenced by a variety of factors, including the effects of the focal activity). Certainly, such an approach might provide a powerful way to assess responses of biological systems to interventions (Carpenter 1990, Jassby and Powell 1990), but currently it remains limited due to the paucity of detailed knowledge about dynamics of ecological systems. In cases where data from an unaffected Control site are available, we may be able to incorporate them into such time series analyses to compensate for the sparseness and complexity of ecological data (Stewart-Oaten, Chapter 7: see below).

Before-After-Control-Impact (BACI) Designs

One possible solution to the problems with the Control-Impact and Before-After designs is to combine them into a single design that simultaneously attempts to separate the effect of the activity from other sources of spatial and temporal variability. There are a variety of such designs. In the first, which we refer to simply as BACI (Before-After-Control-Impact), a Control site and an Impact site are sampled one time Before and one time After the activity (Green 1979). The test of an impact looks for an interaction between Time and Location effects, using variability among samples taken within a site (on a single date) as the error term. Data from our studies of a wastewater outfall (Figure 1.1c) demonstrate such an interaction; the decline in the density of seapens (*Acanthoptilum* sp.) at the Impact site relative to the Control site suggests that the wastewater had a negative effect on density of seapens. However, discharge of wastewater at this site was delayed several years, and did not occur when first anticipated. Thus, the observed changes were due to other sources of variability and were not effects of the wastewater. This design confounds effects of the impact with other types of unique fluctuations that occur at one site but not at the other (i.e., Time × Location interactions). Unless the two sites track one another perfectly through time, this design will yield erroneous indications that an impact has occurred.

To circumvent this limitation of Green's BACI design, Stewart-Oaten et al. (1986; see also Campbell and Stanley 1966, Eberhardt 1976, Skalski and McKenzie 1982) proposed a design based on a time series of differences between the Control and Impact sites that could be compared Before and After the activity begins. We refer to this design as the Before-After-Control-Impact Paired Series (BACIPS) design to highlight the added feature of this scheme (see Stewart-Oaten, Chapter 7 and Bence et al., Chapter 8). In the original derivation of this design (e.g., Stewart-Oaten et al. 1986), the test of an impact rested on a comparison of the Before differences with the After differences. Each difference in the Before period is assumed to provide an independent estimate of the underlying spatial variation between the two sites in the absence of an impact. Thus, the mean Before difference added to the average state of the Control site in the After period yields an estimate of the expected state of the Impact site in the absence of an impact during the After period: i.e., the null hypothesis. If there were no impact, the mean difference in the Before and After periods should be the "same" (ignoring sampling error). The difference between the Before and After differences thus provides an estimate of the magnitude of the environmental impact (and the variability in the time series of differences can be used to obtain confidence intervals: Stewart-Oaten, Chapter 7 and Bence et al., Chapter 8).

The BACIPS design is not without its limitations, for it also makes a set of assumptions, which if violated can lead to erroneous interpretations (e.g., due to nonadditivity of Time and Location effects or serial correlation in the time series of differences). Indeed, one of the fundamental contributions of Stewart-Oaten's work (Stewart-Oaten et al. 1986, 1992, Stewart-Oaten, Chapter 7)has been to make explicit the assumptions that underlie the BACIPS design, pointing out the importance of using the Before period to generate and test models of the behavior of the Control and Impact sites, and to suggest possible solutions if some of the assumptions are violated. Importantly, many of these assumptions can be directly tested. Of course, it is still possible that a natural source of Time $\times$ Location interaction may operate on the same time scale as the study, and thus confound interpretation of an impact. However, this problem is far less likely than those inherent to the other designs (e.g., that variation among Times and Locations be absent and that there be no Time $\times$ Location interaction).

In this volume, Stewart-Oaten (Chapter 7) and Bence et al. (Chapter 8) elaborate upon and apply a more flexible BACIPS design based on the use of the Control site as a "covariate" or predictor of the Impact state, which might have even greater applicability than the original design (which was based on the "constancy" of the differences in the Before and After periods). Underwood (1991, Chapter 9) has suggested a "beyond-BACI" approach, which incorporates multiple Controls, as well as random sampling of the study sites (thus, the "Paired Series" aspect of the BACIPS design is not present in Underwood's beyond-BACI design). Underwood suggests that the beyond-BACI design is able to detect a greater variety of impacts than the BACIPS design (e.g., detection of pulse responses as well as sustained perturbations); however, he also notes that his design is not able to deal explicitly with problems of serial correlation. By contrast, the presence of serial correlation can be directly assessed, and appropriate action taken, when applying the BACIPS design (Stewart-Oaten et al. 1986, 1992, Stewart-Oaten, Chapter 7). In a variety of important ways, Underwood's approach differs from Stewart-Oaten's and some others represented in this book (e.g., Osenberg et al., Chapter 6, Bence et al., Chapter 8). While both schools-of-thought advocate the advantages of using more than one Control site, they do not agree on the ways in which this added information should be incorporated into the analyses. We expect that debate on these issues is far from over, and hope that this book serves to further the discussion and facilitate advancements in the design and application of BACI-type studies.

The Organization of This Book

The issues highlighted above are tackled directly in the second section of this book, which is devoted to elaboration of the application and design of BACI-type studies. However, prior to the implementation of any Field Assessment a number of initial issues must be considered, and some of these are highlighted in the book's first section. For example, the general goal and purpose of the study is critical, and Stewart-Oaten's first chapter (Chapter 2) tackles the standard "P-value culture" that places undo emphasis on the detection of impacts, rather than estimation of their magnitude or importance. Ecological parameters must also be selected for study, and although the use of bio-indicators has been often criticized, Jones and Kaly (Chapter 3) point out that any study necessarily must select a limited number of parameters from the myriad available (thus necessitating the selection of a subset of "bio-indicators"). Once appropriate species (or parameters) are selected, sampling error can constrain our ability to discern the temporal dynamics of populations, and thus impair our ability to use time series analyses to assess ecological change (Thrush, Hewitt, and Pridmore, Chapter 4). Variability also limits the power of statistical tests of impacts, and Mapstone (Chapter 5) suggests a novel way to incorporate such a constraint directly into the permitting process by simultaneously weighting Type I and Type II errors in assessment studies.

The second section of the book (*Improving Field Assessments of Local Impacts: Before-After-Control-Impact Designs*), provides the core of the book and elaborates on the theory and application of BACI-type designs. Osenberg, Schmitt, Holbrook, Abu-Saba and Flegal (Chapter 6) provide a segue from Mapstone's discussion of statistical power by evaluating sources of error in BACIPS designs and specifically evaluating the power to detect impacts on chemical-physical vs. biological (individual-based vs. population-based) parameters. Stewart-Oaten (Chapter 7) follows with a theoretical treatment of BACI-type designs, which extends and generalizes much of the earlier research on BACI(PS). Bence, Stewart-Oaten and Schroeter (Chapter 8) apply this more general and flexible BACIPS design to data derived from an intensive study of the impacts of a nuclear power plant. In the final chapter in this section (Chapter 9), Underwood offers an alternative approach in which multiple Control sites are used to detect impacts in a different way than proposed by the other authors, and which potentially can detect a greater variety of impacts (e.g., pulses as well as sustained impacts).

While BACI-type designs offer great potential to detect impacts of local perturbations, they require sampling of a Control site(s) (a site(s) sufficiently close to the Impact site(s) to be influenced by similar environmental fluctuations, but sufficiently distant to be relatively unaffected by the disturbance). In many cases, such a control does not exist, or the significant biological effects of interest are dispersed over large spatial scales and therefore are difficult (if not impossible) to detect. In such cases, a BACI-type design or other empirical measure of impact is unlikely to be able to quantify the effect with the desired level of precision. Therefore, we must be able to extrapolate results obtained from localized or smaller scale effects to those arising on larger spatial scales. This is the theme for the book's third section. Raimondi and Reed (Chapter 10) discuss how the spatial scales of impacts on chemical–physical parameters might differ from the scale of impacts on ecological parameters based on life-history features of the affected organisms. Larval dispersal is central to many of their points. Understanding the coupling between larval pools and benthic populations and their response to impacts will require integration of oceanographic models and ecological studies (Keough and Black, Chapter 11). Oceanographic processes may also "collect" larvae and pollutants in particular sites (e.g., along linear oceanographic features), and this aggregation of larvae in high concentrations of pollutants might amplify deleterious effects of many types of discharge (Kingsford and Gray, Chapter 12). Ultimately however, effects on larvae need to be translated into consequences at the population level, and in Chapter 13, Nisbet, Murdoch, and Stewart-Oaten provide an approach intended to provide an estimate of how local larval mortality, induced by a nuclear power plant, may impact the abundance of adults assessed at a regional level. Their work points out the critical need to better understand the role of compensation in the dynamics of fishes and other marine organisms affected by anthropogenic activities.

In the final section of the book, we return to the issue of Predictive vs. Postdictive approaches, emphasizing how and why the Predictive phase (which typically yields an Environmental Impact Report or Statement) should be integrated with the Postdictive phase (which yields the Field Assessment) to improve the overall quality of the entire process. The introductory chapter in this section (Chapter 14: Schmitt, Osenberg, Douros, and Chesson) provides a brief summary of the current state of the EIR/S process in the United States and Australia, and Carney (Chapter 15) evaluates the biological data that have been collected with regard to EIR/S studies (as well as Field Assessments). He concludes that numerous problems (including taxonomic errors, design flaws, statistical inaccuracies) plague even the most extensive studies. Piltz (Chapter 16) then elaborates on how institutional constraints can impede sound scientific investigations that require long-term monitoring. His chapter encourages both scientists and administrators to find solutions that ensure the research continuity that is required to obtain the most defensible Field Assessments. If EIR/S consistently fail to yield consensus, or if too few data exist to determine the accuracy of such studies, considerable debate can ensue. This often leads to judicial involvement in the EIR/S process, which in turn leads to tremendous effort expended on documentation during the EIR/S process, but with little additional clarity regarding the likely or actual environmental impacts of a project (Lester, Chapter 17).

Ultimately, the interplay between the Predictive process (EIR/S or Risk Assessment) and the Postdictive process (the Field Assessment) is critical to help guide our development of frameworks used to predict and understand anthropogenic impacts: are our predictions accurate, and if not how might we modify our approach? Ambrose, Schmitt and Osenberg (Chapter 18), provide an audit of an intensive study of the San Onofre Nuclear Generating Station (SONGS) and compare effects that were predicted during the EIR/S process with those subsequently observed. Their findings reveal that the EIR/S process (even in such a major project) revealed little of the actual impacts; even a detailed scientific study conducted by an independent committee erred in a number of crucial ways. Their results demonstrate the need for continued vigilance in conducting well-designed monitoring studies, such as those using BACI-type analyses.

At times the tone of many of these contributions is rather critical of existing approaches and even the new approaches outlined in other chapters. It is only through healthy debate of the merits of alternative designs, and by better integration of administrative and scientific goals, that improvements in the EIR/S and Field Assessment processes occur. Indeed, despite this body of criticism, it is undeniable that refinements in our scientific tools have led to recent improvements in our understanding of anthropogenic impacts. This can only continue by avoiding complacency and by continuing to develop and refine new tools that can be used to tackle these important issues. Of course, realization of our ultimate goal depends upon expanding our basic knowledge of the dynamics and functioning of ecological systems, understanding the mechanisms by which anthropogenic activities impact these systems, and incorporating this information into models and theory to permit us to predict the occurrences of future impacts. But first, we must be better able to quantify the actual impacts that specific activities have induced in ecological systems. This simple goal is neither trivial nor commonly realized, but it is fundamental and achievable. Our hope is that this book helps to further advance understanding of the interactions between human activities and our impacts on our environment.

Acknowledgments

We gratefully acknowledge the helpful comments of S. Holbrook, C. St. Mary, and A. Stewart-Oaten. Research that led to the production of this chapter was funded by the Minerals Management Service, U.S. Department of Interior under MMS Agreement No. 14-35-001-3071, and by the UC Coastal Toxicology Program. The views and conclusions in this chapter are those of the authors and should not be interpreted as necessarily representing the official policies, either expressed or implied, of the U.S. Government.

References

Bartell, S. M., R. H. Gardner, and R. V. O'Neill. 1992. Ecological risk estimation. Lewis Publishers, Chelsea, Michigan.

Box, G. E. P., and G. C. Tiao. 1975. Intervention analysis with applications to economic and environmental problems. Journal of the American Statistical Association 70:70–79.

Buckley, R. 1991a. Auditing the precision and accuracy of environmental impact predictions in Australia. Environmental Monitoring and Assessment 18:1–24.

Buckley, R. 1991b. How accurate are environmental impact predictions? Ambio 20:161–162.

Campbell, D. T., and J. C. Stanley. 1966. Experimental and quasi-experimental designs for research. Rand McNally, Chicago, Illinois.

Carpenter, S. R. 1990. Large-scale perturbations: opportunites for innovation. Ecology 71:453–463.

Culhane, P. J. 1987. The precision and accuracy of U.S. Environmental Impact Statements. Environmental Monitoring and Assessment 8:217–238.

DeMartini, E. 1987. Final report to the Marine Review Committee. The effects of operations of the San Onofre Nuclear Generating Station on fish. Marine Science Institute, University of California, Santa Barbara.

Eberhardt, L. L. 1976. Quantitative ecology and impact assessment. Journal of Environmental Management 4:27–70.

Erickson, P. A. 1994. A practical guide to environmental impact assessment. Academic Press, San Diego, California.

Gilpin, A. 1994. Environmental impact assessment (EIA). Cambridge University Press, Cambridge, England.

Green, R. H. 1979. Sampling design and statistical methods for environmental biologists. Wiley and Sons, New York, New York.

Hurlbert, S. J. 1984. Pseudoreplication and the design of ecological field experiments. Ecological Monographs 54:187–211.

Jassby, A. D., and T. M. Powell. 1990. Detecting changes in ecological time series. Ecology 71:2044–2052.

Kastendiek, J. and K. R. Parker. 1988. Interim technical report to the California Coastal Commission. 3. Midwater and benthic fish. Marine Review Committee, Inc.

NRC (National Research Council). 1990. Managing troubled waters: the role of marine environmental monitoring. National Academy Press, Washington, DC.

NRC (National Research Council). 1992. Assessment of the U.S. outer continental shelf environmental studies program. II. ecology. National Academy Press, Washington DC.

Osenberg, C. W., S. J. Holbrook, and R. J. Schmitt. 1992. Implications for the design of environmental impact studies. Pages 75–89 in P. M. Griffman and S. E. Yoder, editors. Perspectives on the marine environment of southern California. USC Sea Grant Program, Los Angeles, California.

Petts, J., and G. Eduljee. 1994. Envionmental impact assessment for waste treatment and disposal facilities. John Wiley and Sons, Chichester, England.

Skalski, J. R., and D. H. McKenzie. 1982. A design for aquatic monitoring systems. Journal of Environmental Management 14:237–251.

Spellerberg, I. F. 1991. Monitoring ecological change. Cambridge University Press, Cambridge, England.

Stewart-Oaten, A., W. W. Murdoch, and K. R. Parker. 1986. Environmental impact assessment: "psuedoreplication" in time? Ecology 67:929–940.

Stewart-Oaten, A., J. R. Bence, and C. W. Osenberg. 1992. Assessing effects of unreplicated perturbations: no simple solutions. Ecology 73:1396–1404.

Suter, G. W., II, editor. 1993. Ecological risk assessment. Lewis Publishers, Boca Raton, Florida.

Tomlinson, P., and S. F. Atkinson. 1987. Environmental audits: A literature review. Environmental Monitoring and Assessment 8:239–261.

Underwood, A. J. 1991. Beyond BACI: experimental designs for detecting human environmental impacts on temporal variations in natural populations. Australian Journal of Marine and Freshwater Research 42:569–587.

Wathern, P. 1988. Environmental impact assessment. Unwin Hyman Ltd., London, England.

Westman, W. E. 1985. Ecology, impact assessment, and environmental planning. John Wiley and Sons, New York, New York.

CHAPTER 2

GOALS IN ENVIRONMENTAL MONITORING

Allan Stewart-Oaten

The goals of analyses of data on human environmental impacts ("interventions") vary among investigators. Many analyses aim only at description. The data are manipulated to display some suggestive patterns, and the patterns are taken as demonstrating the effects of interest or concern. Such "survey and explain" studies make little or no effort to distinguish the observed patterns from patterns that could have arisen from natural fluctuations or from sampling error (Carney 1987).

Several authors, notably Green (1979) and Carney (1987), stress the importance of formal "confirmatory" statistical methods, such as tests of null hypotheses, or confidence intervals and regions. These differ from "exploratory" methods by supplying objective rules for assessing uncertainty in the results, using procedures whose long-run properties (e.g., the probability of false rejection, or of covering the true value) are known (at least approximately) under plausible assumptions.

Measuring the reliability of conclusions is not the only benefit of confirmatory methods. They improve sampling design by forcing investigators to define "impact" and account for natural variation, and promote clarity by forcing them to organize the data in standardized ways, to state explicitly the models underlying the analyses, and to assess whether these models are appropriate. These requirements may help later workers by facilitating the development of a standardized cumulative data base, helping focus future studies on the more likely effects, improving sampling designs and analytical models, and reducing problems of gross errors (e.g., in data entry).

For many biologists, "formal statistical methods" means hypothesis tests, which are often thought of as a rigorous, objective way of making decisions. I argue here that hypothesis tests are usually poor ways to make decisions, that the aim of formal analyses of monitoring data should not be decisions but descriptions with allowance for error, and that this is best accomplished by confidence intervals or regions, not hypothesis tests.

Detecting Ecological Impacts: Concepts and Applications in Coastal Habitats, edited by R. J. Schmitt and C. W. Osenberg

A Case for Confidence Intervals

Tukey (1960) makes a useful distinction between conclusions and decisions. Roughly, this is the distinction between choosing what to believe and choosing how to act. Conclusions are statements we accept until unusually strong contradictory evidence appears: they are "subject to future rejection" and are "judged by their long-run effects, by their 'truth', not by specific consequences of specific actions." Further distinctions are made between statistical and experimenter's conclusions (the latter involving more uncertainty, because of the possibility of systematic error) and between qualitative and quantitative conclusions (the former judging the truth of a single assertion about a parameter, the latter presenting a region where its value is believed to lie).

Decisions are choices of actions, determined by our assessment of their probable consequences in a specific situation. They may mimic conclusions—that is, we may choose to act as if a particular conclusion is true—but we make some decisions without any evidence at all, and we may reasonably make decisions implying opposite conclusions (e.g., to carry life insurance and also to save for retirement).

In the subsections below, I argue that (a) hypothesis tests are not well suited to decision making; (b) what is really wanted from biologists, ecologists and biological data analysts in environmental monitoring is not decisions but conclusions (including the allowance to be made for error) which can become part of the basis for decision making, usually by others; (c) the conclusions should not be results of significance tests because these carry too little information, e.g., about effect size, and are confusing; and (d) confidence intervals provide clear and informative conclusions, though they may need to be augmented to allow for uncertainty not considered in the formal framework.

Hypothesis Tests and Statistical Decisions

The inadequacy of statistical hypothesis testing for making decisions was recognized by the founders of the predominant methodology:

> The sum total of the reasons which will weigh with the investigator in accepting or rejecting the hypothesis can very rarely be expressed in numerical terms. All that is possible ... is to balance the results of a mathematical summary ... against other less precise impressions. ... The tests themselves give no final verdict, but as tools help the worker who is using them to form his final decision ... (Neyman and Pearson 1928).

Among these less precise impressions are "*a priori* or *a posteriori* considerations," as stressed by Bayesians. A later paper recognizes the costs of wrong decisions:

> If we reject H_0, we may reject it when it is true; if we accept H_0, we may be accepting it when it is false ... in some cases it is more important to avoid the first, in others the second. ... Is it more serious to convict an innocent man or to acquit a guilty? That will depend on the consequences of the error. ... The use of these statistical tools in any given case, in

determining just how the balance should be struck, must be left to the investigator (Neyman and Pearson 1933).

Despite these cautions, the Neyman-Pearson theory does seem to have been intended as a decision-making system, as evidenced by the use of acceptance sampling, clearly a decision setup, as a paradigm (Neyman and Pearson 1936) and by Neyman's later advocacy of "inductive behavior" (e.g., Neyman 1957, 1962). Together with game theory, it laid the groundwork for the more comprehensive theory of decision making under uncertainty developed by Wald (1950). This posed the problem of choosing from among a set of actions, A, when the gain of choosing act a is $G(a,s)$, where s is the "state of Nature," known only to be a member of a set, S. Data analysis enters if we are allowed to observe a variable, X, whose distribution, P_s, depends on s. The task then is to choose a decision function, δ, which selects an action, $\delta(X)$, for each X, so as to maximize the expected gain, $\int G(\delta(X),s)P_s(dX)$. To pose hypothesis testing as a decision problem, S is the set of possible values of the true parameter (e.g., the mean of a population), H_0 asserts that this parameter is in some subset of S, the set of actions is $A = \{$"accept H_0", "reject H_0"$\}$, and the gain may be 0 if we choose correctly but some negative number (depending on the true s) if we do not; X may be the values in a sample from the population whose mean is s.

For all but the simplest problems (e.g., testing a simple hypothesis against a simple alternative), there is no best solution: even with the observed X, the gain depends on the unknown value of s. There are various types of "good" solutions, such as minimax (the solution whose worst result is least bad) and admissible (a solution such that no other solution does at least as well for every s, and better for at least one s). An important type of admissible solution is the Bayesian: a solution whose expected gain, averaged over S according to some distribution p, i.e., $\int \int G(\delta(X),s)P_s(dX)p(ds)$, is largest. If p is known, then Bayes' theorem can be used to rewrite this integral as $\int \int G(\delta(X),s)p(ds|X)P(dX)$, where $p(.|X)$ is the conditional distribution of s given X, and P is the unconditional distribution of X; the "best" $\delta(X)$ can then be chosen to be the action that maximizes the inner integral. Controversy arises here over the determination, or even the existence, of p, the "prior" distribution on the states. It is usually not possible to interpret this as a distribution in the frequentist sense, i.e., as given by the relative frequency of various outcomes in a long run of trials. But this "objectivist" interpretation of probability is not the only one. One alternative is the "personalistic" view, in which probability measures the confidence that an individual has in some proposition. From this basis, Savage (1954) developed and persuasively advocated a Bayesian approach to statistical inference and decision making. Shorter or more gentle accounts of this work are given by Edwards et al. (1963) and Pratt et al. (1965a, 1965b).

Neyman-Pearson inference, Wald's decision theory, and Savage's Bayesianism have all spawned large literatures and many useful insights and techniques. There is no consensus on a "best" approach: many statisticians will mix fields, e.g., using Bayesian methods to make inferences with good Neyman-Pearson properties. A minority would add Fisher's (1937, 1956) fiducial inference to the list. Some feeling for the variety of positions, the main arguments, and perhaps the areas of general agreement can be found in the special issue of *Synthese* (1977).

Tukey (1960) sees the Neyman-Pearson theory of hypothesis testing as a step toward a theory of decision making but "if this view is correct, Wald's decision theory now does much more nearly what tests of hypotheses were intended to do. Indeed, there are three ways in which it does better": focusing on gains or losses (rather than error probabilities), considering a wider range of setups with less stringent assumptions, and showing that there will not be "a single best procedure but rather an assortment of good procedures ... from which judgment and insight ... (perhaps best expressed in the form of an *a priori* distribution) must be used to select the 'best' procedure." However, aspects of the Neyman-Pearson theory, such as the power function and confidence procedures, remain valuable, along with tests of significance, in "conclusion theory".

These views seem broadly accepted by statisticians. They are not unchallenged, but in most disciplines involving data-based decision making (e.g., statistics, economics, business and engineering), the dissenters would give the Neyman-Pearson theory a smaller role, not a larger one. They would apply some form of decision theory, most often Bayesian, to conclusions as well as decisions.

However, biology is a holdout of the "*P*-value culture" (Nelder 1991). Hypothesis testing is presented as a decision problem and treated as the only way to deal quantitatively with uncertainty, whether of conclusions or decisions. Conversely, decision problems are frequently forced, with great effort and ingenuity, into the hypothesis testing mold. The achievements of 50 years of decision theory are not rejected but simply ignored altogether: like trying to use an ax to do fine woodwork, while ignoring the band saw.

Conclusions, Not Decisions

In pure research, we usually need conclusions, not decisions. Environmental monitoring is not pure research, and its ultimate aim is decision making, but here, too, the role of scientific investigators and data analysts is to present conclusions which become part (and only part) of the basis for decision making.

The ultimate decisions are usually not made by the investigator. In pure research, investigators decide what to study and how to study it, but the result is a set of conclusions; others (referees, editors, funding agencies and readers) decide whether to accept the conclusions, or to publish them, or to make or fund further studies because of them. Their main function is to "reduce the spread of the bundle of working hypotheses which are regarded as still consistent with the observations" (Tukey 1960).

In monitoring, investigators often have even less discretion. Managers and review boards often decide what to study, considering not only scientific interest and feasibility but also commerce, aesthetics, public interest, and the law. More important, they also make final decisions, e.g., concerning shutdown, redesign, operational changes, mitigation, or penalties, except when these decisions are removed even further from the investigators, into negotiations, the courts, or the legislature.

These decisions rarely depend on a single variable. Biological effects can be of many kinds (abundance, average size, demographic or sex ratios, etc.) on many species. None of these is likely to be decisive, except for rare instances of dramatic reduction of an important or popular species over a wide range. Indeed, the totality of all biological effects may not be decisive: final decisions will also depend on an array of economic, legal, political, and social goals and requirements, many of them unwritten.

Even for a single variable, the analysis of the monitoring data gives only partial information. Decision makers are usually choosing among many possible actions, but the monitoring data apply directly only to some of these: e.g., "do nothing," "impose a penalty," and "shut it down." (Even this assumes that shutting it down would return the environment to the "Before" condition.) These data may be indirectly informative about the consequences of other actions, such as redesign, but usually only when supplemented by other information: theory, modeling, experiments, and general biological knowledge.

Even if a decision were to be made entirely on the basis of the biological monitoring data, it would be too complicated a function of them to be specified in advance. With n parameters of concern (e.g., the changes in the mean abundances of n species), the data summary is likely to contain at least $2n$ values (e.g., the confidence bounds, or estimates and P-values, for the n changes). Thus, a decision procedure would need to divide the $2n$-dimensional space of possible data summaries into subregions corresponding to the different possible actions. But real decisions will involve far more than the monitoring data, including some factors which, while knowable (e.g., models of the future of the local economy from various starting points), would not be worth determining until we know they are needed, and probably other factors which we cannot anticipate, since decision makers cannot be expected to specify every possible contingency, and their corresponding decisions, in advance. Thus the aim of monitoring should not be decisions but conclusions: succinct descriptions of the biological effects, with the allowances to be made for uncertainty, in as clear a form as possible.

Hypothesis Tests: Meager Information and Unnecessary Confusion

Assessment decisions will rarely depend on the existence of an effect. Almost any intervention big enough to be worth studying will have effects on most of the local environmental parameters studied, whether we "detect" them or not. Knowing an effect exists is useless for decision making: it is the direction and size that matter. Even this question may be too narrow: some effects might be positive under some conditions (e.g., in winter, or when currents flow north) and negative under others (Eberhardt 1976, p. 34, Murdoch et al. 1989, p. 94, Reitzel et al. 1994). Neyman and Pearson's conviction/acquittal analogy is false here: the question is not "is he guilty?" but "what is he guilty of?" and (for decision makers) "what should be the sentence?"

The null hypothesis of "no change" is a straw man, and "detection" of changes is irrelevant. "Nonsignificance" or "failure to detect" an effect means merely that our data or analyses are insufficient to allow us to make an assertion about the change's direction, at a significance level of no demonstrated relevance. It does not mean we have no information: the evidence may point to a large change, but be highly uncertain when taken in isolation. To report it only as "NS" on the basis of the 0.05 cutoff is to engage in self-censorship. "Detection" or a P-value is better but still inadequate; it conveys information about direction, but not about size.

Thus hypothesis testing provides too little information for most decision making. At its best (the P-value), it uses an implausible model of "no effect" to compute the probability of observing data more unfavorable to this model than ours are. An accept/reject "decision" conveys even less. An estimate is far more informative, but a test result and even a power curve (or power evaluated at some arbitrary alternative) adds virtually nothing to it. (That is, nothing directly; from the estimate and the P-value, one can often compute a measure of the estimate's reliability, such as a standard deviation or confidence interval—but this justification applies also to handing over the raw data, unanalyzed.)

In addition, test results are misleading or confusing for many people. Hypothesis testing is awash in jargon, and its logic and the meaning of its results are not simple. Berger and Sellke (1987) argue that the P-value "gives a very misleading impression as to the validity of H_0, from almost any evidentiary viewpoint," mainly by showing how different it is from any reasonable calculation of $\Pr(H_0|x)$, the conditional probability of H_0 given the data x. They justify this by claiming that "Most nonspecialists interpret (the P-value) precisely as $\Pr(H_0|x)$"—an unproven claim, but many statisticians believe it. Perhaps a more striking justification is that Neyman himself once made this error (Good 1984).

A far more damaging misinterpretation is that a "significant" result (or a small P-value) indicates a large, important effect, while a "nonsignificant" effect is nonexistent or unimportant (Yoccoz 1991). This also seems common to "most nonspecialists"—and to specialists not on their toes. Indeed, it is unclear why a test of "no effect" would be proposed for decision making in impact assessment except on the basis of this error.

This confusion can do practical damage. For example, the California Ocean Plan forbids "significant declines in light transmittance," and defines "significant" to mean "statistically significant at the 95% level." (This is interpreted to correspond to a 0.05 level test.) But a large enough monitoring program will eventually find a "significant" change, and the law seems to attribute no relevance to the question of its *biological* significance. A power company or municipality can best satisfy the law by restricting the monitoring program (e.g., on the grounds of expense) so as to ensure a high level of uncertainty and low power of "detection" (Mapstone, Chapter 5).

When many possible effects are studied, multiple testing adds additional confusion. Mead (1988) presents published examples of results, which biologically motivated plots would have made beautifully clear, rendered incomprehensible by multiple testing methods. In assessment, it has been claimed that not only should estimated changes be statistically significant to be taken seriously, but also the *number* of statistically significant changes should itself be statistically significant (Patton 1991).

These artificial complexities are harmful. Managers, nonscientist review boards, and some investigators are very likely to misunderstand the meaning of test results, especially confusing statistical and practical significance. Also, tests focus attention too strongly on only part of the evidence. No standard formal method provides a complete assessment of the reliability and practical significance of a field result. All are affected by model uncertainty, and the multiple testing problem is real, even though the methods are usually unhelpful. Biological understanding is needed to relate the conclusions to each other, to auxiliary experiments and observations, and to mechanisms and processes which are known or plausible consequences either of the intervention or of alternative explanations. Struggling with a variety of tests and their interpretations is not the best use of biologists' time and skills.

Confidence Intervals: Quantitative Conclusions

Final reports must present both descriptions of changes and measures of the uncertainty of these descriptions. One way to do this is to focus on parameter estimation, especially confidence intervals:

> The greatest ultimate importance, among all types of statistical procedures we now know, belongs to *confidence procedures* which, by making interval estimates, attempt to reach as strong conclusions as are reasonable by pointing out ... whole classes (intervals, regions, etc.) of *possible* values, so chosen that there can be high confidence that the 'true' value is *somewhere among them*. Such procedures are clearly quantitative conclusion procedures. They make clear the essential 'smudginess' of experimental knowledge (Tukey 1960).

These descriptions must be easily understood by decision makers who are not trained in statistics. This is especially important if changes are expected to vary with seasons or other environmental conditions, so that estimated changes will have both varying values and varying uncertainties. Confidence intervals satisfy this too. Complex results can be presented clearly, without oversimplifying, to an audience of nonscientist decision makers: parameter estimates and confidence regions not only have obvious relevance to decisions but also are natural candidates for graphing. Even "power" is portrayed, in the length of the interval rather than in a welter of α's, β's, Δ's and noncentral t and F distributions.

Estimates of direction and size are needed for two reasons. One is obvious: these are the main determinants of whether an effect is harmful. The other has to do with the reliability of conclusions. The assessment of a given field result should consider not only the formal statistical summary of the "internal" evidence of the data directly pertaining to it, but also the "external" evidence of the agreement of the conclusion with our understanding of the mechanisms involved, and with other data or conclusions from the study (e.g., concerning changes in similar species). This is particularly the case when the conclusion concerns not only whether a given change has occurred but also whether the intervention caused it, i.e., whether it is an "effect." For these external judgments, the estimated sizes of changes, when combined with measures of these estimates' reliability, are more useful summaries of the internal evidence than are measures of how strongly it indicates the changes' existence (Hill 1965).

A final advantage of presenting confidence intervals and regions rather than hypothesis tests can only be outlined here. If impact assessment is seen as a statistical decision problem, it is difficult to avoid Bayesian formulations and solutions, at least as an ideal (Pratt et al. 1965a, 1965b). This ideal may be unattainable. It often needs detailed specification and quantification of all aspects of the problem: "states of nature" (possible impacts—but also, ultimately, economic, political, and aesthetic parameters), possible actions, the cost or gain function, and "personal" prior distributions on the states. The last risks having debates over assessment degenerate into arguments about prior distributions and the qualifications of their proponents. At present, it seems safer to present "objective" assessments of uncertainty, based on the field observations, separately from assessments based on compatibility with prior information, other results, and auxiliary experiments. However, if the Bayesian logic is accepted, then a non-Bayesian approach adopted for practical reasons should approximate it as well as possible. In fact, confidence intervals based on approximately Normal point estimates do approximate their Bayesian equivalents (posterior probability or "credible" intervals), at least for the diffuse priors one would expect when many poorly known processes are at work, while hypothesis tests do not approximate theirs (Edwards et al. 1963, Pratt 1965, Lindley 1965, Berger and Sellke 1987, Casella and Berger 1987).

Discussion

I have argued that confidence intervals are preferable to hypothesis tests of "no effect" because they directly assess the main concern (effect size), are easy to understand, display "power" automatically, are more relevant to an overall or causal assessment, and correspond reasonably well to the Bayesian ideal.

Some counter arguments should be noted. Perhaps the weakest is the unfounded claim that "the majority" prefers tests. This is a surprising argument for scientists to use: the majority once preferred phlogiston, and thought Copernicus was wrong. It is a poor reason for using a bad procedure, especially since many users do *not* understand the tests they use.

Tests may be more familiar to regulators, who frequently test whether a pollutant is below a threshold. However, this does address effect size: the threshold is not usually zero but a level judged important. The context is also different. Regulation frequently involves ongoing, routine judgments of many pollution sources on the basis of a single variable. Assessment requires a one-time (or few times) judgment of a single project on the basis of many variables. Even so, with moves toward sale of pollution "credits," hypothesis tests may give way to confidence intervals in regulation, too.

There are cases where the law requires hypothesis tests, as for the California Ocean plan discussed above. The tests must then be done. But these laws are written with statistical advice, some of it bad: they can and should be changed with better advice.

Some developers would like decision rules laid out in advance: they don't want the rules to change after the investments have been made. But we all want things we can't have. Sensible decisions will be complex functions of biological and economic data and models, and of factors we can't predict, and we mislead clients by promising what we can't deliver. It may be possible to promise something weaker: e.g., regulators could make a list of key species and promise that, provided none of their abundances has decreased by more than 30%, these species will not be used as the basis for some of the more dramatic possible decisions, such as shutdown or radical redesign. (These species might still be the basis for milder decisions, and the more dramatic decisions might still arise because of other factors.) A small part of the decision-making process could then consist of testing "H_0: Species X has not declined by more than 30%," at the 50% level for each species. This has equal risk at the boundary for developers and for Species X and, assuming abundance estimates have approximately symmetric distributions, is easy to carry out and understand since no variance estimates are needed: H_0 is rejected if, and only if, the estimated decline is >30%. (We would still, presumably, need to describe the estimation method in advance.) With n species all reduced by 30%, the chance of no rejections is only $1/2^n$ for independent estimates; but it would increase with smaller reductions and more accurate estimates, and rejection alone would trigger no penalties, only less restricted decision making.

It can be argued that all this involves no change at all: confidence intervals and tests are interchangeable, the confidence interval containing all values not rejected by the test; 95% is just as arbitrary as 0.05; and one could also test subsets of the data, e.g., winter results only, and convert these to confidence intervals. These arguments are not completely true. The most useful tests may be those for goodness-of-fit (Box 1980), which are not usually invertible. (Nor is the standard test for equality of two Binomial probabilities.) A test demonstrating a change in the mean of a transformed variable, such as $\sqrt{(\text{Impact site abundance} + 0.5)}$ −

$\sqrt{}$(Control site abundance + 0.5), is hard to convert into a confidence interval for the change at the Impact site: confidence intervals force one to focus on the more meaningful parameters (Bence et al., Chapter 8). Even when they are mathematically equivalent, tests and confidence intervals do not convey equivalent messages. A P-value for "no effect" cannot be converted into a confidence interval for effect size unless the size estimate is given. Many readers will not make the conversion, assuming that the P-value summarizes the information, so test results are misleading if confidence intervals (or something similar) are needed. The arbitrariness of 95% is a minor matter: other confidences could be indicated simultaneously on plots. In short, estimates and confidence intervals give the needed information in a usable form; hypothesis tests do not.

Acknowledgments

This chapter owes a great deal to suggestions from J. Bence, M. Keough, R. Schmitt, and especially C. Osenberg. This work was supported in part by the Minerals Management Service, U.S. Department of the Interior, under MMS Agreement No. 14-35-0001-30471 (The Southern California Educational Initiative). The views and interpretation contained in this document are those of the author and should not be interpreted as necessarily representing the official policies, either express or implied, of the U.S. Government.

References

Berger, J. O., and T. Sellke. 1987. Testing a point null hypothesis: The irreconcilability of P-values and evidence (with Discussion). Journal of the American Statistical Association 82:112–139.
Box, G. E. P. 1980. Sampling and Bayes inference in scientific modeling and robustness (with discussion). Journal of the Royal Statistical Society A143:383–430.
Carney, R. S. 1987. A review of study designs for the detection of long-term environmental effects of offshore petroleum activities. Pages 651–696 in D. F. Boesch and N. N. Rabalais, editors. Long-term environmental effects of offshore oil and gas development. Elsevier, New York, New York.
Casella, G., and R. L. Berger. 1987. Reconciling Bayesian and frequentist evidence in the one-sided testing problem. Journal of the American Statistical Association 82:106–111.
Eberhardt, L. L. 1976. Quantitative ecology and impact assessment. Journal of Environmental Management 4:27–70.
Edwards, W., H. Lindman, and L. J. Savage. 1963. Bayesian statistical inference for psychological research. Psychological Review 70:193–242.
Fisher, R. A. 1937. The fiducial argument in statistical inference. Annals of Eugenics 6:391–398.
Fisher, R. A. 1956. Statistical methods and scientific inference. Oliver and Boyd, Edinburgh.
Good, I. J. 1984. An error by Neyman noticed by Dickey. Journal of Statistical Computation and Simulation 20:159–160.
Green, R. H. 1979. Sampling design and statistical methods for environmental biologists. Wiley and Sons, New York, New York.
Hill, A. B. 1965. The environment and disease: association or causation. Proceedings of the Royal Society of Medicine 58:295–300.
Lindley, D. V. 1965. Discussion of Pratt (1965). Journal of the Royal Statistical Society, B 27:192–193.
Mead, R. 1988. The design of experiments. Cambridge University Press, Cambridge, England.

Murdoch, W. W., R. C. Fay, and B. J. Mechalas. 1989. Final report of the Marine Review Committee to the California Coastal Commission. Marine Review Committee, Inc.

Nelder, J. A. 1991. Letter. Biometrics Bulletin 8:2.

Neyman, J. 1957. Inductive behavior as a basic concept of philosophy of science. Review of the International Statistical Institute 25:7–22.

Neyman, J.1962. Two breakthroughs in the theory of statistical decision-making. Review of the International Statistical Institute 30:11–27.

Neyman, J., and E. S. Pearson. 1928. On the use and interpretation of certain test criteria for purposes of statistical inference. Part I. Biometrika 20A:175–240.

Neyman, J., and E. S. Pearson. 1933. On the problem of the most efficient tests of statistical hypotheses. Philosophical Transcripts of the Royal Society of London, A 231:289–337.

Neyman, J. and E. S. Pearson. 1936. Contributions to the theory of testing statistical hypotheses. Statistical Research Memoirs I, 113–137. (Reprinted as pp. 203–239 of "Joint Statistical Papers", J. Neyman and E. S. Pearson, Cambridge University Press.).

Patton, M. L. 1991. Environmental changes in San Onofre Kelp: human impact or natural processes? Chapter 6 in "Marine environmental analysis and interpretation (report on 1990 data)", Southern California Edison Company, document 91-RD-10.

Pratt, J. W. 1965. Bayesian interpretation of standard inference statements. (With discussion). Journal of the Royal Statistical Society, B 27:169–203.

Pratt, J. W., H. R. Raiffa, and R. S. Schlaifer. 1965a. The foundations of decision under uncertainty: an elementary exposition. Journal of the American Statistical Association 59:353-375.

Pratt, J. W., H. R. Raiffa, and R. S. Schlaifer. 1965b. Introduction to statistical decision theory. McGraw-Hill, New York, New York.

Reitzel, J., M. H. S. Elwany, and J. D. Callahan. 1994. Statistical analyses of the effects of a coastal power plant cooling system on underwater irradiance. Applied Ocean Research 16:373–379.

Savage, L. J. 1954. The foundations of statistics. Wiley, New York, New York.

Synthese. 1977. Special issue on foundations of probability and statistics: articles by various authors. Synthese 36:1–269.

Tukey, J. W. 1960. Conclusions vs Decisions. Technometrics 2:423–434.

Wald, A. 1950. Statistical decision functions. Wiley, New York, New York.

Yoccoz, N. G. 1991. Use, overuse and misuse of significance tests in evolutionary biology and ecology. Bulletin of the Ecological Society of America 72:106–111.

CHAPTER 3

CRITERIA FOR SELECTING MARINE ORGANISMS IN BIOMONITORING STUDIES

Geoffrey P. Jones and Ursula L. Kaly

The need to detect human impacts that are detrimental to coastal marine organisms has never been greater. Obvious signs of severe effects can be seen in the shallow estuarine habitats adjacent to the large cities and industrial areas of the world (Root 1990), with habitats such as salt marshes, seagrass beds, and mangroves under particular threat (Hatcher et al. 1989, Thorne-Miller and Catena 1991). However, even for isolated and seemingly pristine habitats such as coral reefs, an alarming array of impacts have been catalogued (Johannes 1975, Brown 1988, Hatcher et al. 1989). The types of effects range in scale from the localized effects of coastal reconstruction or inputs of pollutants, to global effects caused by exploitation or human-induced changes in climatic conditions (Aronson 1990, Bunkley-Williams and Williams 1990, Grigg 1992). The early detection or forecasting of human impacts on the distribution and abundance of organisms will enable us to effectively manage coastal environments and ensure the survival of threatened species.

The onus is on ecologists to carry out appropriate monitoring programs to forecast or detect "unnatural" changes to the structure of coastal marine ecosystems, and determine the scale at which these effects are occurring. Such changes must be distinguished from the natural temporal variation apparent in all these systems, and the variable spatial scales at which these "natural" processes occur (Bernstein and Zalinski 1983). The ecological work must be comprehensive enough to provide unequivocal answers and give a clear recommendation to the regulatory institutions.

An important question that is always asked first in impact assessments is—what organisms should be monitored? Given an unlimited budget, with unlimited people power, time, resources, and energy, those with perseverance would monitor everything. It is only in this way that we could be sure of picking up all possible changes that society might deem unacceptable. However, this is never done. Because we live in a world of constraints, some decision must be made to reduce the number of organisms monitored. Whether it is due to political pressures or lack of money, time, expertise, or sheer laziness, we can be sure that all organisms would not be monitored with equal reliability.

This realization has led to the concept of "indicator organisms" (Soule and Kleppel 1988, Root 1990). These are organisms (a small subset of those available) that have certain characteristics which make them suitable for detecting or forecasting impacts at some level of biological organization, from biochemical to ecosystems (Phillips 1980, Hellawell 1986, Phillips and Segar 1986, Farrington et al. 1987, Wenner 1988, Adams et al. 1989, Cairns and Pratt 1989). More recently, the concept of indicators has been broadened to include monitoring biodiversity (Noss 1990).

The word "indicator" means different things to different people. To some it is an organism that, by measuring bioaccumulation under different conditions, can provide a standardized measure of pollution (and health hazard). Others measure sublethal effects of chemicals or stresses on indicators to provide an early warning of the effects of human impact on the environment, including population and community effects (Cairns and Pratt 1989, Gray 1989). To those focusing on communities in the field, the notion of an indicator may be just a "quick and easy" survey of one or a few species from which the results are extrapolated to the community as a whole (Reese 1981). Some monitor organisms that appear in unhealthy communities, such as the polychaete *Capitella*, which often appears to indicate polluted conditions in marine soft sediments (Henrikkson 1969, Grassle and Grassle 1974). Others monitor organisms that should be present in healthy communities, and which may disappear when the habitat is disturbed, e.g., butterflyfishes on coral reefs (Reese 1981, Hourigan et al. 1988). Still other workers focusing on populations and communities view it important to select a range of key organisms that if affected, will cause drastic changes to the community being examined (Underwood and Peterson 1988). Since no study yet has examined all species in all habitats potentially affected, all EIA studies are to some extent employing an indicator approach, whether this is declared or not.

The different goals of biomonitoring studies, the different levels of organization they focus on (from biochemical to ecosystem), and the different meanings embodied in the term "indicator" have led to a plethora of criteria that have been proposed to characterize the "ideal" species. Many of these criteria represent extremes of biological continua, with one group advocating the selection of organisms at one extreme, and the other groups with different goals, selecting indicators from the other (see Table 3.1). Even within the context of a single goal (e.g., standardized measures of pollution) a wide variety of criteria have been advocated. A situation has now been reached in which the criteria for the perfect indicator may outnumber the species present in some marine habitats. It seems that almost every life history, ecological or behavioral characteristic of a species has a bearing on whether or not it should be monitored. Ecologists whose life work has been devoted to a "pet" taxon, have often championed its use as a bioindicator, sometimes to the exclusion of all others.

Table 3.1.   Double-Ended Criteria for Choosing
Indicator Organisms

Stress		
tolerant ⟵————————⟶ susceptible		
Abundance		
common ⟵——— moderate ———⟶ rare		
Distribution		
cosmopolitan ⟵————————⟶ localized		
Population stability		
stable ⟵————————⟶ unstable		
Life history		
long-lived ⟵————————⟶ short-lived		
Habitat		
specialist ⟵————————⟶ generalist		
Mobility		
sessile ⟵——— sedentary ———⟶ mobile		

In this essay we (1) summarize the proposed criteria for the selection of indicators under the different goals of biomonitoring studies, and (2) challenge the idea that, regardless of the goal, we should have a rigid set of rules with a questionable theoretical basis for choosing our indicators. We focus primarily on detecting effects on population and community parameters, examining the problems with the criteria based on extreme characteristics that do not take into account the context of the habitat or community under threat. We argue against the "shopping list" approach in favor of a much smaller number of criteria stemming from the practical limitations of sampling, and the ecological and social importance of the species present.

What Should an "Indicator" Indicate?

A problem with many studies that use bioindicators is that the aims are not made clear. For example, do the investigators wish to detect the effect of a pollutant/impact only on the species being measured or extrapolate to the whole community or to other species of more interest to humans, but which are harder to monitor directly? Is the interest restricted to one level of organization (e.g., biochemical or ecosystem) or is the ultimate goal to measure effects at one level to predict effects at another? Is the goal to detect an effect present now or predict one in the future? Is the interest because of the effects on the animals themselves or the effects on humans consuming or using the animals? It is possible to identify three main goals that biomonitoring studies normally address, each with a broad set of selection rules, often based on the same biological parameters, but opposite ends of the continuum (see Table 3.1).

A Standardized Measure of Pollution

In many instances we want to rank sites with respect to some standardized pollution index that may "indicate" the relative rates of bioaccumulation of pollutants (e.g., heavy metals). There is a preference for *abundant* species that allow a population to be subsampled over a long period of time without seriously depleting it (Phillips 1980, Wenner 1988). There also seems to be a widely held view that widespread or *cosmopolitan* species make better indicators (Hellawell 1986, Farrington et al. 1987). The same species may then be used at a variety of locations, and one can have a standardized measure of pollution that operates globally and may form the basis of global standards. The Mussel Watch Program was founded on this principle. It was anticipated that the ubiquitous mussel *Mytilus edulis* that occurs on both coasts of the United States and in Europe, could provide a bioassay of pollution levels throughout the northern temperate region (Soule 1988). Many laboratory bioassays that use a single standard species are also based on this philosophy.

Choice of species in the measurement of suborganismal effects usually rely on the species being *resilient* to environmental perturbations (Phillips 1980, Hellawell 1986, Farrington et al. 1987, Wenner 1988). Farrington et al. (1987), for example, recommend the use of bivalves over fish and crustaceans, because their higher tolerance means they can provide a more sensitive test of the presence of toxins. Resilient species will survive through the period of pollution, so the accumulation of toxins and their effects on growth and condition can be measured.

Another set of criteria of significance to many authors are the *life-history* characteristics, e.g., longevity, body size, and mode of reproduction. Phillips (1980), Hellawell (1986), Wenner (1988) and others argue that long-lived species are desirable since they enable samples to be taken over several year classes. Long-lived species also suffer exposure over long periods, and therefore can provide a long-term record of environmental perturbations (e.g., ring spacing on corals, trees, bivalves, fish otoliths, etc.). In suborganismal studies large size ensures that there will be plenty of tissue to sample for prolonged periods (Phillips 1980). Wenner (1988) adds a number of other life-history characteristics that can be desirable particularly for laboratory studies. These include the ability to control reproduction, the absence of a planktonic larval phase, and the ease of assessing egg production.

Mobility is another criterion for which both extremes have been claimed to represent characteristics of good indicators (Table 3.1). Of course those working on sedentary organisms point to the obvious advantage—that these organisms must have been continuously exposed to the local environmental conditions (Gray et al. 1980, Phillips 1980, Christie 1985, Hellawell 1986, Bilyard 1987, Farrington et al. 1987). The Mussel Watch Program is obviously based on this philosophy. If you want an indicator to monitor the long-term health of a particular area then surely you will monitor an organism confined to that area.

Species that are easy to transplant, either to other field sites or into the laboratory, can be desirable for some monitoring purposes (Farrington et al. 1987). It is a common biomonitoring practice to transplant species to polluted locations where they are no longer or not normally found, to monitor sublethal effects. While this may serve some purposes, we stress that if the aim is to detect the impact of pollution of natural populations in an area, then this technique has little value. Indicator species transplanted from healthy areas may not serve as models for different species naturally recruited to the impacted area.

One of the problems with the sentinel organism approach to bioaccumulation is that associated indices lack contextual information. That is, they provide no indication as to whether the dominant local fauna perceive a degraded habitat or are accumulating pollutants in the same way (this is particularly acute when indicators do not occur naturally and must be transplanted). Further, there is little information that extends these sublethal effects to the population level (see next section). Thus, they contribute little to documentation or prediction of impacts on population density or community structure.

Reactive Monitoring for Early Warning Signals

In reactive monitoring programs, a small number of organisms are monitored prior to and during the course of a potentially impacting development. Both sublethal and lethal effects are examined (e.g., coral bleaching and mortality: Brown 1988) for the purpose of the early detection of habitat degradation. If the impact is likely to be severe, the development can be stopped before widespread damage occurs. This differs from the previous goal in that the aim is not to provide a standard, but a site-specific assessment of a pending problem. However, many of the same criteria for the selection of organisms have been applied (e.g., abundant, cosmopolitan, large, and immobile species). Modular organisms that exhibit partial mortality as a first stage of human impact may be particularly useful.

Perhaps the major difference from the last section concerns susceptibility to stress. Here the emphasis clearly shifts to the other end of the scale, to those organisms most susceptible and therefore likely to be the first to respond. Underwood and Peterson (1988) recommend choosing species that are likely to be suffering from stresses under natural conditions in the ecosystem, and that are therefore likely to succumb quickly to new stresses caused by pollutants. Species less tolerant to pollution may provide an early warning system for the rest of the community. They cite the example of bivalves that are more susceptible to smothering by sediment when they are in crowded conditions (Peterson and Black 1988). However, the degree to which natural populations are stressed will not normally be known prior to a study. An indicator may be more representative of the community, in terms of tolerance to stress, if it is chosen at random.

Reactive monitoring programs are usually short term and may not detect more subtle long-term changes to population densities and community structure. If one or two sensitive species are eliminated, the potential impacts on the community at large range from virtually nil to severe, depending on the ecological role of the species. They are valuable when employed in conjunction with longer-term monitoring programs, but cannot serve as a substitute.

It is well known that there are a number of different levels of biological response of organisms to pollutant or environmental stress, from biochemical to ecosystems (Capuzzo 1985). Adams et al. (1989) illustrated the continuum of these responses along gradients of response time and ecological relevance. The choice of suborganismal measurements to forecast population and community changes is suspect (Cairns and Pratt 1989, Gray 1989). Although this may work in theory (e.g., Calow and Sibley, 1990), Underwood and Peterson (1988) have pointed out the difficulties of making these links for marine organisms, and certainly there have been no clear demonstrations of any relationships. As Adams et al. (1989) point out, there are few integrated studies that have examined any species at a variety of levels to make extrapolations possible.

Impacts on Population Density or Community Structure

Here we address only the criteria involved in selecting species for examining effects on distribution and abundance in the field. Our assumption is that an impact assessment should be designed to detect unacceptable increases in the abundance of species, including those that are potentially damaging (pests, disease-causing organisms, etc.) and unacceptable declines in all the rest (whether useful to humans or not). We do not enter into the territory of appropriate sampling designs, which are essential for distinguishing human impacts from natural changes in abundance. There are a number of articles that should and must be consulted in which authors have dealt with sampling strategies for detecting such effects (Bernstein and Zalinski 1983, Stewart-Oaten et al. 1986, Andrew and Mapstone 1987, Clarke and Green 1988, Peterman 1990, Underwood 1991, Chapter 9, Osenberg et al., Chapter 6, Stewart-Oaten, Chapter 7, Bence et al., Chapter 8). We concern ourselves here only with sampling limitations as they might affect the type of species chosen.

Sentinel Organisms or Single Indicators

The idea that a single taxon or habitat may be used to indicate broader impacts on the community is widespread in the marine literature. It seems everyone has been prepared to promote a particular type of organism as the ideal choice (e.g., fish: Cairns and Dickson 1980, Reese 1981, Hourigan et al. 1988, Stephens et al. 1988; bivalves (mussels): Goldberg 1980, Phillips 1980, Martin et al. 1984; polychaetes: Henrikkson 1969, Grassle and Grassle 1974, Maurer et al. 1981, Raman and Ganapati 1983; crustaceans: Wenner 1988; infauna: Bilyard 1987; rocky-reef communities: Christie 1985; sandy beaches: Wenner 1988). The comparative approach can be very useful in identifying the types of organisms that repeatedly respond to particular classes of human impact. For example, coral death appears to occur on coral reefs in response to a wide range of disturbances (Kaly and Jones 1988). Since the very life of the reef is affected, it would be absurd to leave these organisms out of any impact assessment in these habitats.

The question is—should we choose a species just because someone else chose it before? Physiologists and biochemists involved in monitoring sublethal effects have been accused of being shackled to "millstone" species (Underwood and Peterson 1988). That is "one which is chosen for study without consideration of alternatives because of the great weight and past history of invested time and effort that has gone into the development of techniques using it." Are there organisms that have been shown to repeatedly respond in a predictable way to pollution or other disturbances, to the extent that they warrant monitoring all the time?

Unfortunately, there is no general theory of pollution or human impact to aid our choice of indicator species, although there have been three classic attempts to provide one:

Capitella.   Perhaps no other single species in the diffuse history of biological indication has received as much attention as the polychaete worm *Capitella capitata*. Although other species of sedentary, infaunal worms have been considered useful as indicators of environmental stress (Grassle and Grassle 1974, Remani et al. 1983, Blackstock et al. 1986), it was *Capitella* whose apparent *preference* for organically enriched, oxygen starved, polluted habitats inspired a belief that a single species could be used almost universally to herald environmental degradation (see Gilet 1960, Reish and Barnard 1960, Wass 1967, Pearson and Rosenberg 1978, Reish 1979).

The intuitive reasons for selecting polychaetes, particularly those with life histories like *Capitella,* are clear. Although costly to sample and often difficult to identify taxonomically (Bilyard 1987), polychaetes include species that are small but not microscopic, abundant (Grassle and Grassle 1974, Gray and Pearson 1982), sedentary (and therefore forced to experience an impact: Bilyard 1987) and that respond in differing ways to stress with some populations being enhanced and others declining (Remani et al. 1983, Dauer 1984). Polychaetes are also often considered important components of food webs and therefore may accumulate toxins that may be passed on to species that humans directly utilize (Bilyard 1987). In addition to all of these perceived requirements some species, notably *Capitella capitata*, are widely distributed in many parts of the world (Fauchald 1974).

The evidence for using *Capitella* and other polychaetes as indicators has tended to be circumstantial, although a few workers have managed to be more convincing (e.g., the experimental evidence of Tsutsumi 1990). Gray (1989) and others have argued that *Capitella* is the endpoint in the pollution gradient rather than the first indication of change, so it has little predictive value. Many workers have fallen into the impact assessor's trap of comparing only a single control site with a single or several polluted sites (e.g., Remani et al. 1983, Ansari et al. 1986), hence obtaining very little information on the characteristics of "natural" populations (e.g., Wass 1967, Sarala Devi and Venugopal 1989). Even worse, the polluted site often differs markedly from a control site for reasons other than pollution (e.g., Tsutsumi 1990). On occasion, the assumption that a species is a valuable indicator has gone too far. Wass (1967) accepted *Capitella capitata*'s role as an indicator to such an extent that when other data showed low abundances of the worm in an area previously deemed polluted, he discounted it on the grounds of inadequate sampling! When Reish and Winter (1954) concluded that levels of pollution in a bay were low, Wass examined their species list and declared that species present were indicators of enrichment and therefore that the bay *must* be polluted. In cases such as these, it is clear that the concept of indicators is being abused.

Nematode/Copepod Ratios. It has been suggested that nematodes and copepods exhibit differential survivorship in polluted and unpolluted areas, with nematodes favoring polluted areas. Raffaelli and Mason (1981) used the nematode/copepod ratio as an indicator of pollution. Amjad and Gray (1983) found a strong correlation between this ratio and the known degree of pollution at various sampling stations in Oslofjord, Norway.

The use of nematode/copepod ratios as indicators has, however, been subject to considerable criticism (Coull et al. 1981, Platt et al. 1984). The primary concern appears to be that the relationship between the nematode/copepod ratios and organic enrichment is based on correlations established from points along a transect. Along these transects, other factors vary, such as sediment grain size, which may also represent causal factors (Platt et al. 1984). Hence, while these meiofaunal organisms may be important in pollution studies, preconceived "rules"· may lead to erroneous conclusions. The continuing debate (see Gray 1985) would no doubt benefit from experimental procedures involving before/after, treatment/control comparisons. The use of ratios should not replace abundance data on the two categories. It needs to be clear whether it is changes in the abundance of nematodes, copepods, or both that are occurring.

Butterflyfishes. Butterflyfishes (family Chaetodontidae) have been recommended as indicators of the health of coral reefs (Reese 1981, Hourigan et al. 1988, White 1988, Nash 1989). The claim is based on the close relationship between coral cover and butterflyfish diversity and the abundance of obligate coral feeders. Other workers have disputed this, showing that the relationship between representatives of this fish family and habitat structure is not always close (Roberts and Ormond 1987, Roberts et al. 1988).

Our own work, which examines changes in the structure of assemblages of three fish families (Acanthuridae, Chaetodontidae, and Labridae) along a gradient of habitat quality in Tuvalu, indicates that all three contain some member species that increase in relation to coral cover and others that decrease (Figure 3.1). For example, the butterflyfish *Chaetodon trifasciatus* and *C. trifascialis*, the wrasses *Gomphosus varius* and *Labrichthys unilineatus*, and the surgeonfish *Acanthurus nigricans* and *Zebrasoma scopas* increased along a gradient toward higher coral cover at Funafuti lagoon. Other species such as the butterflyfish *Chaetodon auriga* and *C. citrinellus*, the wrasses *Halichoeres trimaculatus* and *Stethojulis bandanensis*, and the surgeonfish *Ctenochaetus striatus* characterize the more degraded conditions near the main settlement. We suspect the same extremes would apply to other taxa and it may be more appropriate to select habitat responders from a range of fish families, rather than just selecting butterflyfish as a single indicator.

We accept that in many cases habitat characteristics may be a poor indicator of fish abundance and vice versa. However, this controversy comes close to totally missing the goal of a rigorous impact assessment. If the aim is to detect changes in coral cover then it is far easier to measure changes in coral cover directly, rather than to monitor the abundance of a mobile organism like a fish. However, it can be very important to detect impacts on fish populations, regardless of how closely they respond to changes in their habitat.

We strongly disagree with limiting the focus of impact assessments to particular species, taxa, or habitats (see also Landres et al. 1988). There are few human impacts that would ever be confined to one of these categories, and we could expect different responses in them all. Obviously you cannot monitor everything. But depending on the scale of the disturbance, different representatives of different habitats, and taxa within habitats will need to be chosen.

Theoretical Considerations and the Selection Criteria for Organisms Studied in Ecological Impact Assessment

Abundance. One can support an argument for selecting as indicators the most abundant species, those that are moderately common and the least abundant species in an assemblage (Table 3.1). Many environmental impact studies focus solely on detecting effects on abundant species, presumably because it is difficult to get reliable quantitative estimates for those of lesser abundance. All else being equal, changes to abundant species are likely to be easier to detect. Also, changes in the abundance of abundant species may have more important consequences for the structure of communities or ecosystems than changes to rare species.

However, as Gray (1985) points out, in communities dominated by one or a few species, this approach can lead a researcher to overlook species exhibiting more "average" density conditions. Gray and coworkers suggest that only by the use of "moderately" common species could an objective assessment be made for the community at large (Gray and Pearson 1982, Gray 1989). However, choosing moderately common species is no more objective than choosing the most abundant. In reality, these workers appear to choose from species with intermediate abundance those that exhibit clear trends away from disturbed sites. Far from being objective, this approach can give you an exceedingly biased view of the overall impact of a point source of pollution. If objectivity is the goal the correct way to choose species is randomly.

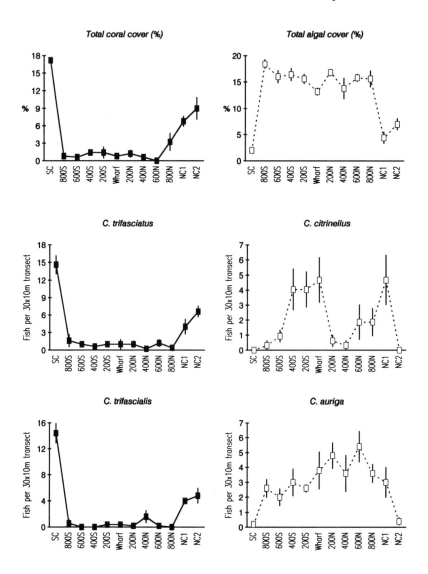

Figure 3.1. Fish-habitat associations at Funafuti Lagoon, Tuvalu. Changes in total coral and algal cover, abundance of four species of Chaetodontidae (*Chaetodon trifasciatus*, *C. trifascialis*, *C. citrinellus*, and *C. auriga*), four species of Labridae (*Gomphosus varius*, *Labrichthys unilineatus*, *Halichoeres trimaculatus,* and *Stethojulis bandanensis*), and three species of Acanthuridae (*Zebrasoma scopas*, *Acanthurus nigricans,* and *Ctenochaetus striatus*), with increasing distance to the north and south of the village wharf. Sites are 200 m apart, except South Control (SC: 2 km south of the wharf), North Control 1 (NC1: 2 km north of the wharf), and North Control 2 (NC2: 3 km north of the wharf).

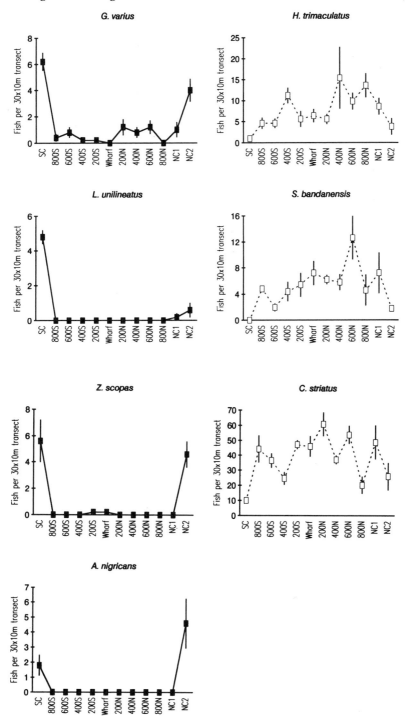

There is a clear case for examining "rare" species, which may become extinct as a consequence of human disturbance. One is intrinsically more worried about reducing a population from 10 to 0, than reducing one from 100,000 to 10,000, even though in numerical terms this is trivial. The problem with sampling in the marine environment is distinguishing low density species from those which are truly endangered. There are not too many documented cases of marine species close to extinction, except for large marine mammals. There are many widely dispersed, low density species for which there is little danger of human-induced extinction, except on a local scale.

Clearly abundance *per se* may or may not mean individuals of a particular species are more or less sensitive to human disturbances. While rare species may be subject to local extinction, abundant species may already be subject to a variety of natural stresses. There is no reason why moderately common species should be more representative in this regard. Since there is no *a priori* way of knowing what situation applies, we feel there is little to be gained from using this as a criterion in selecting organisms. For quantitative field sampling, it is likely that abundant species will be the primary focus, simply by default. However, qualitative "faunal list" type surveys will no doubt emphasize the inclusion of rare species.

Biogeographic Distribution.   In the same way that cosmopolitan species may be useful in providing a standardized measure of pollution, they may aid in the development of simple protocols for the effects of different types of impact on community structure. Such standards may allow impact assessments to be carried out by people with minimal training and may be particularly useful in developing countries. However, since such comparisons do not substitute for local control sites adjacent to impacted areas, we do not really see a great advantage of cosmopolitan species. In fact it is possible to create a perfectly reasonable argument to support the opposite view. Exceedingly localized species may be far more susceptible to negative human disturbances, since they would be far more likely to become globally extinct.

Population Stability. It would be an advantage for statistical reasons if indicator species normally exhibit stable populations, since for these it should be easier to distinguish effects of human disturbance from natural patterns (Hellawell 1986). Farrington et al. (1987) used this reasoning to promote the use of bivalves as indicators, and Stephens et al. (1988) cautioned against the use of temperate reef fish populations on the same grounds. However, there are insufficient data to make gross generalizations about population stability of species in different taxa. Knowledge of the status of any given population prior to a polluting event is almost always lacking. Stability itself is a hotly debated subject and there are few species for which all the required information is available, including the longevity of the indicators (Connell and Sousa 1983). It is not only a measure of how constant (or cyclic) a population is, but how resistant it is to

disturbances, and how likely it is to return to its former level following a disturbance (Dayton et al. 1984). These are variables of particular interest to us in
human impact studies. We should not bias our result toward selecting stable
species that may be more resistant and resilient to human disturbances. Rather
we should be focusing on representative species. There is no *a priori* reason why
constant populations should be more susceptible to disturbance than variable
ones.

Life-History Characteristics: Longevity and Body Size.   Again, it is
possible to erect an argument for the selection of organisms that are either long-
lived and reach large sizes, or small and short-lived. Because long-lived species
may be the slowest to recover from an impact it may be more important to detect
a response. Changes to populations of large, high-profile organisms are usually
obvious and easy to detect (Hourigan et al. 1988; Stephens et al. 1988). Large
organisms are usually readily identified, so you can avoid niggling taxonomic
problems (Hellawell 1986). However, limiting sampling of large, long-lived
organisms is an extremely dangerous proposition. These may be the last to disappear from a disturbed ecosystem, simply because of their longevity or size. Old
mangrove trees, for example, can be all that remains of natural stands of mangroves grazed by cattle in New Zealand (personal observation). Large animals
frequently live the longest and are positioned at the end of food chains. They may
be the last to respond to any breakdown in the structure of a community.

   On the other hand very short-lived, opportunistic species are known to
respond very quickly to environmental perturbations (Grassle and Grassle 1974)
and in doing so may provide a good forecasting method. A negative response of
a short-lived species may led to local extinction very quickly. While some large
organisms may be very important ecologically, it must be remembered that most
of the productivity of shallow marine ecosystems comes from microalgae.
Hence, although working on large organisms may be easier, it may also be
irrelevant.

Habitat Specialists versus Generalists.   The recommendation that butterflyfish are good indicators of the health of coral reefs is based on the assumption that they are habitat specialists and will respond rapidly to changes in their
habitat. The choice of species that are likely to be sensitive to changes in habitat
structure, such as increasing sedimentation and turbidity, or loss of benthic
organisms has been discussed elsewhere. Bilyard (1987), for example, regards
benthic infauna as being particularly susceptible to habitat destruction and organic enrichment. Clearly, some organisms do respond to particular changes in habitat structure and others do not. No single taxon is necessarily sensitive to all types
of anthropogenically induced changes to habitats, so the most suitable indicator
will depend on the type of disturbance expected (Kremen 1992). Also, to choose
only habitat-sensitive organisms assumes that the effect of the impact will occur
only via habitat modification. However, organisms may respond directly to
human disturbance, rather than indirectly through habitat destruction. While

biogenic components of habitat will always be a prime target for impact assessments, particularly those appropriate to a large number of species, we can see no argument for limiting study to these organisms and their associated species. An argument in favor of habitat generalists would be equally valid, particularly for impacts impinging on a wide range of habitat types.

Mobility.   One might favor sedentary organisms simply because they are easier to count and often provide habitat (shelter and food) for mobile organisms. However, since many of the ecologically and economically important species are mobile, this decision would not satisfy everyone. Monitoring extremely mobile species can indicate how far a point source of pollution is likely to spread through the ecosystem. The mobility and sensory perception of fish may allow them to avoid environmental perturbations, thus they can show rapid response to environmental change (Hourigan et al. 1988; Stephens et al. 1988; Nash 1989). Such a response may forecast impacts that will later affect sedentary organisms. Since one can erect an argument to favor immobile and mobile organisms, this criterion is of little practical use.

Ecological Importance.   Ecologists generally believe some species are more important than others in ecological communities. This can include large species ("ecological dominants") that create or modify living habitat for a variety of smaller ones (e.g., kelp forests, corals, mangroves) or species capable of destroying these large species (e.g., sea urchins, starfish). The term "keystone" has been coined for those species that, regardless of the abundance, exert a major impact on the structure of the community of which they are a part (Dayton 1971, Paine 1974). Species of major trophic importance are those that represent a major pathway by which energy is transferred toward top carnivores. Underwood and Peterson (1988) argue that indicator species should be chosen for their direct importance, either as "keystone" species in organizing the community, or because they form an important link in food webs leading to economically important consumers.

We believe that this is a very important concept, and for a number of shallow marine habitats, have listed the types of organisms that should obviously be considered (Table 3.2). A major stumbling block is that for any particular assemblage, the choice must be based on prior knowledge of the system. At present, the selection is based upon extrapolating from basic ecological studies on these systems at some locations. However, if the practice of ecology in the 1980s has taught us anything, it is how much ecological processes can change from place to place and time to time (Underwood and Fairweather 1986). While some things will always be obvious (e.g., corals on coral reefs, kelp in temperate kelp forests, mangroves in mangrove forests, seagrass in seagrass beds), these recommendations should not be applied uncritically. For example, while sea urchins may actively remove kelp forests at some geographic locations, there are large expanses of temperate coastline where sea urchins remain crevice bound and have little effect on community structure (Jones and Andrew 1990).

Table 3.2.   Key Macroorganisms of Shallow Benthic Marine Habitats

Habitat	Organisms
Temperate subtidal kelp forests	Kelp, grazing urchins, microcrustacea, fish
Temperate and tropical intertidal rock platforms	Microalgae, grazing molluscs, sessile invertebrates
Tropical subtidal reefs	Corals, turfing, algae, fish, grazers
Tropical and temperate subtidal soft-sediments	Bivalves, polychaetes
Tropical and temperate sandy beaches	Bivalves, crustaceans
Tropical and temperate seagrass beds	Seagrass, microcrustacea, fish
Tropical mangrove communities	Mangroves, crabs, fish

Economic and Social Importance.   Almost every coastal marine habitat harbors organisms that are of particular significance to the local people. Food, pests, parasites, diseases, endangered species, sacred species, etc. all come into this category. Organisms such as fish, molluscs and crustaceans are commonly consumed, and may represent economically important, recreational or subsistence fisheries (Farrington et al. 1987). Populations of these species may be reduced, or subject to a variety of sublethal effects that affect the quantity, quality or health of the flesh. Populations may be reduced directly, or indirectly as a result of changes to food chains, or the abundance of parasites or disease. For these reasons, such organisms must universally be examined as biological indicators (Hellawell 1986, Underwood and Peterson 1988).

Conclusions

While it is desirable to examine as many species as possible in an EIA, one cannot examine everything before regulatory decisions are made (Soule 1988). Some sort of selection process is always carried out (whether consciously or not) and is necessary (whether you want to or not). This process must be recognized, established, and declared. Considerable lip service has been given to the criteria for choosing marine indicators, to which this commentary will no doubt be added. No single marine taxon and no single criterion for choosing taxa appears to be satisfactory for marine organisms. Similar conclusions have been reached for terrestrial organisms (Landres et al. 1988, Kremen 1992). In contrast, if you took all the advice on the types of species that should be chosen, you would have to examine virtually all the species present in some communities. Such a diverse array of criteria amounts effectively to no direction at all.

We caution against using the double-ended criteria for which an argument for either extreme can be made. These do not take into account the nature of the impact, the particular marine communities in question, or the human uses of the resources in question. If the purpose is to monitor representative species then most of the criteria can be quite safely ignored.

However, we do need to ultimately decide on some subset of species to monitor and detect detrimental changes to our coastal marine systems. In practical terms, how do we do it? We believe the appropriate choice will be based on the context of the particular community and impact in question, rather than an *a priori* "shopping list." Clearly pilot surveys are required to determine the habitats likely to be affected, the species present, and an appropriate sampling design (Underwood and Kennelly 1990). *Abundance* will always be a consideration, given the time and resources available. A quantitative pilot survey will indicate for which species reliable estimates of abundance can be obtained and an effect of a predetermined magnitude can be detected, given the sampling effort at your disposal. Factors such as *ecological* and *sociological* importance must be taken into consideration, which will further extend the list of indicator species. We further stress the need for considering a variety of taxa and habitats in the impacted area, and species from different trophic positions. None of these choices can be preselected from the list of criteria, and many will become apparent during the pilot survey in each habitat type.

Other previously listed criteria, such as biogeographic distribution, population stability, susceptibility to stress, susceptibility to habitat modification, body size, mobility, longevity, etc. can be dispensed with. This is for the simple reason that, in theory, both extremes can be equally important (Table 3.1). In practice, much of this information will not be available. Given a sufficient budget one could justify examining species across the range of any one characteristic, but not confining sampling to those representing one extreme.

One danger of having a rigid set of criteria is that they will be built into the regulatory framework (Hedgpeth 1973). In many instances, it may be impossible to find species that fulfill the legal requirements. Debatable theoretical concepts, such as "stability" and "keystone species" should not be legislated into rigid procedures.

Finally, we get back to the beginning—the question! The literature on indicators is rife with the purposeless accumulation of data. There is nothing intrinsically relevant about examining a variety of biochemical, organismal, population, and community indices. If the question concerns effects on the distribution and abundance, then suborganismal measures are only of significance if they can forecast future population changes. Until it is established that this is possible, recommendations should be based upon detailed ecological studies, in which biological indicators will play an integral part.

Acknowledgments

We thank the New Zealand, Australian, and Tuvalu governments for providing the funding and opportunity for us to explore the role of bioindicators in impact assessments we have done in Tuvalu. J. Caley, K. Clements, R. Cole, B. Jackson, V. Staines, C. Syms, and K. Tricklebank all assisted with field work, the results of which led to this discussion. Special thanks to R. Schmitt and C. Osenberg whose comments improved the structure of this essay.

References

Adams, S. M., K. L. Shepard, M. S. J. Greeley, B. D. Jimenez, M. G. Ryon, L. R. Shugart, and J. F. McCarthy. 1989. The use of bioindicators for assessing the effects of pollutant stress on fish. Marine Environmental Research 28:459–464.

Amjad, S., and J. S. Gray. 1983. Use of the nematode-copepod ratio as an index of organic pollution. Marine Pollution Bulletin 14:178–181.

Andrew, N. L., and B. D. Mapstone. 1987. Sampling and the description of spatial pattern in marine ecology. Oceanography and Marine Biology Annual Review 25:39–90.

Ansari, Z. A., B. S. Ingole, and A. H. Parulekar. 1986. Effect of high organic enrichment of benthic polychaete population in an estuary. Marine Pollution Bulletin 17:361–365.

Aronson, R. 1990. Rise and fall of life at sea. New Scientist:24–27.

Bernstein, B. B., and J. Zalinski. 1983. An optimum sampling design and power tests for environmental biologists. Journal of Environmental Management 16:35–43.

Bilyard, G. R. 1987. The value of benthic infauna in marine pollution monitoring studies. Marine Pollution Bulletin 18:581–585.

Blackstock, J., P. J. Johannessen, and T. H. Pearson. 1986. Use of a sensitive indicator species in the assessment of biological effects of sewage disposal in Fjords near Bergen, Norway. Marine Biology 93:315–322.

Brown, B. E. 1988. Assessing environmental impacts on coral reefs. Proceedings of the 6th International Coral Reefs Symposium 1:71–80.

Bunkley-Williams, L., and E. H. J. Williams. 1990. Global assault on coral reefs. Natural History 4:46–54.

Cairns, J., and K. L. Dickson. 1980. The ABC's of biological monitoring. Pages 1–33 in C. H. Hocutt and J. R. Stauffer, editors. Biological monitoring of fish. Lexington Books, Lexington, Massachusetts.

Cairns, J., and J. R. Pratt. 1989. The scientific basis of bioassays. Hydrobiologia 188/189:5–20.

Calow, P., and R. M. Sibley. 1990. A physiological basis of population processes: ecotoxicological implications. Functional Ecology 4:283–288.

Capuzzo, J. M. 1985. Biological effects of petroleum hydrocarbons on marine organisms: integration of experimental results and predictions of impacts. Marine Environmental Research 17:272–276.

Christie, H. 1985. Ecological monitoring strategy with special reference to a rocky subtidal program. Marine Pollution Bulletin 16:232–235.

Clarke, K. R., and R. H. Green. 1988. Statistical design and analysis for a 'biological effects' study. Marine Ecology Progress Series 46:213–226.

Connell, J. H., and W. P. Sousa. 1983. On the evidence needed to judge ecological stability or persistence. American Naturalist 121:789–824.

Coull, B. C., G. R. F. Hicks, and J. B. J. Wells. 1981. Nematode/copepod ratios for monitoring pollution: a rebuttal. Marine Pollution Bulletin 12:378–381.

Dauer, D. M. 1984. The use of polychaete feeding guilds as biological variables. Marine Pollution Bulletin 15:301–305.

Dayton, P. K. 1971. Competition, disturbance, and community organization: the provision and subsequent utilization of space in a rocky intertidal community. Ecological Monographs 41:351–389.

Dayton, P. K., V. Currie, T. Gerrodette, B. D. Keller, R. Rosenthal, and D. ven Tresca. 1984. Patch dynamics and stability of some California kelp communities. Ecological Monographs 54:253–289.

Farrington, J. W., A. C. Davis, B. W. Tripp, D. K. Phelps, and W. B. Galloway. 1987. "Mussel watch" - measurements of chemical pollutants in bivalves as one indicator of coastal environmental quality. Pages 125–139 in T. P. Boyle, editor. New approaches to monitoring aquatic ecosystems. American Society for Testing and Materials, Philadelphia, Pennsylvania.

Fauchald, K. 1977. The polychaete worms: definitions and keys to the orders, families and genera. Natural History Museum of Los Angeles County, Science Series **28**:1–190.

Gilet, R. 1960. Water pollution in Mareilles and its relation with flora and fauna, in waste disposal in the marine environment. Pages 39–56 *in* E. A. Pearson, editor. Proceedings of the first international conference. Pergamon Press, New York, New York.

Goldberg, E. D. 1980. The international mussel watch (report of a workshop sponsored by the Environmental Studies Board Commission on Natural Resources, National Research Council). National Academy of Sciences, Washington. 148 pp.

Grassle, J. F., and J. P. Grassle. 1974. Opportunistic life histories and genetic systems in marine benthic polychaetes. Journal of Marine Research **32**:253–284.

Gray, J. S. 1985. Ecological theory and marine pollution monitoring. Marine Pollution Bulletin **16**:224–227.

Gray, J. S., 1989. Do bioassays adequately predict ecological effects of pollutants? Hydrobiologia **188-189**:397–402.

Gray, J. S., and T. H. Pearson. 1982. Objective selection of sensitive species indicative of pollution-induced change in benthic communities. Marine Ecology Progress Series **9**:111–119.

Gray, J. S., D. Boesch, C. Heip, A. M. Jones, J. Lassig, R. Vandenhorst, and D. Wolfe. 1980. The role of ecology in marine pollution monitoring (ecology panel report). Rapports et Procès-Verbaux des rèunions Conseil International pour l'exploration de la Mer **179**:237–252.

Grigg, R. W. 1992. Coral reef environmental science: truth versus the Cassandra syndrome. Coral Reefs **11**:183–186.

Hatcher, B. G., R. E. Johannes, and A. I. Robertson. 1989. Review of research relevant to the conservation of shallow tropical marine ecosystems. Oceanography and Marine Biology **27**:337–414.

Hedgpeth, J. W. 1973. The impact of impact studies. Helgoländer Wissenshaft Meersuntersuchungen **24**:436–445.

Hellawell, J. M. 1986. Biological indicators of freshwater pollution and environmental management. Elsevier, New York, New York.

Henrikkson, R. 1969. Influence of pollution on the bottom fauna of the Sound (Oresund). Oikos **20**:507–523.

Hourigan, T. F., T. C. Tricas, and E. S. Reese. 1988. Coral reef fishes as indicators of environmental stress in coral reefs. Pages 107–135 *in* D. Soule and G. S. Kleppel, editors. Marine organisms as indicators. Springer-Verlag, New York, New York.

Johannes, R. E. 1975. Pollution and degradation of coral reef communities. Pages 13–62 *in* E. J. Wood and R. E. Johannes, editors. Tropical marine pollution. Elsevier, Amsterdam, The Netherlands.

Jones, G. P., and N. L. Andrew. 1990. Herbivory and patch dynamics on rocky reefs in temperate Australasia. Australian Journal of Ecology **15**:505–520.

Kaly, U. L., and G. P. Jones. 1988. The construction of boat channels across coral reefs: a preliminary assessment of biological impact and review of related literature. An environmental assessment of the impact of reef channels in the South Pacific. New Zealand Ministry of External Relations and Trade, Report **1**:1–26.

Kremen, C. 1992. Assessing the indicator properties of species assemblages for natural areas monitoring. Ecological Applications **2**:203–217.

Landres, P. B., J. Verner, and J. W. Thomas. 1988. Ecological uses of vertebrate indicator species: a critique. Conservation Biology **2**:316–329.

Martin, M. G., J. Ichikawa, M. Goetzl, M. de los Reyes, and M. D. Stephenson. 1984. Relationship between physiological stress and trace toxic substances in the bay mussel, *Mytilus edulis*, from San Francisco Bay, California. Marine Environmental Research **11**:91–110.

Maurer, D., W. Leathem, and C. Menzie. 1981. The impact of drilling fluid and well cuttings on polychaete feeding guilds from the U.S. Northeastern Continental Shelf. Marine Pollution Bulletin **12**:342–347.

Nash, S. V. 1989. Reef diversity index survey method for nonspecialists. Tropical Coastal Area Management **4**:14–17.

Noss, R. F. 1990. Indicators for monitoring biodiversity: a hierarchical approach. Conservation Biology 44:355–364.

Paine, R. T. 1974. Intertidal community structure: experimental studies on the relationship between a dominant competitor and its principal predator. Oecologia 15:93–120.

Pearson, T. H., and R. Rosenberg. 1978. Macrobenthic succession in relation to organic enrichment and pollution of the marine environment. Oceanography and Marine Biology Annual Review 16:229–311.

Peterman, R. M. 1990. Statistical power analysis can improve fisheries research and management. Canadian Journal of Fisheries and Aquatic Sciences 47:2–15.

Peterson, C. H., and R. Black. 1988. Density-dependent mortality caused by physical stress interacting with biotic history. American Naturalist 131:257–270.

Phillips, D. J. H. 1980. Quantitative aquatic biological indicators. Applied Science Publishers, London, England.

Phillips, D. J. H., and D. A. Segar. 1986. Use of bio-indicators in monitoring conservative contaminants: programme design imperatives. Marine Pollution Bulletin 17:10–17.

Platt, H. M., K. M. Shaw, and P. J. D. Lambshead. 1984. Nematode species abundance patterns and their use in the detection of environmental perturbations. Hydrobiologia 118:59–66.

Raffaelli, D. G., and C. F. Mason. 1981. Pollution monitoring with meiofauna, using the ratio of nematodes to copepods. Marine Pollution Bulletin 12:158–163.

Raman, A. V., and P. N. Ganapati. 1983. Pollution effects on ecobiology of benthic polychaetes in Visakhapathnam (Bay of Bengal). Marine Pollution Bulletin 14:46–52.

Reese, E. S. 1981. Predation on corals by fishes of the family Chaetodonidae: implications for conservation and management of reef ecosystems. Bulletin of Marine Science 31:594-604.

Reish, D. J. 1979. Bristle worms (Annelida: Polychaeta). Pages 77–125 in C. W. Hart and S. H. Fuller, editors. Pollution ecology of estuarine invertebrates. Academic Press, New York, New York.

Reish, D. J., and J. L. Barnard. 1960. Field toxicity tests in marine waters utilizing the polychaetous annelid Capitella capitata (Fabricius). Pacific Naturalist 1:1–8.

Reish, D. J., and H. A. Winter. 1954. The ecology of Alamitos Bay, California, with special reference to pollution. California Fish and Game 40:105–121.

Remani, K. N., K. Sarala Devi, P. Venugopal, and R. V. Unnithan. 1983. Indicator organisms of pollution in Cochin backwaters. Mahasagar-Bulletin of the National Institute of Oceanography 16:199–207.

Roberts, C. M., and R. F. G. Ormond. 1987. Habitat complexity and coral reef fish diversity and abundance on Red Sea fringing reefs. Marine Ecology Progress Series 41:1–8.

Roberts, C. M., R. F. G. Ormond, and A. R. D. Shepherd. 1988. The usefulness of butterflyfishes as environmental indicators on coral reefs. Proceedings of the 6th International Coral Reefs Symposium 2:331–336.

Root, M. 1990. Biological monitors of pollution. BioScience 40:83–86.

Sarala Devi, K., and P. Venugopal. 1989. Benthos of Cochin backwaters receiving industrial effluents. Indian Journal of Marine Science 18:165–169.

Soule, D. F. 1988. Marine organisms as indicators: reality or wishful thinking? Pages 1–11 in D. F. Soule and G. S. Kleppel, editors. Marine organisms as indicators. Springer-Verlag, New York, New York.

Soule, D. F., and G. S. Kleppel. 1988. Marine organisms as indicators. Springer-Verlag, New York, New York.

Stephens, J. S. J., J. E. Hose, and M. S. Love. 1988. Fish assemblages as indicators of environmental change in nearshore environments. Pages 91–105 in D. F. Soule and G. S. Kleppel, editors. Marine organisms as indicators. Springer-Verlag, New York, New York.

Stewart-Oaten, A., W. W. Murdoch, and K. R. Parker. 1986. Environmental impact assessment: "psuedoreplication" in time? Ecology 67:929–940.

Thorne-Miller, B., and J. Catena. 1991. The living ocean: understanding and protecting marine biodiversity. Island Press, Washington, D.C.

Tsutsumi, H., S. Fukunaga, N. Fujita, and M. Sumida. 1990. Relationship between growth of *Capitella* sp. and organic enrichment of the sediment. Marine Ecology Progress Series **63**:157–162.

Underwood, A. J. 1991. Beyond BACI: experimental designs for detecting human environmental impacts on temporal variations in natural populations. Australian Journal of Marine and Freshwater Research **42**:569–587.

Underwood, A. J., and P. G. Fairweather. 1986. Intertidal communities: do they have different ecologies or different ecologists? Proceedings of the Ecological Society of Australia **14**:7–16.

Underwood, A. J., and S. J. Kennelly. 1990. Pilot studies for designs of human disturbance of intertidal habitats in New South Wales. Australian Journal of Marine and Freshwater Research **41**:165–173.

Underwood, A. J., and C. H. Peterson. 1988. Towards an ecological framework for investigating pollution. Marine Ecology Progress Series **46**:227–234.

Wass, M. L. 1967. Biological and physiological basis of indicator organisms and communities. Section II - Indicators of pollution. Pages 271–283 *in* T. A. Olson and F. J. Burgess, editors. Pollution and marine ecology. Interscience, New York, New York.

Wenner, A. M. 1988. Crustaceans and other invertebrates as indicators of beach pollution. Pages 199–229 *in* D. F. Soule and G. S. Kleppel, editors. Marine organisms as indicators. Springer-Verlag, New York, New York.

White, A. T. 1988. The effect of community-managed marine reserves in the Philippines on their associated coral reef fish populations. Asian Fisheries Science **2**:27–41.

CHAPTER 4

IMPACTS ON SOFT-SEDIMENT MACROFAUNA

The Effects of Spatial Variation on Temporal Trends[1]

**Simon F. Thrush, Richard D. Pridmore,
and Judi E. Hewitt**

Many aspects of environmental degradation can only be detected and accurately assessed when there are sufficient data to reveal long-term trends compared to short-term fluctuations. Without a long-term perspective natural fluctuations may be mistakenly attributed to human impacts. Impact studies should ideally be able to document density changes and demonstrate causality (Underwood and Peterson 1988), and treating a proposed impact as a large experiment (e.g., using BACI-type designs) will be appropriate in some situations (Hilborn and Walters 1981; Underwood 1991). However, an experimental approach is often impractical when considering diffuse and complex impacts that operate over large spatial and temporal scales (e.g., urban runoff, habitat disturbance by fishing), especially where preimpact assessment is not possible (Livingston 1987, Parker 1989, Tilman 1989). In these situations the use of time-series data is particularly relevant to identify both trends and changes in cyclic data together with broad correlative relationships. Well-developed statistical tests exist for time-series analysis (Chatfield 1980) and are frequently used to assess trends in pollution monitoring (e.g., special issue of Water Resources Bulletin, Vol. 21(4) 1985) and to provide ecological insights (Jassby and Powell 1990). To obtain the most from time-series analyses, the data used should be as unbiased and precise as possible.

---

[1]This chapter was previously published in a modified form in Ecological Applications 4:31–41. © 1994 by the Ecological Society of America.

One factor that can markedly influence time-series analyses in marine macrofaunal studies is spatial variation caused by aggregated distribution patterns. It has long been known that the scale of sampling relative to the distributional pattern of the organisms to be sampled can influence both the precision and the interpretation of the data (Greig-Smith 1983). However, even if sampling is conducted on an appropriate scale, poor estimates of abundance can be obtained if insufficient samples have been collected to account for patchiness. To obtain accurate density estimates, some knowledge of the spatial arrangement of the organisms to be sampled must be acquired. Identifying spatial patterns in marine soft-sediment habitats is difficult as the density of many species is not apparent from the sediment surface. Techniques are available, however, that enable patterns to be described from sediment samples taken from known locations (e.g., Legendre and Fortin 1989, Legendre 1993). Where patterns have been quantitatively assessed in soft-sediment environments, homogeneous density patches ranging from 0.01 to 100 m radius have been described (Eckman 1979, McArdle and Blackwell 1989, Thrush et al. 1989, Thrush 1991, Trueblood 1991).

The potential for spatial variation to confound assessments is often an important consideration in short-term studies where the precision of within-site sampling is emphasized (see Osenberg and Schmitt, Chapter 1). In contrast, an examination of the methods used in a variety of long-term macrofaunal studies (Table 4.1) reveals that typically less than five samples are collected from within a site. Sites are usually situated kilometers apart and/or in different habitats, indicating that in most cases each site was considered to be representative of a relatively large area. In many of the studies, less emphasis has been placed on the number of within-site samples, possibly under the belief that increasing the number of sampling occasions increases the power and conservativeness of future statistical tests. Unfortunately, temporal autocorrelation, which is common in environmental data, can lead to frequently collected samples not being independent (e.g., Edwards and Coull 1987). This raises the question of how to best apportion effort in accounting for spatial and temporal variability within a site. For many long-term macrofaunal studies effort may best be spent obtaining accurate density estimates at each time to prevent spatial variation from confounding the temporal sequence.

In this chapter, we use data collected from a monitoring program of intertidal sandflat communities to show how spatial variation can influence observed time series. We illustrate some of the problems likely to occur when analyzing and interpreting time-series data based on sampling strategies that do not account for spatial variation. No attempt has been made to interpret the ecological significance of the temporal sequences presented. In many of the examples given, such interpretations would be inappropriate until a much longer time series has been collected. Although all our examples are based on density estimates of macrofaunal populations, similar problems will occur with any other environmental variable that is patchily distributed.

TABLE 4.1.   Sampling Strategies Used in Long-Term Studies of
Soft-Bottom Macrofauna[a]

No. of sites (No. of replicates × sample size)	Comment on sites	Time period	Reference
8 sites (5 × 0.05 m²)	located km's apart	3 monthly over 1 year	Lincoln Smith 1991
14 sites (3 × 0.1 m²)	located km's apart	twice over 10 years	Josefson and Rosenberg 1988
12 sites (4 × 0.05 m²)	located km's apart	3–6 monthly over 4 years	Jones 1990
2 sites (5 × 104 cm²)	different localities	monthly over 10 years	Barnett and Watson 1986
3 sites (5 or 10 or 20 × 0.1m²)	different depths	20 times over 4 years	Persson 1983
12–21 sites (3–4 dredges or 0.01 m² grabs or 0.02 m² cores	located km's apart and often different habitats	4 times a year for 13 years with some sites 10 times a year for 4 years	Holland et al. 1987
4 sites (5–10 U × 0.01 m²)	located kms apart and range in habitat from mud and silt to sand	3 monthly for 6 years	Lie and Evans 1973
2 sites (4 × 3.5 cm diam)	mud and sand habitats	monthly for 3 years	Coull and Fleeger 1977
3 sites (3 × 0.09 m²)	located km's apart	monthly for 11 years	Flint 1985
2 sites (3–4 × 2.7 cm diam) reduced to (2 × 2.7 cm diam)	mud and sand habitats	monthly for 8 years increased to fortnightly for 13 years	Coull 1985
5 sites (10 × 0.025 m² or 0.08 m²)	located km's apart	mainly bimonthly for 5.5 years	Hines et al. 1987
60–74 sites (4–5 × 0.01 m²)	located km's apart	erratically for 5 years	Govaere et al. 1980
3 sites or 6 sites (12 × 7.62 cm diam)	located km's apart	weekly for 3 years monthly for 12 years	Livingston 1987
1 site (4 × 0.03 m² or 10 × 0.01 m²)		2–3 monthly for 10 years	Ibanez and Dauvin 1988

[a]Presented in a haphazard selection of material from peer-reviewed publications.

Methods

The Monitoring Program

Manukau Harbor is a large (340 km^2) shallow inlet, adjacent to Auckland, on the west coast of the North Island of New Zealand. Intertidal sandflats constitute about 40% of the area of the Harbor and are commonly used for recreation and food gathering. The monitoring program was established in 1987 to provide managers with some stocktaking of the resource under stewardship and to create a framework against which more process-oriented studies could be conducted (e.g., Pridmore et al. 1991, Thrush et al. 1991).

Data on macrofaunal populations are collected bimonthly from six sites, each 9000 m^2, situated on midtide sandflats around the Harbor. The six sites (AA, CB, CH, EB, KP and PS) are situated kilometers apart; their geographic location and a description of the macrobenthic community structure is given in Pridmore et al. (1990). Each site is divided into 12 equal sectors. On the first sampling occasion (October 1987 at all sites except CB which was first sampled on December 1987), 36 cores (13 cm dia., 15 cm depth) were collected from each site (3 randomly located within each sector) to establish the optimum sampling intensity for future visits. On all subsequent occasions 12 cores have been collected from each site (1 randomly from each sector). The decision to collect 12 cores from each site was determined by a process of balancing effort against precision (Bros and Cowell 1987, Hewitt et al. 1993). To date, two sampling occasions have been missed (October and December 1988). After collection, samples are sieved (500 μm mesh) and the residue fixed in 5% formalin and 0.1% rose-bengal in seawater. In the laboratory, macrofauna are sorted, identified to the lowest possible taxonomic level, counted and preserved in 70% alcohol.

Data Analyses

For the purposes of this chapter each site is treated separately. The term population is thus used to refer to an individual species at an individual site. Analyses were only performed on populations that exhibited mean abundances greater than or equal to one per core on most sampling occasions (i.e., >16 of the 21).

Spatial patterns were investigated using the program SAAP (Wartenberg 1989) to construct correlograms and test both for overall significance of the correlogram and its individual points. Correlograms were constructed from values of Moran's I spatial autocorrelation coefficient calculated for each 5 m distance class, i.e., those samples lying within 0 to 5 m of each other, 5 to 10 m of each other, etc. Significant coefficients indicate that a spatial pattern exists and allow estimates of the mean patch size to be made. To aid interpretation of the correlograms, three-dimensional plots of species abundance at a site were generated. Only some of the plots that illustrate particular patterns are presented as this analysis is simply used to illustrate the types of patterns that can affect spatial variability.

In order to provide generalizations of how spatial variation can confound temporal sequences we used the ratio of spatial to temporal variation to allocate individual populations to three categories (i.e., spatial variation less than temporal, equal to temporal, or greater than temporal variability). Allocation to the first or last category was based on spatial variability less than half or greater than twice the temporal variability respectively; populations not falling within these two categories were considered to exhibit spatial variation roughly equivalent to temporal variability. Two techniques were utilized to create the categories. First, we compared the magnitude of the range in abundance exhibited on each sampling occasion, relative to the range in the mean abundances over time. While this technique made no assumptions about the data, problems may have arisen if the range was not a good description of variation or if estimates of temporal variability were confounded by utilizing means from each sampling occasion. To assess the importance of these problems, a second technique was utilized; it involved estimating and comparing between-group variance (temporal variability) with within-group variance (spatial variability) from one-way ANOVA's for each population (log-transformed where appropriate). Approximately the same number of populations were found in each category using the two methods. A few populations close to the category boundaries switched depending on which technique was used.

To determine how our interpretation of temporal patterns may have been affected if fewer samples were collected on each occasion, we generated 100 unique time-series plots for each population based on a sample size of 4, using randomization techniques without replacement from the samples available on each occasion. A sample size of 4 was chosen as 3 to 5 samples are commonly collected to quantify macrofaunal abundance in long-term studies (Table 4.1). We used these time-series plots to illustrate the changes in temporal patterns (e.g., timing and height of peaks) that can occur with decreased precision on each sampling occasion. This is done visually because it is not simply the increased level of variation with decreased sample size that is important but also the inability to identify features of the temporal sequence that can be attributed to biological/environmental events (e.g., recruitment periods). We also assessed the increased variation that can occur with decreased precision on each sampling occasion by testing for linear trends in the time series.

Results

Spatial patterns of the macrobenthic populations were investigated at sites on the first sampling occasion only. For most of the populations studied (i.e., 45 of the 58 with mean densities greater than 1 per core), significant spatial correlograms were obtained. Figure 4.1 illustrates the variety and complexity of spatial patterns that contribute to variations in density estimates within the 9000 m^2 sites.

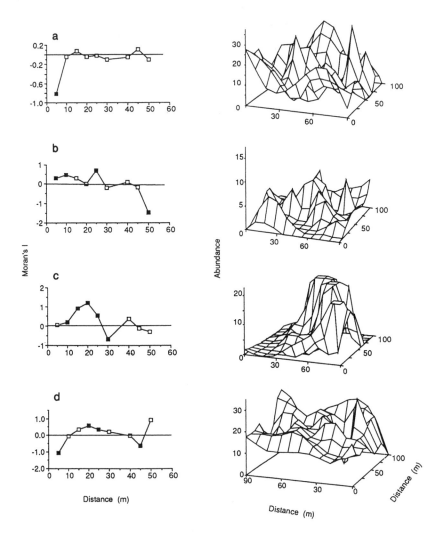

Figure 4.1. Spatial autocorrelograms and corresponding 3-D plots (based on 36 samples from the first sampling occasion) showing: (a) small-scale heterogeneity (*Heteromastus filiformis* at EB); (b) homogeneous density patches of 5–10 m radius (*Goniada emerita* at CH); (c) large patches conferring a gradient (*Nucula hartvigiana* at PS); (d) gradient superimposed on small-scale heterogeneity within patches of 20–25 m radius (*Tellina liliana* at AA). Filled squares represent significant Moran's I.

At each site, the number of individuals of a given species in a core commonly varied by an order of magnitude. Coefficients of variation on the first sampling occasion frequently exceeded 0.75 (Table 4.2). On subsequent sampling occasions, coefficients of variation for many (55%) of the populations varied little

Table 4.2. Distribution of Coefficients of Variation Calculated from the First Sampling Occasion for Populations from Individual Sites

Coefficient of variation	No. of population
<0.5	4
0.5–0.75	11
0.75–1.0	11
1.0–1.25	10
1.25–1.5	4
>1.5	5

from those initially determined (e.g., Figure 4.2, Table 4.3). Of the remaining populations, most (85%) exhibited wide variations in the coefficient of variation because of low (<1 per core) densities on one or more occasions. Approximately 86% of the populations studied have exhibited spatial variation that was equal to or greater than the observed temporal fluctuations in mean density (Table 4.4). The ratios of temporal to spatial variation show no relationship with the coefficient of variation from the first sampling occasion (Figure 4.3), indicating that the magnitude of the ratios of spatial to temporal variation observed is not a product of the precision achieved by our sampling strategy.

Most populations exhibited density variations in time similar to or less than those in space. Populations that exhibited greater temporal than spatial variation usually showed distinct annual fluctuations in density (e.g., species with high levels of annual recruitment). Density variation in space was greater than that in time for only 33% of the populations. These populations did not form any distinct functional grouping based on mobility, size or life-history characteristics; however, bivalves were not represented. All exhibited very low levels of temporal variation and the density distributions at any one time usually exhibited high maximum ranges relative to the 75th percentile ranges, i.e., spatial patterns that include a small number of very high or low density patches.

Figure 4.4 shows some of the time sequences generated for a population that to date has exhibited similar scales of spatial and temporal variation. Although data from only one population are used to illustrate our point, similar results were obtained for all other populations that exhibited similar scales of spatial and temporal variation. Routine monitoring with a sample size of 12 has established that the polychaete *Heteromastus filiformis* at Site CB commonly attains a mean density of 17 $\pm$ 10% individuals per core during winter (June–October) (Figure 4.4a). In 3 of the 4 years sampled, the observed population declined by at least 70% to less than 5 individuals per core in February/March. Of the 100 time sequences generated using a sample size of four, only 28 closely mimicked this pattern with abundances in winter in the range of 17 $\pm$ 20% and a decline of at least 50% in February/March (e.g., Figure 4.4b). An additional 20 adequately

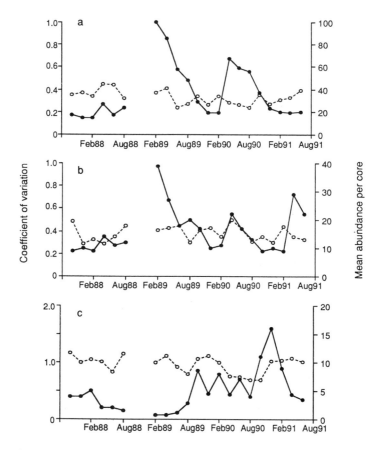

Figure 4.2. Plots of mean abundance (solid line) and coefficient of variation (dashed line) over time: (a) *Tellina liliana* at AA; (b) *Tellina liliana* at PS; (c) *Nucula hartvigiana* at PS.

described the three declines in population abundance but showed highly variable (i.e., $> \pm 20\%$) densities during the winter months, (e.g., Figure 4.4c). Fifty two of the time sequences generated (e.g., Figure 4.4d) showed little sign of the temporal patterns observed using a sample size of 12.

Fewer discrepancies in the generated time sequences were noted for the 7 populations in which spatial variation was much smaller than the observed temporal fluctuations. For example, Figure 4.5 shows some of the time sequences generated for the bivalve *Tellina liliana* at Site AA. Routine monitoring with a sample size of 12 indicated that little or no recruitment occurred during 1988 (Figure 4.5a). Two periods of high recruitment were observed in February 1989 (or earlier) and April 1990, with the latter period being at least 30% less than the preceding year. After each recruitment period, densities gradually declined to

Table 4.3.   Maximum Change in Coefficients of Variation Over Time (as a Percentage Deviation from the Coefficient of Variation Calculated on the First Sampling Occasion) for Populations from Individual Sites

% deviation of coefficient of variation	No. of populations	Cumulative % of populations
<50	7	16
50–60	10	38
60–70	8	55
70–80	7	71
80–90	2	75
90–100	2	79
100–200	6	92
200–300	1	94
>300	3	100

about 25 individuals per core. Of the 100 time sequences generated, 44 closely mimicked the observed time sequence (e.g., Figure 4.5b): that is, no peak or gradual decline in abundance in 1988, maximal densities occurring in February 1989 and April 1990 followed on each occasion by a slow decline to 15–35 per core, and the 1990 maximal density being 20–40% less than that estimated the preceding year. The remaining 56 generated time sequences each depicted little or no recruitment in 1988 with periods of recruitment in 1989 and 1990, but either the timing (e.g., Figure 4.5c) or magnitude (e.g., Figure 4.5d) of recruitment was markedly different.

As expected, the time sequences generated for populations with relatively high spatial variation compared to temporal variation were the most divergent. Figure 4.6, for example, shows some of the time sequences generated for the isopod *Exosphaeroma falcatum* at Site PS. Routine monitoring with a sample size of 12 indicated that from February 1989 to October 1990 the mean density of this population was 40% lower than that observed at the beginning (October 1987–August 1989) and end (October 1990–June 1991) of the time series (Figure 4.6a). Few (<15%) of the generated time sequences, however, depicted a 30–50% decrease in mean density during this period (e.g., Figure 4.6b). Most (70%) of the generated time series showed no decline in abundance at all or if a pattern was apparent it bore little resemblance to that obtained with a sample size of 12 (e.g., Figure 4.6c). This increasing divergence of that generated from the observed time sequence with increasing spatial relative to temporal variability occurred irrespective of the size of the coefficient of variation of the observed population.

Spatial variation was also found to influence our ability to quantify step-changes in abundance over time and to detect temporal trends. For example, routine monitoring with a sample size of 12 established that *Heteromastus filiformis* declined in abundance by 71% between August 1989 and February

Table 4.4.   Details of Spatial and Temporal Variability of Populations Used to Illustrate the Effect of Spatial Variation on Observed Temporal Sequencing

Population and site	Spatial pattern October, 1987 (n=36)	Spatial range October, 1987 (n=36)	Temporal range of means October, 1987–June, 1991	Type of population	% of populations	Range of temporal/ spatial variability
Tellina liliana AA	Small scale hetero-geneity with patches of 20–25 m radius	8–36 (12–20)	15–100 (18–58)	Spatial variation less than temporal	14	2.0–3.2
Heteromastus filiformis CB	No pattern at scale of sam-pling; Signif-icant variance/ mean ratio	3–45 (9–20)	2–21 (10–19)	Spatial variation equal to temporal	53	0.50–2.0
Exosphaeroma fulcatum PS	Patches of 10 m radius	0–22 (2–8)	0–10 (1–5)	Spatial variation greater than temporal	33	0.10–0.50

Note: Spatial and temporal ranges are given; interquartiles are noted parenthetically. The percentages of similar types of populations found in the study are also given together with the range of the ratios of temporal/spatial variability calculated from ANOVA.

1990 (Figure 4.4a). Of the 100 time sequences generated, only 31 indicated a drop in abundance of 65–75% during this period; 17 indicated a change of less than 60%, while 21 indicated a decrease of more than 80%. For populations such as *Tellina liliana* at Site AA that to date have exhibited low spatial relative to temporal variability, the quantification of step-changes was much less affected by sample size. For example, routine monitoring has indicated that between February 1989 and February 1990 *Tellina liliana* declined in abundance by 80% (Figure 4.5a). A total of 71 of the 100 time sequences generated indicated a decrease in abundance of 75–85% during this period; only 6% indicated a change of less than 70% or greater than 90%.

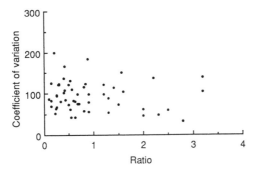

Figure 4.3. Plot of coefficient of variation for the first sampling occasion versus the ratio of temporal to spatial variability from one-way ANOVA's for individual populations.

Although we do not recommend that analysis for cyclic patterns or trends be conducted with such a short time series we illustrate how the analysis of trends may be affected by spatial variation. Few of the populations sampled have shown a general increase or decrease in abundance through the short time span of the monitoring program. Data for the isopod *Exosphaeroma falcatum* at Site PS (Figure 4.6a), however, can be used as an illustration. In the case of *Exosphaeroma falcatum* an apparent decline occurred between October 1987 and August 1990. Using autocorrelation tests (Chatfield 1980), no significant first order temporal autocorrelation was found for this population. Using linear regression, a significant decline ($r = -0.71$, $n = 17$, $P = 0.0021$) can be detected using the routine monitoring data. In only 19 of the time sequences generated for this highly patchily distributed population could a significant linear trend be detected during this period.

Discussion

Macrofaunal populations within the 9000 m^2 intertidal sandflat sites mostly exhibited patchiness on various spatial scales. However, most populations that exhibited density variations in time similar to or greater than those in space usually showed distinct annual fluctuations in density. When we reduced sample size to a level similar to that frequently used in other long-term studies (Table 4.1) we were able to illustrate various anomalies. The timing and size of annual density changes often shifted, and trends in density were often missed. However, temporal patterns of populations exhibiting strong annual cycles frequently remained consistent even when the precision achieved on each sampling occasion was reduced by decreasing the sample size. In contrast, the largest anomalies were apparent for populations where spatial variation was much greater than temporal.

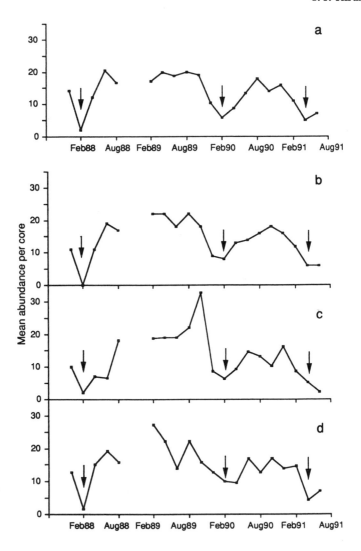

Figure 4.4. Plots of the observed time series (a) and selected examples of randomly generated time series (b–d) of *Heteromastus filiformis* at site CB. Arrows indicate events discussed in the text.

We have illustrated potential problems using data from within single large sites (9000 m^2) and from a short time series. However, the problems mentioned above are unlikely to disappear in more complex sampling programs which encompass different scales of variation.

A crucial part of the design of a monitoring program, once the appropriate time scale has been determined, is choosing the sample size. This is usually done

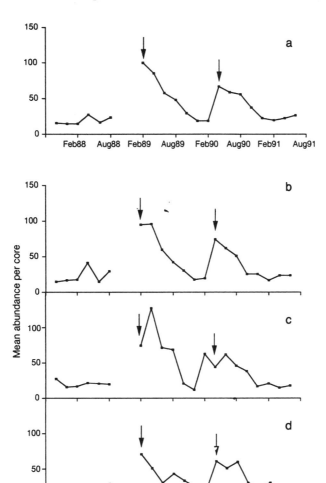

Figure 4.5. Plots of the observed time series (a) and selected examples of randomly generated time series (b-d) of *Tellina liliana* at site AA. Arrows indicate events discussed in the text.

by examining the relationship between spatial variation and sampling intensity from data collected in a pilot survey. This approach contains two potential problems. First, it assumes that the true variance of a population will not change over time (e.g., with recruitment). The consistency in the coefficients of variation of many of the populations from Manukau Harbor (Table 4.3) indicates that information gained during pilot sampling is likely to be relevant over time, even for populations that show large annual density fluctuations (Figure 4.2). Second,

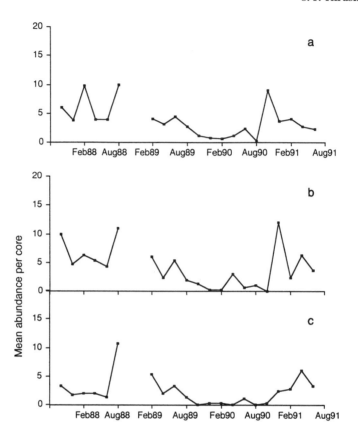

Figure 4.6. Plots of the observed time series (a) and selected examples of randomly generated time series (b–c) of *Exosphaeroma falcatum* at Site PS. See text for further explanation.

commonly cited methods for the determination of sample size require specification of a level of difference that needs to be detected (e.g., Andrew and Mapstone 1987). But it may be difficult to decide whether 20, 50, or 100% changes in mean density will be ecologically significant. In making these kinds of decisions the relationship between spatial and temporal variance comes into play; a standard error of 20% might be sufficient if the population varies temporally by much more than this or may be quite insufficient if the temporal variation is small. When designing a time-series program the question then becomes—what precision in density estimates at a site is required to prevent spatial variation from confounding the temporal sequence?—and the magnitude of the ratio of spatial/temporal variability becomes a dominant factor. Unfortunately, the magnitude of this ratio is unknown at the start of the time series and many areas of the world will not have background information on temporal patterns of species available. Thus, methods for determining sample size are needed that allow recognition of when the increase in precision is low relative to the increase in sampling effort. Once the time-series program is under way, modifications can be made to the sampling strategy in light of the developing temporal sequence.

Unpredictability in density estimates over time suggests the role of stochastic processes, at least at the level of observation (Frost et al. 1988, Rahel 1990). Where spatial variation contributes to apparent unpredictability this can influence environmental management decisions. For example, populations or communities may be considered too unpredictable for effective management or, for species with variable spatial and temporal dynamics apparently driven by stochastic processes, it may be difficult to identify localized extinctions and their relation to impact effects. Moreover, direct analysis of variability is itself useful in, for example, assessing the relative importance of processes operating on different scales. If abundance varies among sites but is consistent at each site over time, then a major influence on processes is associated with the sites. Conversely, if abundance shows strong variability among years at sites but a high degree of consistency in each year across sites, then a major influence of a large-scale factor (e.g., weather) is implicated (Kratz et al. 1987). Time-series analysis is useful in evaluating the influence of impacts on population and community characteristics (e.g., persistence) because of the effect that observational scales in space and time have on both these characteristics and our ability to describe them (Connell and Sousa 1983, Dayton and Tegner 1984, Weins 1989).

Separating trends from greater than annual cycles will always be limited by the length of the time series (e.g., Loftis et al. 1991). However, confidence in the biological interpretation of events and the recognition of unusual events can yield potentially useful ecological insights (Carpenter 1988). Monitoring is inherently retrospective; therefore it is important to identify events that are likely to warrant remedial action before a critical and potentially irreversible level is reached. The ability, which a time series gives, to identify patterns can be useful in this and in the design of associated experimental studies. For example, periods of bivalve recruitment identified in Manukau Harbor have been used to time studies on pesticide impacts (Pridmore et al. 1991). The ability to confidently identify repeatable density fluctuations is particularly important if the determination of temporal changes (e.g., level of recruitment) are deemed a sensitive measure of impact (Underwood 1991).

Ultimately it is important that the design of a time-series or impact study defines specific and answerable questions relevant to information users and environmental managers, such that appropriate data are collected within logistic and cost constraints. In the case of generating a time series, design is particularly important because of the long-term commitment and cost. If time-series data are used, either to provide background information or for impact assessment, it is important that patterns and trends can be inferred confidently. It may be possible to reduce the overall cost of collecting time-series data by pooling within-site samples and subsampling the aggregate. While this approach would not allow estimates of spatial variability to be generated, good estimates of mean density may be achieved with reduced effort. Whether this approach is appropriate will depend on the study objectives. It is likely to be relevant only when common species are to be studied and errors due to subsampling are small. While estimates of spatial variability may not be important in tests for temporal trends, the information may still be useful (e.g., demonstrating the consistency of coefficients of variation over time). Moreover, spatial variation should not be considered as noise, but as an important component of population dynamics. Information on the consistency or change in spatial patterns may provide useful ecological insights.

We have illustrated the problems that can arise in inferring patterns or testing for trends in time series when insufficient effort is expended in accounting for spatial variation. Our analysis indicates the crucial feature is the relative magnitude of spatial and temporal variation. In Manukau Harbor this often varies for the same species at different sites or for different species at the same site. Thus, it is not appropriate to generalize on the levels of variation that may be expected for particular species or groups of species. Sample sizes for each site must be chosen such that low biased and highly precise estimates are obtained on each sampling occasion. In general, we do not recommend that any trade-off occur between the number of times a site is sampled and the number of within-site replicates collected. When sufficient data have been collected, analysis of the relative magnitude of spatial and temporal variation may indicate possible modifications to the sampling strategy.

Acknowledgments

The monitoring program that provided data for this study was funded by the Auckland Regional Council—Auckland Regional Water Board. We thank C. Osenberg, R. Schmitt, B. Whitlatch and an anonymous reviewer for comments on early drafts of this manuscript.

References

Andrew, N. L., and B. D. Mapstone. 1987. Sampling and the description of spatial pattern in marine ecology. Oceanography and Marine Biology Annual Review 25:39–90.

Barnett, P. R. O., and J. Watson. 1986. Long-term changes in some benthic species in the Firth of Clyde, with particular reference to *Tellina tenuis* (da Costa). Proceeding of the Royal Society of Edinburgh 90B:287–302.

Bros, W. E., and B. C. Cowell. 1987. A technique for optimizing sample size (replication). Journal of Experimental Marine Biology and Ecology 114:63–71.

Carpenter, S. R. 1988. Transmission of variance through lake food webs. Pages 119–135 *in* S. R. Carpenter, editor. Complex interactions in lake communities. Springer-Verlag, New York, New York.

Chatfield, C. 1980. The analysis of time series: an introduction, 2nd Edition. Chapman and Hall, London, England.

Connell, J. H., and W. P. Sousa. 1983. On the evidence needed to judge ecological stability or persistence. American Naturalist 121:789–824.

Coull, B. C. 1985. The use of long-term biological data to generate testable hypotheses. Estuaries 8:84–92.

Coull, B. C., and J. W. Fleeger. 1977. Long-term variation and community dynamics of meiobenthic copepods. Ecology 58:1136–1143.

Dayton, P. K., and M. J. Tegner. 1984. The importance of scale in community ecology: a kelp forest example with terrestrial analogs. Pages 457–483 in P. W. Price, C. N. Slobodchikoff and W. S. Gaud, editors. A new ecology: novel approaches to interactive systems. J. Wiley and Sons, New York, New York.

Eckman, J. E. 1979. Small-scale patterns and processes in a soft substratum intertidal community. Journal of Marine Research 37:437–457.

Edwards, D., and B. C. Coull. 1987. Autoregressive trend analysis: an example using long-term ecological data. Oikos 50:95–102.

Flint, R. W. 1985. Long-term estuarine variability and associated biological response. Estuaries 8:158–169.

Frost, T. M., D. L. DeAngelis, S. M. Bartell, D. J. Hall, and S. H. Hurlbert. 1988. Scale in the design and interpretation of aquatic community research. Pages 229–258 in S. R. Carpenter, editor. Complex interaction in lake communities. Springer-Verlag, New York, New York.

Govaere, J. C. R., D. Van Damme, C. Heip, and L. A. P. De Coninck. 1980. Benthic communities in the Southern Bight of the North Sea and their use in ecological monitoring. Helgoländer Meeresuntersuchungen 33:507–521.

Greig-Smith, P. 1983. Quantitative plant ecology, 2nd Edition. Butterworths, London, England.

Hewitt, J. E., G. B. McBride, R. D. Pridmore, and S. F. Thrush. 1993. Patchy distributions: optimising sample size. Environmental Monitoring and Assessment 27:95–105.

Hilborn, R., and C. J. Walters. 1981. Pitfalls of environmental baseline and process studies. Environmental Impact Assessment Review 2:265–278.

Hines, A. H., P. J. Haddon, J. J. Miklas, L. A. Wiechert, and A. M. Haddon. 1987. Estuarine invertebrates and fish: sampling design and constraints for long-term measurements of population dynamics. Pages 140–164 in T. P. Boyle, editor. New approaches to monitoring aquatic ecosystems, ASTM STP 940. American society for testing and materials, Philadelphia, Pennsylvania.

Holland, A. F., A. T. Shaughnessy, and M. H. Hiegel. 1987. Long-term variation in mesohaline Chesapeake Bay Macrobenthos: spatial and temporal patterns. Estuaries 10:227–245.

Ibanez, F., and J. C. Dauvin. 1988. Long-term changes (1977–1987) in a muddy fine sand Abra alba–Melinna palmata community from the Western English Channel: multivariate time-series analysis. Marine Ecology Progress Series 49:65–81.

Jassby, A. D., and T. M. Powell. 1990. Detecting changes in ecological time series. Ecology 71:2044–2052.

Jones, A. R. 1990. Zoobenthic variability associated with a flood and drought in the Hawkesbury Estuary, New South Wales: some consequences for environmental monitoring. Environmental Monitoring and Assessment 14:185–195.

Josefson, A. B., and R. Rosenberg. 1988. Long-term soft-bottom faunal changes in three shallow fjords, West Sweden. Netherlands Journal of Sea Research 22:149–159.

Kratz, T. K., T. M. Frost, and J. Magnuson. 1987. Inferences from spatial and temporal variability in ecosystems: long-term zooplankton from lakes. American Naturalist 129:830–846.

Legendre, P. 1993. Spatial autocorrelation: trouble or new paradigm? Ecology 74:1659–1673.

Legendre, P., and M. J. Fortin. 1989. Spatial pattern and ecological analysis. Vegetatio 80:107–138.

Lie, U., and R. A. Evans. 1973. Long-term variability in the structure of subtidal benthic communities in Puget Sound, Washington, USA. Marine Biology 21:122–126.

Lincoln Smith, M. P. 1991. Environmental impact assessment: the roles of predicting and monitoring the extent of impacts. Australian Journal of Marine and Freshwater Research 42:603–614.

Livingston, R. J. 1987. Field sampling in estuaries: the relationship of scale to variability. Estuaries 10:194–207.

Loftis, J. C., G. B. McBride, and J. C. Ellis. 1991. Considerations of scale in water quality monitoring and data analysis. Water Resources Bulletin 27:255–264.

McArdle, B. H., and R. G. Blackwell. 1989. Measurement of density variability in the bivalve *Chione stutchburyi* using spatial autocorrelation. Marine Ecology Progress Series **52**:245–252.

Parker, G. G. 1989. Are currently available statistical methods adequate for long-term studies? Pages 199–201 *in* G. E. Likens, editor. Long-term studies in ecology, approaches and alternatives. Springer-Verlag, New York, New York.

Persson, L.-E. 1983. Temporal and spatial variation in coastal macrobenthic community structure, Hano Bay (Southern Baltic). Journal of Experimental Marine Biology and Ecology **68**:277–293.

Pridmore, R. D., S. F. Thrush, J. E. Hewitt, and D. S. Roper. 1990. Macrobenthic community composition of six intertidal sandflats in Manukau Harbour, New Zealand. New Zealand Journal of Marine and Freshwater Research **24**:81–96.

Pridmore, R. D., S. F. Thrush, R. J. Wilcock, T. J. Smith, J. E. Hewitt, and V. J. Cummings. 1991. Effect of the organochlorine pesticide technical chlordane on the population structure of suspension and deposit feeding bivalves. Marine Ecology Progress Series **76**:261–271.

Rahel, F. 1990. The hierarchical nature of community persistence: a problem of scale. American Naturalist **136**:328–344.

Thrush, S. F. 1991. Spatial patterns in soft-bottom communities. Trends in Ecology and Evolution **6**:75–79.

Thrush, S. F., J. E. Hewitt, and R. D. Pridmore. 1989. Patterns in the spatial arrangement of polychaetes and bivalves in intertidal sandflats. Marine Biology **102**:529–536.

Thrush, S. F., R. D. Pridmore, J. E. Hewitt, and V. J. Cummings. 1991. Impact of ray feeding disturbances on sandflat macrobenthos: do communities dominated by polychaetes or shellfish respond differently? Marine Ecology Progress Series **69**:245–252.

Tilman, D. 1989. Ecological experimentation: strengths and conceptual problems. Pages 136–157 *in* G. E. Likens, editor. Long-term studies in ecology, approaches and alternatives. Springer-Verlag, New York, New York.

Trueblood, D. D. 1991. Spatial and temporal effects of terebellid polychaete tubes on soft-bottom community structure in Phosphorescent Bay, Pureto Rico. Journal of Experimental Marine Biology and Ecology **149**:139–159.

Underwood, A. J. 1991. Beyond BACI: experimental designs for detecting human environmental impacts on temporal variations in natural populations. Australian Journal of Marine and Freshwater Research **42**:569–587.

Underwood, A. J., and C. H. Peterson. 1988. Towards an ecological framework for investigating pollution. Marine Ecology Progress Series **46**:227–234.

Wartenberg, D. 1989. SAAP - A spatial autocorrelation analysis program, version 4.3, Department of Environmental and Community Medicine. Robert Wood Johnson Medical School, University of Medicine and Dentistry of New Jersey, Piscataway, NJ USA.

Weins, J. A. 1989. Spatial scale in ecology. Functional Ecology **3**:385–397.

CHAPTER 5

SCALABLE DECISION CRITERIA FOR ENVIRONMENTAL IMPACT ASSESSMENT

Effect Size, Type I, and Type II Errors[1]

Bruce D. Mapstone

Large-scale habitat destruction, local species extinctions, and the possibility of global environmental change have emphasized that anthropogenic impacts can exceed the environment's capacity to absorb them (Soulé 1991). Further, human activities that change the status of the environment from that which is considered productive or undisturbed impinge on the economic viability of many industries, ranging from fishing and agriculture to eco-tourism. The potential for environmental degradation as a result of development, and the subsequent economic, social and political ramifications, are now routinely juxtaposed with the mainly economic and political consequences of impeding development and economic growth.

Historically, Environmental Impact Assessment (EIA) simply involved the prediction of impacts of development, and wrangling over approval to proceed with development often focused on the soundness of those predictions (e.g., Buckley 1989a, 1989b, 1990, Fairweather 1989, Lincoln Smith 1991). An increasing tendency to regulate development, however, carries the implicit expectation that such predictions will be tested, that impacts will be measured, and that regulations will be enforced. Permission to develop is now often contingent on funding an Environmental Impact Monitoring (EIM) program to measure (potential) environmental impacts, with the implicit expectation that real impacts can and will be detected. Decisions about environmental impacts are made empirically as well as prophetically, and inevitably the statistical and inferential bases for those decisions will be subject to increasingly thorough legal and scientific scrutiny (Millard 1987, Hoverman 1989, Christie 1990, Beder 1991, Buckley and McDonald 1991, Fairweather 1991, Jarrett 1991, MacDonald 1991, Martin and Berman 1991, Lester, Chapter 17).

[1]This chapter was previously published in a modified form in Ecological Applications 5:401–410. © 1995 by the Ecological Society of America.

Decisions about environmental impacts have important consequences, whether they are correct or in error (Bernstein and Zalinski 1983, Andrew and Mapstone 1987, Millard 1987, Peterman 1990, Fairweather 1991, Faith et al. 1991, Lincoln Smith 1991). For example, concluding that an impact has occurred may result in the cessation of work, the closing of a factory, or the imposition of fishing quotas. Alternatively, concluding that no deleterious impact has occurred usually provides tacit support for continued development, possibly resulting in the collapse of a fishery, catastrophic pollution, or extinction of species. If the conclusion that an impact had occurred was wrong (a Type I error, α), the curtailment of development would have been unwarranted and the local economic and social hardship likely would have been unnecessary. On the other hand, an incorrect conclusion of "no impact" (a Type II error, β), might mean that serious environmental degradation occurred before the real impact was noticed. Fairweather (1991) and Peterman (1990) have suggested that a Type II error would be more costly than a Type I error, since in addition to a more severe environmental impact, the same hardships and economic costs would eventually be incurred.

It follows that *any* erroneous conclusion (either a Type I or a Type II error) would be cause for concern, yet traditional inferential statistical decision making has been preoccupied with only Type I errors. The acceptable (or critical) level of Type I error (α_c) has been dictated by convention, with the result that α_c is treated as a constant. In this chapter, I suggest specific steps for evaluating new critical levels of α, and procedures for making decisions based on a variable α_c and consideration of the probability of Type II error (β).

Components of a Decision

Under the hypothetico-deductive paradigm, empirical decisions typically rest on statistical tests of null hypotheses. In EIM, as elsewhere, the most common null hypothesis (H_0) is one of "no difference" or "no effect". Having collected our data, a statistical analysis provides us with a probability (α_0) of observing those data if the H_0 were in fact true. If this probability is less than some arbitrary critical level (α_c), almost always 0.05 in ecology, we reject H_0 in the belief that there is only a small probability (<0.05) of having done so incorrectly (a Type I error). If our associated probability is greater than the critical 0.05, we do not reject H_0. Such nonrejection typically results in a conclusion of "no impact". Here, it is the probability of Type II error (β)—the likelihood of the data if an alternative hypothesis (H_a) was true—that is important (Table 5.1).

Evaluation of β, however, cannot proceed without first stipulating (i) the critical Type I error rate (α_c) by which we failed to reject H_0, (ii) the departure from H_0 represented by H_a that we would wish to detect if it were true, and (iii) the sampling design and statistical model on which the test was based. The difference between H_0 and H_a has been termed the Effect Size (ES; Winer 1971, Cohen 1988) and in EIM would be defined by the magnitude and form (Cohen

Table 5.1.   The Four Alternative Outcomes in Tests of Hypotheses, Illustrating the Relationships between Rejection of H_0 and Type I Errors and Nonrejection of H_0 and Type II Errors[a]

	Reality	
Decision	H_0 True (No Impact)	H_0 False (? Impact)
Reject H_0		
(? $\Rightarrow$Impact)	Type I error (α)	No error: Power ($1 - \beta$)
Not Reject H_0		
($\Rightarrow$No Impact Detected)	No Error ($1 - \alpha$)	Type II Error (β)

Note: Also shown are the functional analogues of these outcomes in decisions about hypothesized environmental impacts. The "?"s are used to emphasize that untruth and/or rejection of an H_0 is not always indicative of an impact, though it would be prerequisite to inferring an impact from hypothesis-testing procedures.
[a]Modified after Toft and Shea (1983), Peterman (1990).

1988; $\approx$ "type" and "size," Bernstein and Zalinski 1983) of the *maximum* environmental impact that we would be prepared to tolerate in a particular case. This might roughly equate to the "limit of acceptable change" due to impacts of development. Sampling designs and statistical models for environmental impact studies have been discussed recently in considerable detail, and I will not discuss these issues here (see Winer 1971, Green 1979, 1989, Bernstein and Zalinski 1983, Millard and Lettenmaier 1986, Stewart-Oaten et al. 1986, Millard 1987, Underwood and Peterson 1988, Warwick 1988, Faith et al. 1990, 1991, Underwood 1990, 1991a, 1991b, Chapter 9, Keough and Quinn 1991, Warwick and Clarke 1991, Stewart-Oaten, Chapter 7; for further detail about the relations among sample size, variance, ES, α, and β see Bernstein and Zalinski 1983, Millard and Lettenmaier 1986, Andrew and Mapstone 1987, Cohen 1988, Peterman 1990, Osenberg et al., Chapter 6). My emphasis is on the criteria by which we make empirical decisions with the data from carefully designed EIM programs.

Problems with Traditional Decisions

Statistical decisions based only on Type I error rates are essentially decisions by a singular rule: Reject H_0 if $\alpha_0 < \alpha_c$. The singularity is emphasized by the tyranny of convention—everyone (in ecology) uses $\alpha_c = 0.05$! Being stipulated arbitrarily by us, the critical Type I error rate is not subject to other components of our research such as sample size and error variance. In rejecting H_0 by this rule, it is not incumbent upon us to worry about the magnitude of "statistically significant" differences (e.g., among means), even though that is perhaps the most interesting facet of our data: statistical significance has come to be treated almost synonymously with biological importance, even though no such relationship exists. Neither α nor β have any intrinsic meaning in terms of the biological variables we measure or the biological importance of an outcome.

Toft and Shea (1983) and Peterman (1990) have pointed out that such an approach to decisions embraces an implicit, but unacknowledged, weighting of the importance attached to errors of Type I and Type II, a thesis supported by the weighting given to the two decision outcomes. In research, a "significant" result is cause for excitement, confidence, a manuscript, etc.; a nonsignificant result leads to despondency, a reevaluation of research direction, a conclusion of a nonresult at best, or a failed experiment at worst. It follows that many authors are (perhaps unknowingly) content to consider Type I errors of far greater concern than Type II errors. This bias is echoed in editorial lore that we all know and endure. In issues of environmental impacts, a "significant" result is cause for concern, action, litigation, controversy—no matter how small or inconsequential the impact; nonsignificance is sometimes controversial, often leads to complacency, but rarely precipitates action, concern, or litigation—even though large impacts might have been missed. Given the routine stipulation of small values of α_c, this constitutes a tacit prioritizing of development over environment and dictates a fundamental constraint that permeates the entire design and decision-making process.

Further, with α_c fixed, negotiations over the costs of monitoring programs involve bartering the risks of failing to detect an impact (β) and/or the size of an impact that might be detectable with confidence against increased costs of monitoring. Except where adherence to standards is mandatory (e.g., by legislation), better monitoring explicitly or implicitly equates with increased probability (statistical Power, ρ) of detecting critical impacts (ES), if they occur. This equation provides little leverage in arguments with environmentally indifferent development proponents concerned over increased costs of monitoring: their interests would be served best by a cheap and insensitive EIM program. Here, notions of the importance of a detectable impact become entwined in a debate over costs and probabilities of error—with the result that the question of "how big an impact is tolerable?" often is translated to "how small an impact can the developer afford (economically) to worry about?"

The apparent immutability of the critical Type I error rate ($\alpha_c = 0.05$) might not present a problem if the Type II error rates typically realized in EIM were at least close to the α_c that drives decisions, or if small ESs were detectable with reasonable certainty. In my experience, however, this is rarely the case—a view consistent with those of others who have addressed the neglect of statistical power in decision making (Andrew and Mapstone 1987, Fairweather 1991, Peterman 1990 and references therein, Toft and Shea 1983, Hayes 1987). In reviewing the unpublished reports of many EIM programs, I have found many in which the H_0 of no impact was not rejected, but very few where the likelihood of Type II error in doing so was less than 0.4 for an impact that would constitute an 80–100% change in the measured variables.

New Decision Rules

Resolving the above problems requires a substantial shift in approach to statistical decision making. An environmentally conservative approach would be to stipulate a low rate of Type II error, perhaps—though not necessarily—at the expense of elevated Type I error rates. Such an emphasis, however, leads to the consideration of α_c as a variable. Although other authors have suggested that the value of α_c should be carefully considered, implying that it should be considered as a variable at some stage (Bernstein and Zalinski 1983, Millard and Lettenmaier 1986, Oakes 1986, Andrew and Mapstone 1987, Millard 1987, Cohen 1988, Peterman 1990, Fairweather 1991), few have suggested an alternative strategy for decision making that incorporates a variable α_c (Cohen 1988, Peterman 1990) and few have adopted a "non-standard" α_c (e.g., Holt et al. 1987, Mapstone 1988, Mapstone et al. 1989).

If α_c is to be "liberated", the singular decision rule currently in use must be replaced with a sensible alternative. It is important, however, that we do not simply replace historic dogma ($\alpha_c = 0.05$) with contemporary chaos (anything goes) or the foundations of future dogma (e.g., $\beta = 0.05$). It is imperative also that any alternative decision rule(s) be specified and agreed *a priori* for any EIM (or research) program—placing bets after the race has been run has always been illegal.

I suggest that changes to decision-making procedures must occur both before and after data collection. Two parallel, but initially independent, procedures should be followed prior to the collection of data. The first involves the choice of the level of impact (= critical ES) that we want to detect (if it really occurs). The stipulation of an Effect Size is a biological (or chemical, physical, aesthetic, etc.) decision, not simply a statistical or procedural decision, and involves a raft of judgments about the biological importance of an effect of that magnitude. As such, specification of ESs is perhaps the most critical aspect of environmental impact decisions, and possibly also of decisions in fundamental research, since this is where the importance of potential impacts (or alternative hypothesized outcomes) must be evaluated. Such evaluation should proceed in parallel with, but not subject to, negotiations over the costs of EIM and risks of erroneous decisions.

The second *a priori* requisite of the amended decision rules involves the derivation of critical levels of α and β. Evaluation of the relative consequences of Type I and Type II errors should play an instrumental part in determining the critical level of α that would lead to the rejection of a null hypothesis (see also Peterman 1990, Fairweather 1991). Setting a desired α_c in relation to the consequences of decision outcomes involves four steps:

(i) Establish the relative importance of consequences (economic, political, environmental, social costs) of Type I and Type II errors.
(ii) Designate the potential costs of Type I and Type II errors as C_I and C_{II}

respectively and calculate $k = C_{II}/C_I$. Set the *ratio* of critical Type I and Type II errors according to the relative costs of committing those errors: i.e., $\alpha_c/\beta = k = C_{II}/C_I$ (see also Peterman 1990), or more conveniently, $\alpha_c = k\beta$ (see later). This is an important and difficult task and I suggest that in the absence of sufficient information to relate the costs of errors, C_I and C_{II} should be weighted equally (i.e., $k = C_{II}/C_I = 1$, $\alpha_c = \beta$) rather than the traditional strategy of tying down α_c and letting β roam (see also Kmietowicz and Pearman 1981 and references therein; Peterman 1990). Note that thus far the absolute level of neither potential error rate has been specified.

(iii) Specify: (a) the maximum risk of development being unnecessarily interrupted because of a Type I error with which the development proponent(s) would be prepared to work; and (b) the maximum risk that real and important impacts would go undetected because of a Type II error with which managers would be prepared to live. From these starting values, the values for risks of errors are negotiated with reference to the agreed relation $\alpha_c = k\beta$. These negotiations determine *a priori* the desired values of both α_c and β (denoted α'_c and β') that all parties would be prepared to accept as criteria for the design of a monitoring program. For example, if it is decided that $k = 1$ and a risk of Type I error of more than 0.05 is unacceptable, then this procedure will dictate that the risk of Type II error should also be 0.05 or less.

(iv) Design an EIM program that would be likely to realize the desirable Type II error rate (β'), given the effect size specified previously (and independently), and the desired critical Type I error rate, α'_c. Sample size and design would be the dependent variables in these calculations (e.g., see Millard and Lettenmaier 1986).

After collecting the data from such a program, statistical decisions would be driven by the agreed value(s) of k. Two procedures might be used to make those decisions. The first amounts to a decision against a critical α; the second entails a decision against the relative realized values of α and β.

In the first decision procedure, given the critical ES and the design of the program from which the data were collected, the actual critical Type I error rate (α_c) would be set by iteration, as follows (Figure 5.1):

   (i) Set α_c to that desired *a priori*: i.e., $\alpha_c = \alpha'_c$;
   (ii) Calculate the Type II error rate expected (β_0) if the null hypothesis was not rejected at that α_c;
   (iii) If $\alpha_c < k\beta_0$ or $\alpha_c > k\beta_0$, adjust α_c (higher or lower respectively) to reduce the inequality, and then repeat step (ii) with this new α_c;
   (iv) Iterate steps (ii) and (iii) until $\alpha_c = k\beta_0$;
   (v) Compare the observed probability of the data if H_0 was true (α_0) with the final value of α_c and reject H_0 if $\alpha_0 \le \alpha_c$.

An alternative decision procedure differs from the first in that the decision rests on direct comparison of α_0 and $k\beta_0$, as follows:

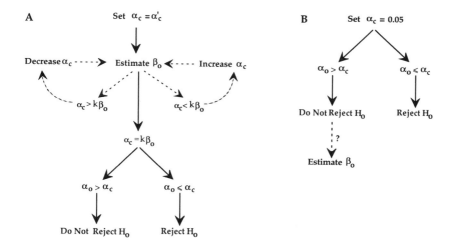

Figure 5.1. Flow diagrams of the proposed decision procedures (A) and conventional procedures (B). Solid lines indicate the central components of a decision, i.e., choices or steps that must always be taken. Dashed lines in A indicate iterative steps that may be necessary in refining decision criteria. The ? in B indicates that the estimation of β is optional under conventional practice.

(i) Calculate the probability of the data if H_0 true (α_0);

(ii) Calculate β_0 on the assumption that the H_0 was not rejected against a critical Type I error rate marginally less than α_0: i.e., $\alpha_c = \alpha_0 - \delta$ (δ very small);

(iii) If $\alpha_0 < k\beta_0$, reject H_0, otherwise, do not reject H_0.

This decision is based on the argument that had α_c been set at $\alpha_0 + \delta$, the null hypothesis would have been rejected (since $\alpha_0 < \alpha_c$) and the probability of Type I error would be α_c or less. If $k\beta_0$ (as calculated above) $> \alpha_c = \alpha_0 + \delta$, the probability of error in rejecting H_0 would be less than the probability of error in nonrejection against a critical value of $\alpha_c = \alpha_0 - \delta$. If $k\beta_0 < \alpha_0$, then non rejection would be the least error prone option. Ambiguity would arise if $k\beta_0 = \alpha_0$ exactly. This is unlikely, but in that event either decision would carry the same probability of error.

In both of the above alternatives, the decision rule is based on the evaluation *a priori* of the relative importance of consequences of Type I and Type II errors, and adherence to that evaluation in the final decision. Both error rates are considered variable, and consequently the critical probabilities upon which decisions will be made might be expected to vary from case to case, even though the decision rule might be constant. It is assumed here that the critical ES has been stipulated already, and is not subject to negotiation at the time of data analysis or in view of the realized rates of α or β.

In both cases it is essential to derive an absolute value for β' because it is the desire to perpetuate that value to the final decision that provides the main criterion for the design of the EIM program. In the first procedure, the value of α_c used in the eventual hypothesis test will also depend on how successfully the EIM program has been designed to realize $\beta_0 = \beta'$, because $\alpha_c = k\beta_0$. Hence, minimizing the risk of frequent intervention in development is contingent upon minimizing the risk of failing to recognize real (important) impacts. In the second procedure, although the level of α_c is not set by the level of β_0, the decisions about the likely presence of impacts will depend on the relative values of α_0 and β_0. Frequent intervention will be avoided, therefore, by that EIM design that maximizes the potential for $k\beta_0 < \alpha_0$ for moderate and small values of α_0: i.e., a design which minimizes β_0.

It is important to note, however, that the values of β_0 and α_c ultimately realized are likely to differ from those considered desirable (β', α'_c) because of the likelihood of estimation errors in the design of the EIM. Although we may design an EIM program to realize a specified $\beta_0 = \beta'$, such a design will be based on estimates of variance etc. that will be unlikely to be realized exactly in the final data set (e.g., see McArdle et al. 1990). The consequences of not realizing the *a priori* desired probability of errors, however, are shared by both proponents and managers. Had the traditional strategy of decision against a fixed α_c been followed, the consequences of inexact design would have reflected exclusively on the value of β_0, and been borne by the environment (Figure 5.2).

Advantages of Liberating α

The liberation of α_c and adoption of a scalable decision rule has several advantages. Firstly, trading a variable α'_c against a variable β' means that attention is clearly focused on the consequences of both decision outcomes. This forces the explicit *a priori* evaluation of the relative importance of Type I errors and Type II errors, and hence direct comparison of the risks the interested parties are prepared to take in proceeding with a development (given a nominated level of EIM). Here, better monitoring means a lower probability of Type II error *and* of Type I error. Both are essentially arbitrary limits of confidence in an outcome with which we are prepared to live, and in this process they are evaluated as such.

Secondly, cost savings in monitoring will be traded against costs of frequent intervention in development that might arise from a high α_c, tied to a high β_0 by the relation $\alpha_c = k\beta_0$. Lower risk of erroneous interference is achieved only by realizing a lower value for β_0: i.e., by better monitoring (Figure 5.2). Further, given that it is likely to be more difficult to prove an error of Type I (for which a developer might seek compensation) than an error of Type II (for which a developer might have to pay reparation), it is clearly in the developer's best interest to promote rigorous monitoring.

Thirdly, the stipulation of effect size, the crucial ecological variable, is removed from the bargaining table when the scope and costs of EIM are being

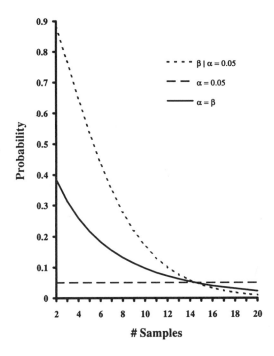

Figure 5.2. Illustrative example of the effects of increasing sample size on the values of α_c and β_0 under the conventional and proposed statistical decision rules. Under conventional practice, the value of α_c (– – – –) does not change with increasing sample size, and all benefits of increased sample size reflect on the value of β_0 (- - - -). With the proposed rules, small samples mean equally high values of both α_c and β_0 (———), and increasing sample sizes reduces both. Effect size is constant throughout and $k = 1$ (see text).

negotiated. By having effect size stipulated independently on the basis of ecological principles or empirical evidence, responsibility for this decision is vested with biologists (and managers?), who should be best placed to make it! Argument over whether a detected impact is important will be dealt with before monitoring rather than afterward, as is often the case now. At no point should the critical ES in a chosen variable be adjusted because of the cost of designing an EIM program that would be likely to detect that effect. Such logistic and financial considerations, however, might influence the choice of variables to monitor. If it is impossible to stipulate what level of perturbation would be considered important for particular variables, then monitoring those variables should be seen as information gathering, exploration, provision of data for the future - not as vehicles for testing hypotheses about impacts. The above procedures precipitate these issues *a priori*, whereas conventional procedures may never stimulate consideration of them.

Fourthly, the suggested procedures allow different decision criteria (α_c, β) to be attached to different scales of impact and/or different variables measured in EIM. Where EIM incorporates assessment of impacts at a hierarchy of spatial or temporal scales (Mapstone et al. 1989, Underwood 1991b), the consequences of impacts at each scale can be evaluated independently and decision criteria set accordingly. For example, it might be considered unimportant that an impact of specified magnitude was missed on a very small scale (such as a single small site, or for a period of only 1 week), and the maximum acceptable level of β might be set at a relatively large value (e.g., 0.3). An impact of similar, or different, magnitude (the ES might also be scale dependent) at a larger scale, such as over an entire bay or effective for a month, might be considered far more important and, accordingly, the maximum acceptable level of β might be set to a very low level. Thus, the significance criteria by which H_0s were rejected might vary among the terms within a single multifactorial ANOVA. Similarly, it might be considered crucial that a real impact for one variable (e.g., levels of organochlorines) was detected, but of only marginal concern that an impact on another variable (e.g., water clarity) was missed. Different decision criteria would be applied to different variables, and their sampling intensities determined accordingly.

Fifthly, the above procedures explicitly identify a set of priorities in assessing environmental impacts. The independent establishment of a critical Effect Size—the maximum "acceptable" impact—takes highest priority. Such limits to impacts might vary among cases and should be determined in advance by reference to the local environment, but should not be dictated by political or economic expediency. Establishing the relative importance of Type I and Type II errors, and the levels of certainty ($1-\alpha$ or $1-\beta$) desired in decisions, are also high priorities and should be assessed relative to stipulated critical ESs. Lastly, the costs of monitoring are evaluated against the risks involved in assessments of impact, again, without impinging on the size of impact that is considered ecologically (or by some other independent criterion) important. I suggest that these priorities provide a fair basis for balancing the burden of proof (Peterman 1990) between developer and environmental manager.

Some Problems with a Variable α

The Determination of Critical Effect Size

Effect size is defined by two components: the form(s) and magnitude(s) (Cohen 1988) of the impact(s) we seek to detect if they occur. The form of impact involves deciding whether we are concerned with changes in means and/or variances at impact sites relative to control sites (Green 1979, 1989, Stewart-Oaten et al. 1986, Underwood 1991a, 1991b, Chapter 9, Stewart-Oaten, Chapter 7), deciding at what scales we expect impacts might occur (Raimondi and Reed, Chapter 10), and specifying which means (or groups of means) differ from which others. The magnitude of an impact is a measure of the amount by which means or variances change. Cohen (1988) discusses the need to specify exactly both components of an effect to sensibly construct alternative hypotheses. Specifying both aspects of ES will be difficult for complex EIM designs, particularly when impacts of interest are measured by interaction terms in analyses (e.g., Underwood, Chapter 9; but see Stewart-Oaten, Chapter 7).

For each variable being monitored, we must answer the questions "how much anthropogenic disturbance is acceptable?", and "what amount of development-related change should precipitate management action?" The two answers will not always be similar, since with some variables (e.g., concentrations of persistent or toxic pollutants) we might wish to initiate management action at levels of change well below those that would be considered the limits for environmental or human well being. For such variables we might wish EIM to provide the basis for proactive rather than reactive management. Even when limits of acceptable change are specified by regulations (e.g., EPA regulations), we will still be faced with deciding at what point we would wish to know if that level were being approached, in the interests of ensuring that it was never reached. The above procedures highlight the shortcomings of our current understanding of environmental impacts, their importance, and how to manage them. Clearly, more research in this field is required.

Two further points should be noted here. Firstly, although I have routinely referred to ES in terms of "ecological" importance, other criteria for critical ESs might be important. For example, aesthetic or economic consequences of environmental impacts might result in the stipulation of more stringent critical ESs than would arise from consideration of ecological consequences alone. Two examples illustrate this point: (i) in areas where eco-tourism is important (e.g., the Great Barrier Reef), impacts that might be considered relatively minor ecologically might represent important degradations in aesthetic quality and reduced tourism; and (ii) where environmental impacts impinge on a fishery, either by pollution or stock reduction, the magnitude of impacts that precipitated economic losses to the fishing industry might be smaller and considered more critical than those which would cause substantive ecological effects.

Secondly, it must be emphasized that the critical ES is not a categorical distinction between "impact" and "no impact," but simply provides a cut-off point on a continuous relation between ES and β (or statistical power). A monitoring program will have lower power to detect smaller than "critical" impacts that have slighter consequences. The critical ES merely specifies where the (increasing) consequences of impact become unacceptable.

Weighting α_c and β

The weighting of Type I and Type II errors is unfamiliar to most scientists, managers, and development proponents, although the need to do so has been discussed for some time (see Kmietowicz and Pearman 1981, Oakes 1986, Peterman 1990). If α_c becomes large as a consequence of liberating α_c and demanding low values for β, there is likely to be increased intervention in development, and more frequent litigation of decisions from EIM programs. Avoiding such an action will entail greater costs of EIM, but those costs will become increasingly critical (for the proponent) as the economic scale of development decreases. Further difficulty arises when the costs of Type I and Type II errors are measured in different currencies—e.g., money versus genetic diversity. It is likely that economic rationalist attempts to value all costs in monetary terms will favor short-term development interests, especially where entire communities are dependent on a nominated development. Such issues make the evaluation of α'_c/β' particularly critical. Again, these issues can be comfortably ignored within traditional decision-making practice, but are critical elements of the procedures I suggest above.

Administrative Reality

Pragmatists will say that the above procedures involve considerably more work than existing practice and that it will be difficult to agree upon and sustain decision criteria throughout the life of an EIM exercise. The vagaries of political priorities, management changes, and new information are likely to mean that what was considered appropriate initially will be repeatedly challenged, debated, and possibly changed—particularly where developments and EIM span several years. The above procedures precipitate such debate, but the issues are not new. In current practice, issues such as the "importance" of an observed impact, the adequacy of monitoring, or the basis of a decision are fought after the event, when revision, adjustment and improvement are impossible. The likelihood that these debates would be brought forward is an advantage of the above procedures, but will require diligent legal management.

Summary

Conventional decision-making practices in respect of environmental impacts perpetuate an inherently one-sided perspective of "significance." Masked by the security of a well established convention of statistical decision making is a suite of difficult inferential and epistemological problems that have real, tangible implications. The revised approach to statistical decision making I suggest here does not solve these problems, but focuses attention on them and highlights the need for urgent attention to them. More importantly, the procedures I suggest provide a mechanism by which the burden of limitations of current practice is shared between potential environmental assailant and environmental defender rather than being borne solely by the latter.

Acknowledgments

The ideas expressed here benefited from discussions with N. Andrew, J. H. Choat, G. Jones, S. McNeill, H. Marsh, J. Oliver, K. Shurcliff, and A. J. Underwood. I am grateful to N. Andrew, S. Holbrook, G. Jones, C. Osenberg, R. Schmitt and two anonymous reviewers for comments on earlier manuscripts. The main theme of the chapter was presented at the 2nd International Temperate Reef Symposium (Auckland 1992). I acknowledge the support of the Australian Research Council (for a National Research Fellowship), the Great Barrier Reef Marine Park Authority, and the CSIRO (Pulp and Paper Research Project).

References

Andrew, N. L., and B. D. Mapstone. 1987. Sampling and the description of spatial pattern in marine ecology. Oceanography and Marine Biology Annual Review **25**:39–90.

Beder, S. 1991. The many meanings of means. Search **22**:88–90.

Bernstein, B. B., and J. Zalinski. 1983. An optimum sampling design and power tests for environmental biologists. Journal of Environmental Management **16**:35–43.

Buckley, R. 1989a. Precision in environmental impact predictions: first national environmental audit, Australia. Center for Resource and Environmental Studies, Australian National University, Canberra, Australia.

Buckley, R. 1989b. What's wrong with EIA? Search **20**:146–147.

Buckley, R. 1990. Environmental science and environmental management. Search **21**:14–16.

Buckley, R., and J. McDonald. 1991. Science and law—the nature of evidence. Search **22**:94–95.

Christie, E. 1990. Science, law and environmental litigation. Search **21**:258–260.

Cohen, J. 1988. Statistical power analysis for the behavioural sciences, 2nd Edition. Lawrence Erlbaum Associates, Hillsdale, New Jersey.

Fairweather, P. G. 1989. Environmental impact assessment—where is the science in EIA? Search **20**:141–144.

Fairweather, P. G. 1991. Statistical power and design requirements for environmental monitoring. Australian Journal of Marine and Freshwater Research **42**:555–567.

Faith, D. P., C. L. Humphrey, and P. L. Dostine. 1990. Requirements for effective biological monitoring of aquatic ecosystems in the area of the Kakadu Conservation Zone. Submission No. KA 90/047 to the Inquiry into the Kakadu Conservation Zone, Resource Assessment Commission, Canberra, Australia.

Faith, D. P., C. L. Humphrey, and P. L. Dostine. 1991. Statistical power and BACI designs in biological monitoring: comparative evaluation of measures of community dissimilarity based on benthic macroinvertebrate communities in Rockhole Mine Creek, Northern Territory, Australia. Australian Journal of Marine and Freshwater Research **42**:589–602.

Green, R. H. 1979. Sampling design and statistical methods for environmental biologists. Wiley and Sons, New York, New York.

Green, R. H. 1989. Power analysis and practical strategies for environmental monitoring. Environmental Research **50**:195–205.

Hayes, J. P. 1987. The positive approach to negative results in toxicology studies. Ecotoxicology and Environmental Safety **14**:73–77.

Holt, R. S., T. Gerrodette, and J. B. Cologne. 1987. Research vessel survey design for monitoring dolphin abundance in the eastern tropical Pacific. Fisheries Bulletin of the U.S. **85**:435–446.

Hoverman, J. S. 1989. Scientists must punt or be damned. Search **20**:147–148.

Jarrett, R. G. 1991. Means to an end: a statistical adjudication. Search **22**:92–93.

Keough, M. J., and G. P. Quinn. 1991. Causality and the choice of measurements for detecting human impacts in marine environments. Australian Journal of Marine and Freshwater Research **42**:539–554.

Kmietowicz, Z. W., and A. D. Pearman. 1981. Decision theory and incomplete knowledge. Gower, Hampshire, United Kingdom.

Lincoln Smith, M. P. 1991. Environmental impact assessment: the roles of predicting and monitoring the extent of impacts. Australian Journal of Marine and Freshwater Research **42**:603–614.

MacDonald, R. 1991. Statistics in the real world. Search **22**:91–92.

Mapstone, B. D. 1988. The determination of patterns in the abundance of *Pomacentrus moluccensis* Bleeker on the southern Great Barrier Reef. Ph.D. Thesis. University of Sydney, Australia.

Mapstone, B. D., J. H. Choat, R. L. Cumming, and W. G. Oxley. 1989. The fringing reefs of Magnetic Island: benthic biota and sedimentation. Great Barrier Reef Marine Park Authority Publication No. 13, Townsville, Australia.

Martin, R., and P. A. Berman. 1991. Environmental research in Australia. Search 22:95–96.

McArdle, B. H., K. J. Gaston, and J. H. Lawton. 1990. Variation in the size of animal populations: patterns, problems, and artefacts. Journal of Animal Ecology 59:439–454.

Millard, S. P. 1987. Environmental monitoring, statistics, and the law: room for improvement. The American Statistician 41:249–253.

Millard, S. P., and D. P. Lettenmaier. 1986. Optimal design of biological sampling programs using the analysis of variance. Estuarine, Coastal and Shelf Science 22:637–656.

Oakes, M. 1986. Statistical inference: a commentary for the social and behavioural sciences. John Wiley and Sons, Chichester, UK.

Peterman, R. M. 1990. Statistical power analysis can improve fisheries research and management. Canadian Journal of Fisheries and Aquatic Sciences 47:2–15.

Soule, M. E. 1991. Conservation: tactics for a constant crisis. Science 253:744–750.

Stewart-Oaten, A., W. W. Murdoch, and K. R. Parker. 1986. Environmental impact assessment: "psuedoreplication" in time? Ecology 67:929–940.

Toft, C. A., and P. J. Shea. 1983. Detecting community wide patterns: estimating power strengthens statistical inference. American Naturalist 122:618–625.

Underwood, A. J. 1990. Experiments in ecology and management: their logics, functions and interpretations. Australian Journal of Ecology 15:365–389.

Underwood, A. J. 1991a. Beyond BACI: experimental designs for detecting human environmental impacts on temporal variations in natural populations. Australian Journal of Marine and Freshwater Research 42:569–587.

Underwood, A. J. 1991b. Biological monitoring for human impact: how little it can achieve. Pages 105–223 in R. V. Hyne, editor. Proceedings of the 29th Congress of the Australian Society of Limnology, Jabiru, NT, 1990. Office of the Supervising Scientist Alligator Rivers Region, Australian Government Publishing Service, Canberra, Australia.

Underwood, A. J., and C. H. Peterson. 1988. Towards an ecological framework for investigating pollution. Marine Ecology Progress Series 46:227–234.

Warwick, R. M. 1988. The level of taxonomic discrimination required to detect pollution effects on marine benthic communities. Marine Pollution Bulletin 19:259–268.

Warwick, R. M., and K. R. Clarke. 1991. A comparison of some methods for analysing changes in benthic community structure. Journal of the Marine Biology Association of the United Kingdom 71:225–244.

Winer, B. J. 1971. Statistical principles in experimental design, 2nd Edition. McGraw-Hill, New York, New York.

IMPROVING FIELD ASSESSMENTS
OF LOCAL IMPACTS
Before-After-Control-Impact Designs

CHAPTER 6

DETECTION OF ENVIRONMENTAL IMPACTS

Natural Variability, Effect Size, and Power Analysis[1]

Craig W. Osenberg, Russell J. Schmitt, Sally J. Holbrook, Khalil E. Abu-Saba, and A. Russell Flegal

A principal challenge posed in field assessments of environmental impacts is to isolate the effect of interest from noise introduced by natural spatial and temporal variability. If the size of an impact from a human disturbance is small relative to natural variability, it will be difficult to detect with any degree of confidence. Therefore, it is critical to consider statistical power in planning and interpreting environmental impact assessment studies (Green 1989, Fairweather 1991, Faith et al. 1991, Osenberg et al. 1992a, Mapstone, Chapter 5; see also Peterman 1990, Cooper and Barmuta 1993). Consideration of power can also guide the selection of environmental parameters and sampling intensity. These are important design criteria because time and financial constraints typically limit the number of parameters that can be measured and the number of samples that can be collected.

Calculation of statistical power, which is the probability of rejecting the null hypothesis of "no effect" when it is false and a specified alternative is true, requires specification of the number of replicates as well as the ratio between the size of the "true" effect and the variability among the replicates (Cohen 1977). Because there are many assessment designs, each of which makes different assumptions about the meaning of "effect", "variability" and "replicate" (Green 1979, Stewart-Oaten et al. 1986, Eberhardt and Thomas 1991, Underwood 1991, Chapter 9, Osenberg and Schmitt, Chapter 1), the general assessment design must be specified before power can be discussed unambiguously. In assessing the environmental impacts of a particular anthropogenic activity, we typically

---

[1]This chapter was previously published in a modified form in Ecological Applications 4:16–30. © 1994 by the Ecological Society of America.

require a design that explicitly deals with the lack of spatial replication and randomization (e.g., nuclear power plants are not replicated and placed at random sites along the U.S. coastline: Stewart-Oaten et al. 1986). The Before-After-Control-Impact Paired Series design (BACIPS: Stewart-Oaten et al. 1986, Schroeter et al. 1993, Stewart-Oaten, Chapter 7; see also Campbell and Stanley 1966, Eberhardt 1976, Skalski and McKenzie 1982, Bernstein and Zalinksi 1983, Carpenter et al. 1989) meets this criterion, and is the focus of our analyses and discussion.

In its simplest formulation, BACIPS requires simultaneous (*Paired*) sampling several times *Before* and *After* the perturbation at a *Control* and an *Impact* site. The measure of interest is the difference (hereafter referred to as "delta", Δ) in a parameter value (in its raw or transformed state) between the Control and Impact sites as assessed on each sampling date (e.g., $\Delta_{Pi} = \log(C_{Pi}) - \log(I_{Pi})$, where C_{Pi} and I_{Pi} are estimates of the parameter at the Control and Impact sites on the ith date of the Period P: i.e., Before or After). The average delta in the Before period is an estimate of the average spatial variation between the two sites, which provides an estimate of the expected delta that should exist in the After period in the absence of an environmental impact (i.e., the null hypothesis). The difference between the average Before and After deltas ($\Delta_{B.} - \Delta_{A.}$) provides an estimate of the magnitude of the environmental impact. Confidence in this estimate is determined by the variation in deltas (among sampling dates within a period, S_{Δ}), as well as the number of sampling dates (i.e., replicates) in each of the Before and After periods ($n_B + n_A = n$). For the purposes of this study, we define

$$\text{Effect Size} = \Delta_{B.} - \Delta_{A.} \tag{1}$$

$$\text{Variability} = S_{\Delta} = [\Sigma \, (\Delta_{Pi} - \Delta_{P.})^2]^{1/2}/(n_P - 1)^{1/2} \tag{2}$$

$$\text{Standardized Effect Size} = |\, \Delta_{B.} - \Delta_{A.} \,|/(2S_{\Delta}) \tag{3}$$

We assume for convenience that variability (S_{Δ}), as well as sample size (n_P), are equal in the Before and After periods (but see Stewart-Oaten et al. 1992), and we double the standard deviation of deltas (S_{Δ}) in the denominator of Equation 3 based on the assumption that the resulting test will be two-tailed (Gill 1978).

Note that the standardized effect size (Equation 3), which consists of the components defined by Equations 1 and 2, expresses the effect size in standard deviation units and enters directly into conventional calculations of power (Cohen 1977). Note also that our terminology differs from that used by some authors (e.g., Cohen 1977): we use "effect size" to refer to the absolute magnitude of the effect and "standardized effect size" to refer to the standardized measure (which is often simply labelled "effect size").

Unlike other designs, the variability of interest, S_{Δ}, is not a simple measure of within-site sampling variability. Rather, it is a measure of the actual temporal variation in deltas, as well as within-site sampling error (which contributes to error in estimating the actual delta on any date). Figure 6.1 illustrates how this variability of deltas can be altered without any change in the average temporal

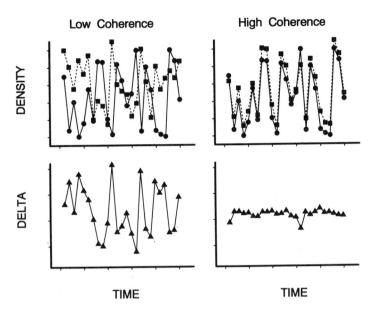

Figure 6.1. Patterns of spatial and temporal variation in population densities that lead to high and low variation in deltas. Simulated data (top panels) are from two pairs of sites. In both panels temporal variation in density (at a site) and the average difference between the sites are similar. The panels differ in the degree to which the estimated densities at the paired sites track one another through time. On the left, poor tracking (i.e., low coherence: Magnuson et al. 1990) leads to a low correlation between densities at the two sites ($r = -0.25$), while on the right, good tracking (i.e., high coherence) leads to a stronger correlation in densities ($r = 0.98$). The bottom graphs show the resulting differences in density (deltas). Low temporal coherence in densities (or any other parameter of interest) leads to high variability in deltas (i.e., relatively low power), while high coherence leads to low variability in deltas (i.e., relatively high power).

variability of a parameter (e.g., density), or in the amount of within-site sampling error. The critical feature in determining the variability among deltas is the extent to which estimates of parameters at the two sites track one another though time; Magnuson et al. (1990) refer to this as temporal coherence.

To aid in the planning of a BACIPS study, it would be helpful to find previous BACIPS studies conducted in a comparable situation (e.g., similar perturbation in a similar environment) and review the results for variability and effect size. This would permit estimation of the number of sampling dates needed to achieve a given level of power (e.g., Bernstein and Zalinski 1983) or a given amount of confidence in estimates of the effect size (e.g., Bence et al., Chapter 8, Stewart-Oaten, Chapter 2). For example, parameters with large standardized effect size (i.e., relatively large effect size and small variability) will yield more powerful assessments with fewer sampling events than parameters with low standardized effect size. Obtaining an adequate number of sampling events in the Before period is crucial in a BACIPS assessment since once the perturbation begins, it is no longer possible to obtain additional Before samples. Unfortunately, there are few existing BACIPS studies that permit this type of analysis.

In the absence of this information, other data could be used to guide the design of BACIPS studies. Two types of non-BACIPS studies are more common and can offer insight. The first are long-term studies that document natural spatial and temporal variability, and therefore can provide estimates of S_Δ (Equation 2). The second are "After-only" (or Control-Impact) studies that assess impacts using a postimpact survey of sites that vary in proximity to the perturbation. After-only studies are a common type of field assessment approach, but they confound effects of the perturbation with natural spatial variability. Still, After-only studies can suggest the size of effects that might occur in response to a particular perturbation (Equation 1).

In this chapter, we illustrate how information from long-term studies and After-only studies can be combined to help plan BACIPS studies. We show how this information can be used to guide the selection of parameters and determine sampling schedules given constraints of time and funding. Our presentation consists of four analytical steps: (i) estimation of temporal variability of deltas using results from a long-term study; (ii) estimation of the likely magnitude of impacts using results from an After-only study; (iii) determination of the number of sampling dates required to detect the estimated impact given the background variability (at a specified level of power); and (iv) exploration of serial correlation, using the long-term data set, to assess the time necessary to achieve the required number of independent sampling dates. We contrast results for chemical–physical (e.g., chemical concentrations, sediment characteristics), individual-based biological (e.g., body size, growth), and population-based biological (e.g., density) parameters, and conclude there is a critical need to increase the use of individual-based parameters in field studies of environmental impacts.

Methods

Background

To help guide the planning of a BACIPS study of a particular planned intervention, it would be best to examine results of several preexisting BACIPS studies that examined impacts on many parameters in response to the same intervention in identical environments. Of course, such studies do not (and cannot) exist, but the congruence between this ideal and the realized match serves as a guide to the potential accuracy of the general guidelines that emerge.

The first step in this process is to define the intervention. To illustrate our approach, we focus on the nearshore discharge of an aqueous waste called produced water. Produced water is a complex wastewater generated from the production of oil and contains a variety of petroleum hydrocarbons, heavy metals, and other potential pollutants (Middleditch 1984, Higashi et al. 1992). Although concerns have been raised about possible environmental effects of produced water in marine environments (Neff 1987, Neff et al. 1987, Osenberg et al. 1992b, Raimondi and Schmitt 1992), there have been no field assessments with sufficient Before data to allow separation of impacts from other sources of spatial and temporal variability (Carney 1987; also see Underwood 1991, Osenberg and Schmitt, Chapter 1).

We explore results from a long-term study of natural spatial and temporal variability and an After-only study to substitute for the absence of existing BACIPS studies. The two studies were both conducted in nearshore habitats along the coast of Santa Barbara County in southern California. The benthic environments are both dominated by soft-bottom habitats, and the studies used many of the same methods and quantified many of the same parameters. In each study, parameters had been selected based upon their perceived relevance to the impacts of produced water (e.g., Boesch and Rabalais 1987). (Because the long-term study is actually part of the "Before" sampling of a BACIPS study of produced water impacts, even the parameters examined in this study were selected with respect to produced water discharge.) However, these parameters, which include chemical, physical and biological characteristics (Table 6.1), are commonly measured in field assessments of other impacts in marine environments. We next review the two studies, the methods that were used and the parameters that were measured.

Natural Variability Assessed from a Long-Term Study

The two sites that comprise the long-term study are located approximately 1.6 km apart offshore of Gaviota, CA (ca. 34°27'29"N, 120°12'43'W) at a water depth of approximately 27 m. Various biological and chemical–physical parameters (Table 6.1) were sampled at the sites for periods ranging from 1.5 to just over 3 years beginning in February, 1988. For a given sampling date, a single value was obtained for each parameter at each site, and a delta was calculated as the difference between the log-transformed values:

$$\Delta_i = \log(X_{1i}) - \log(X_{2i}), \tag{4}$$

where X_{1i} and X_{2i} are the values of parameter X at each of the two sites (1 and 2) on the ith date. Original parameter values were log-transformed to better satisfy assumptions of additivity required by BACIPS (Stewart-Oaten et al. 1986) and to facilitate comparison of deltas for parameters measured in different units (the transformed deltas are dimensionless). For each parameter, variability was quantified as the standard deviation of the deltas (S_Δ) calculated over all available sampling dates (Equation 2).

Population-Based Parameters. Densities of infaunal organisms were estimated approximately eight times per year. On each sampling date, 12 cores

Table 6.1.   List of the Types of Parameters Used to Explore Natural Temporal Variability in Deltas (from the Long-Term Study) and to Obtain Estimates of Effect Size from an Existing Perturbation (from the After-Only Study)

Parameter type	Source	
	Long-term study (*Variability*)	After-only study (*Effect size*)
Chemical–Physical		
Water temperature (No. depths)	2	2
Seston characteristics	3	0
Sediment quality	2	2
Sediment elements	11	9
Water column elements	12	8
Individual-based: Field collections		
Urchin size and condition	5	0
Cumacean body size	2	0
Individual-based: Transplants		
Mussel performance	(10)	12
Abalone performance	0	4
Population-based (No. taxa)		
Band transects	6	0
Infaunal cores	11	10
Quadrats	1	0
Emergence traps	4	0
Re-entry traps	3	0

Note: For each parameter type, we give the number of parameters quantified at each site (e.g., for infaunal density, 11 taxonomic groups yielded sufficient data for analysis in the long-term study). Details on parameters are given in the methods section. The ten estimates of variability for mussel performance, in parentheses, were collected as part of the After-only study but analyzed in the same manner as data from the long-term study.

(each 78 cm^2 × 10 cm deep) were collected. Samples were preserved in 10% buffered formalin and sieved through a 0.5 mm mesh sieve. Organisms were identified and counted from at least 4 of these cores per site per sampling date. Because this community is extremely speciose, with many species represented by only a few organisms or by zero counts on particular dates, and because zeros can cause difficulties in BACIPS analyses (Stewart-Oaten et al. 1986), infaunal organisms were grouped into broad taxonomic units, such as families and classes (see discussions on aggregation in Herman and Heip 1988, Warwick 1988, Frost et al. 1992, Carney, Chapter 15).

Numbers of infaunal organisms that migrated from the sediments into the overlying water (i.e., demersal zooplankton) were estimated using two emergence funnel traps (each covering a bottom area of 0.23 m²) and three reentry traps (each 0.05 m² in area), which were deployed at both sites approximately eight times per year (for more detail on trap designs and function, see Alldredge and King 1980, Stretch 1983). Traps were set out for a 24-hr period. Following retrieval, contents were preserved, sieved through a 0.5 mm mesh sieve, and organisms were identified and counted as with the infaunal cores (Table 6.1).

Densities of larger epifaunal and demersal organisms (e.g., fish, sea stars, tube anemones) were estimated visually along band transects by divers. Two band transects (each 40 m by 1 m) were established along the 27 m isobath at both sites on each sampling date, and all large organisms within the transect were counted. Most were identified to species, although we grouped many of them into larger taxonomic units for these analyses. Due to their greater maximum density, white sea urchins (*Lytechinus anamesus*) were counted in 5 nonpermanent quadrats, each 1 m² in area, at both sites on all dates. Densities of urchins and other epifaunal and demersal organisms were estimated 8–12 times per year.

Individual-Based Parameters. The size (length of metasome) of two cumacean species was measured from samples obtained from the emergence traps. Other individual-based parameters (Table 6.1), were calculated from samples of the white sea urchin, *Lytechinus anamesus*: test diameter, gonad mass, somatic tissue mass, and gonadal-somatic index. In addition, the condition of urchins was estimated by calculating an adjusted mean for each site and date based on ANCOVA using each collection as a group, log(test diameter) as the covariate, and log(total tissue mass) as the response parameter. Urchins were sampled for these analyses 11 times during the study. As part of the After-only study, we also obtained estimates of variability for several other individual-based parameters derived from study of the mussel, *Mytilus californianus* (see below: *Combining Results on Effect Size and Natural Variability*).

Chemical–Physical Parameters. Chemical and physical parameters were examined that were thought to be indicative of the future plume's chemistry (e.g., elevated levels of certain heavy metals) or of the discharge's physical effects (e.g., altered sediment traits due to scouring of substrate or altered sedimentation rates and temperature due to local oceanographic effects) (Table 6.1). Seston flux was estimated by particulate accumulation in two sediment traps (5.1 cm diameter) that were filled with a mixture of seawater, formalin, and salt; the dense preservative remained in the sediment traps during the deployment and had an initial salinity of approximately 65 ppt and a formalin concentration of 5%. Sediment traps were deployed approximately 3 m above the sediments and retrieved by divers after 3–7 days. Traps were deployed approximately 8 times per year. Prior to analysis, large invertebrates were removed (aided by a dissecting microscope), following which the dry mass and ash free dry mass (AFDM) of the particles were determined. Sedimentation rate was calculated as the mass of material (on a dry mass or AFDM basis) per cm^2d^{-1}. The percentage organic matter in the seston was estimated as the ratio of AFDM to dry mass.

Sediment grain size and percent organic matter were characterized from two sediment cores (20.3 cm^2/core, 5 cm deep) collected from both sites approximately 8 times per year. Sediment organic matter (SOM) was estimated based on combustion (for 4 hr at $450^{\circ}C$) of subsamples from one core. The fine sediment fraction (percent) was estimated from the other core as the percent (by dry mass) of the sample that passed through a 0.063 mm mesh sieve.

Water temperature was recorded approximately monthly at 3 m depth intervals. Here we use data for the 6 m and 21 m depths.

Surficial sediments (approximately the top 1 cm) were collected 4 times per year for analyses of trace and bulk elements. Three samples were collected at each site in acid-cleaned polyethylene containers by divers using trace metal clean sampling techniques. Any overlying water was decanted and samples were frozen. Sediments were later thawed and extractions performed by leaching 2 g sediment in 20 ml of 0.5 N HCl for 24 hr. The leachate was then filtered through a 0.45 μm mesh teflon filter using procedures reported previously (Oakden et al. 1984). This extraction is considered to be relatively selective for the biologically available concentrations of many metals, such as Pb, Cu, and Ag (Luoma et al. 1991). Leachates were analyzed for bulk elements (Al, Ca, Fe, Mg, Mn, P) and trace elements (Ba and Zn) by inductively coupled plasma-atomic emission spectrometry (ICP-AES). Other trace element (Cr, Cd, and Pb) concentrations were determined by graphite furnace atomic absorption spectrometry (GFAAS). Environment Canada reference sediments (BCSS-1, MESS-1, PACS-1) were analyzed concurrently to quantify the extraction efficiency for each element. All analyses were normalized to sediment dry weight.

Unfiltered water samples were collected two times per year from each site at two depths (surface and 21 m). The samples were extracted using the ammonium 1-pyrrolidinedithiocarbamate/diethylammonium diethyldithiocarbamate (APDC/DDC) extraction method described by Bruland et al. (1985). Trace element concentrations (Ag, Cd, Co, Cu, Fe, Ni, Pb, Zn) were measured by GFAAS. Procedural blanks were measured in each sample set. Each set of samples was analyzed in duplicate after a series of intercalibrations with Environment Canada reference seawater (CASS-1). These analyses were conducted concurrently with analyses of sea water from San Francisco Bay, and details of the procedural blanks and intercalibrations are provided in a report on those data (Flegal et al. 1991).

Effect Size Estimated from an After-Only Study

The After-only study was conducted at a produced water outfall located near Carpinteria, CA (34°23'10"N, 119°30'31"W) that was the subject of recent investigations of potential environmental impacts (Higashi et al. 1992, Krause et al. 1992, Osenberg et al. 1992b, Raimondi and Schmitt 1992, Raimondi and Reed, Chapter 10). The Carpinteria sites are approximately 50 km from the Gaviota sites. Although the two locations (Carpinteria and Gaviota) are both open coast, soft-bottom environments in the Santa Barbara Channel and have many species in common, the bottom depths sampled differed between the Carpinteria (11 m) and Gaviota (27 m) sites.

An intensive spatial survey of infauna was conducted along the 11 m isobath at the Carpinteria study area in 1990, approximately 12 years after produced water was first discharged at this location (Osenberg et al. 1992b). In a single survey, 20 sites were sampled along a spatial gradient from 2 to 1000 m upcoast (West) and downcoast (East) of the diffusers. Infaunal densities were estimated at each site by collecting eight cores (78 cm^2/core to a depth of 10 cm). These were processed as described for the long-term study, and a mean density was calculated for each taxon at each of the 20 Carpinteria sites.

All chemical–physical parameters examined as part of the long-term study at Gaviota were also estimated at the Carpinteria sites, except those related to seston quality and deposition and several elements. Methods were identical to those used at Gaviota (described above).

Individual-based biological data were obtained by transplanting individuals of known size and/or age to several of the sites. Mussels (*Mytilus californianus* and *M. edulis*) were transplanted to six sites to determine if proximity to the outfall influenced their individual growth and condition (Osenberg et al. 1992b). Forty individuals from a uniform size distribution (range 20–60 mm shell length) of a mussel species were put into a bag of 1.25 mm oyster netting, and one bag of *M. californianus* and one of *M. edulis* were attached to buoy lines approximately 3 m above the sediments. Mussels were retrieved and frozen after 3–4 months in the field. Final shell length, initial shell length, dry gonadal tissue mass, and somatic tissue mass were then measured for each mussel. Site-specific estimates of average gonadal condition (gonad mass at a given size), somatic condition, total condition, and gonadal-somatic index were obtained by running analyses of covariance (ANCOVA) for each parameter for each mussel species using log(final shell length) as the covariate. Average shell growth and tissue production were estimated using log(initial shell length) as the covariate. Adjusted means were obtained for each parameter at each of the 6 sites.

Abalone larvae were raised in the lab and transplanted in small flow-through cages to 6–8 sites located 5–1000 m from the diffuser (Raimondi and Schmitt 1992). Three measures of per capita settlement and metamorphosis were derived from transplants that lasted about four days: (i) the proportion of late-stage larvae that successfully settled in the field, (ii) the proportion of late-stage larvae that successfully metamorphosed in the field and (iii) the proportion of early-stage larvae that subsequently settled in the lab after addition of a chemical inducer (for details, see Raimondi and Schmitt 1992). An additional measure of individual performance was obtained from a short-term transplant: the proportion of early-stage larvae still swimming after 6 hr in the field.

To obtain estimates of the magnitude of impacts due to produced water, we calculated means (e.g., of density or performance) for three distance categories: Near (sites < 25 m of the diffuser), Far (25–200 m), and Control (> 200 m). We then calculated a near-field and far-field effect size as the difference between log(Mean Near or Mean Far) and log(Mean Control). This is equivalent to the impact size (expressed in log units) of a BACIPS study (Equation 1) assuming no natural spatial variation between the sites (i.e., $E(\Delta_B) = 0$). While this assumption cannot be tested without Before data (and is certainly false), available evidence suggests that natural spatial gradients are small relative to the impacts of produced water (Osenberg et al. 1992b, Raimondi and Schmitt 1992).

Combining Results on Effect Size and Natural Variability

For parameters that were common to both the After-only study and the long-term study, the standardized effect size was calculated as the ratio between the absolute value of the effect size, which was obtained from the After-only study, and twice the standard deviation of deltas, which was obtained from the long-term study: Equation 3. In some cases, however, the same parameters were not measured in both studies, and other steps were required before proceeding with the power analyses.

For example, there were four chemical–physical parameters that provided estimates of effect size but not variability. All four parameters were elemental concentrations (i.e., Cu in sediments and Co, Ag, and Pb in the water column), so we used the average standard deviation for other elements (in either the sediments or water column) in the calculation of the standardized effect size.

Conversely, there were chemical–physical and population-based parameters that provided estimates of variability but not effect size (i.e., parameters estimated from sediment traps, band transects, emergence traps, reentry traps, and quadrats in addition to several elemental concentrations: Table 6.1). For these parameters we calculated standardized effect sizes using the average effect size for similar parameters that were measured as part of the After-only study.

Estimating standardized effect sizes for individual-based parameters posed a more difficult analytical problem because the individual-based data from the long-term study were derived from field collections of organisms, whereas the transplants conducted in the After-only study used organisms of known size, or cohorts of known number and age. Therefore, the transplants removed several sources of potential variability present in estimates from the long-term study. Because mussels had been transplanted during four different periods (spread over a total of 14 months), we were able to obtain estimates of variability for the mussel parameters. The standard deviation of differences between log-transformed parameters measured at the 1000 m and 100 m sites was calculated for ten of the mussel parameters over the four periods. Because the 100 m site is probably influenced by the discharge of produced water (Osenberg et al. 1992b, Raimondi and Schmitt 1992), this approach will overestimate S_Δ if there is temporal variation in the effects of produced water.

Standardized effect size was then calculated as explained above using these new estimates of variability for all mussel parameters except tissue production (for which we had only one survey and therefore could not estimate S_Δ). The mean standard deviation of deltas for the mussel parameters was used to estimate the standardized effect sizes for mussel tissue production and abalone performance parameters, which lacked estimates of S_Δ. The standardized effect sizes for the individual-based parameters derived from the long-term study were calculated using the mean effect sizes based on the mussel and abalone transplants.

For each parameter, we estimated the sample size (total number of sampling dates in the Before and After periods) needed to have an 80% chance of detecting ($\alpha = 0.05$) an impact characterized by the parameter's standardized effect size. All power analyses were based on two-tailed t-tests as provided in Gill (1978). The number of sampling dates in the Before and After periods was assumed to be equal.

Serial Correlation

The power analyses yield the number of independent sampling events (i.e., dates) needed for a given level of power (e.g., 80%). The time scale over which those samples must be collected will depend on the amount of serial correlation in the time series of deltas for each parameter (Stewart-Oaten et al. 1986). Serial correlation can be directly incorporated into the analyses of BACIPS data (Stewart-Oaten et al. 1992, Stewart-Oaten, Chapter 7), but power is greatest when serial correlation is absent. Therefore, we tried to determine the most intensive sampling schedule that would avoid substantial amounts of serial correlation. By doing so, we could roughly translate the number of independent sampling events into an estimate of the minimum amount of time required by the BACIPS study.

Because rigorous analyses of serial correlation require long time series of data, and because the approach we outline here is imprecise to begin with (i.e., extrapolating from two different studies to the design of a future one), we used a simpler approach to provide a general guide to sampling frequency. For each parameter sampled as part of the long-term study, we examined the correlation between the delta measured on one sampling date (Δ_i) and the delta measured on the next date on which sampling for that parameter was conducted (Δ_{i+1}). Only parameters with data from 8 or more dates were included in the analyses.

Results

Natural Variability Assessed from a Long-Term Study

Data from the long-term study revealed that the variation in deltas (i.e., in the difference in parameter values between sites) was lowest for chemical-physical parameters, intermediate for individual-based parameters, and greatest for population-based parameters (Figure 6.2). Most (28 of 30) of the chemical-physical parameters exhibited less variation in deltas than did the least variable population-based parameter. Almost all of the population-based parameters (24 of 25) were more variable than the most variable of the 7 individual-based parameters. Within a parameter group, no systematic differences were apparent among data collected using different techniques (e.g., densities based on infaunal cores versus band transects, or water column elements versus sediment elements), and there were no apparent trends among the population-based parameters related to the level of taxonomic aggregation (see Frost et al. 1992). All else being equal, these data suggest that chemical–physical parameters will provide more reliable indicators of environmental impacts than population-based parameters due to their smaller variability.

Effect Size Estimated from an After-Only Study

The After-only study provided estimates of effect sizes, which varied with proximity of the sampled sites to the produced water diffuser. In general, sizes of

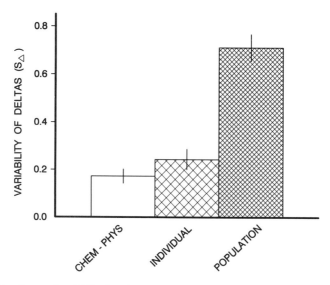

Figure 6.2. Temporal variability in estimates of the deltas (S_Δ) for chemical-physical, individual-based, and population-based parameters. Data were derived from the long-term study. For each parameter on each sampling date, a delta was estimated based on the difference between the log-transformed means at two sites [e.g., Log(mean density at Site 1 on date i) − Log(mean density at Site 2 on date i]. Shown are the standard deviations of deltas (mean ± SE) for parameters in each of the three groups. Means are based on 30, 7 and 25 different parameters for chemical-physical, individual, and population groups respectively. Here all individual-based data are derived from field collections.

effects were correlated ($r = 0.62$, $n = 47$) for sites near to and far from the diffuser (Figure 6.3), and the magnitudes of effects consistently were greatest nearer the diffuser. This pattern suggests that impacts diminished with distance away from the disturbance.

Both positive and negative changes in parameter values with distance from the diffuser were observed, and the sign depended on the particular parameter or parameter group examined. For example, concentrations of water column metals were higher nearer the diffuser, whereas most measures of individual performance were lower. Similarly, some taxa were more abundant closer to the diffuser, while others were less abundant. These two patterns in density probably reflect positive responses to organic enrichment (from oil constituents) and negative responses to toxicants present in produced water (e.g., Spies and DesMarais 1983, Osenberg et al. 1992b, Steichen 1994, see also Pearson and Rosenberg 1978, Ferris and Ferris 1979).

In evaluating power, the crucial factor is the absolute size of the change and not the sign (i.e., a positive or negative response). Although quite variable, the population-based parameters had absolute values of effect sizes that were about twice those for individual-based parameters, and four times larger than effect sizes for chemical-physical parameters (Figure 6.4). This pattern was similar for both Near and Far sites ($r = 0.72$, $n = 47$), although the overall magnitude of effects was lower at the Far sites (Figure 6.4). For simplicity, we focus on results from the Near sites in the following sections.

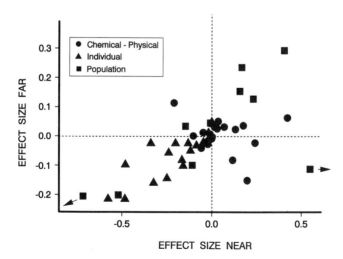

Figure 6.3.   Effect sizes estimated from sites near and far from an operating produced water diffuser (an After-only study). Positive values indicate larger parameter values near (or far from) the diffuser relative to control sites, while negative values indicate the opposite. The two population-based parameters next to the arrows have effect sizes that are off the scale: (-0.92, -0.85) and (0.917, -0.13).

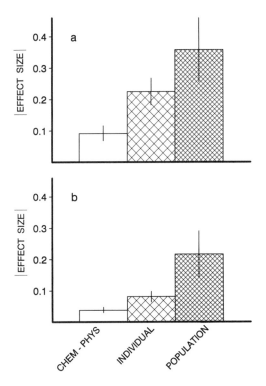

Figure 6.4. Absolute effect sizes ($\pm$ SE) for chemical-physical, individual-based, and population-based parameters based on sites (a) near and (b) far from the diffuser. Sample sizes (number of parameters) were 21, 16 and 10 for the chemical-physical, individual, and population groups respectively.

Combining Results on Effect Size and Natural Variability

Estimates of natural variability in individual-based parameters were derived from field collections, whereas those for effect size were obtained from transplants. To make the estimates more comparable, we calculated variability of deltas for individual performance of mussels from four separate transplants in the After-only study. The results show that all ten indices of mussel performance were relatively invariable over time (Mean S_Δ = 0.080, S.E. = $\pm$ 0.20, Range 0.007 − 0.220). Indeed, most (70%) of these estimates of mussel performance were less variable than almost all (94%) of the parameters measured in the long-term study.

The results from the long-term study and the After-only study yielded the opposite conclusions about the power associated with different parameter groups. On one hand, the population-based (and individual-based) parameters should be the most powerful due to their larger average effect sizes (Figure 6.4), whereas the chemical–physical (and individual-based) parameters should be more powerful due to their smaller average variability (Figure 6.2). Ultimately, the more powerful parameters will be those with the greatest standardized effect size (i.e., signal to noise ratio: Equation 3). Due to their relatively large effect sizes but low variability, individual-based parameters (particularly those derived from transplants) had larger standardized effect sizes than either the chemical-physical or population-based parameters (Figure 6.5). With respect to the individual-based parameters, the transplants yielded standardized effect sizes that were over three times larger than those derived from field collections.

The standardized effect sizes for both chemical-physical and population-based parameters were low and quite similar (Figure 6.5), due to the lower variability associated with chemical–physical parameters (Figure 6.2) and the greater effect sizes associated with population-based parameters (Figure 6.4). The standardized effect sizes for these two groups of parameters were 1/2 and 1/7 the magnitude of those for individual-based parameters derived from field collections and transplants respectively (Figure 6.5).

These results indicate that power to detect changes from exposure to produced water should be greatest for individual-based parameters derived from transplants, and next greatest for individual-based parameters obtained from field collections. For an equivalent number of estimates (i.e., sampling dates), power

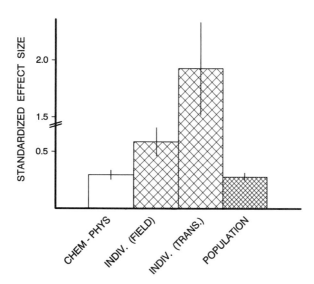

Figure 6.5. Standardized effect size (|Effect size| / 2 × S_Δ) for each parameter group; the measure is the ratio of effect size to twice the standard deviation of delta. Shown are means (± SE), based on 34, 7, 16, and 26 parameters (from left to right). Individual-based parameters are divided into estimates derived from field collections and those derived from transplants of marked individuals or caged cohorts. Note the break in scale between 0.7 and 1.5.

should be considerably lower for chemical–physical and for population-based parameters. For example, based upon average standardized effect sizes (Figure 6.5) and a Type I error rate of 0.05, the numbers of independent sampling dates needed to achieve power of 80% are approximately four for individual-based parameters from transplants, 24 for individual-based parameters from field collections, 90 for chemical-physical parameters, and 95 for population-based parameters.

Most individual-based parameters required <20 (and typically < 10) sampling dates to achieve 80% power (Figure 6.6). Over half of the chemical–physical and population-based parameters required 100 or more sampling dates to reach 80% power (Figure 6.6). To provide an idea of how many parameters would have high power for a logistically reasonable number of surveys that would also permit model development and testing (Stewart-Oaten, Chapter 7), we determined the fraction of parameters in each group with a sufficiently large standardized effect size (> 0.52) to yield power of at least 80% with 30 sampling dates ($n_B = n_A = 15$). Using this guideline, 81% (13/16) of individual-based parameters from transplants and 43% (3/7) of those from field collections had power that exceeded 80%. By contrast, only 18% (6/34) of the chemical–physical and 4% (1/26) of the population-based parameters achieved this level of power from 30 surveys.

The preceding analyses were based on effect sizes estimated from sites near the produced water diffuser. Repeating the analyses using data from the Far sites yielded similar patterns, although as expected the overall power was much lower

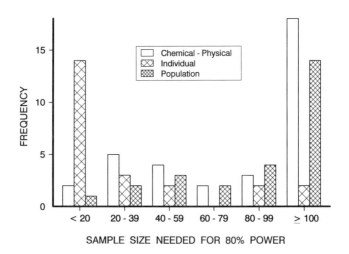

Figure 6.6. Frequency distribution of the sample size (number of independent sampling dates) for parameters in each group that is required for 80% power. Power analyses are based on standardized effect sizes (Figure 6.5).

or number of sampling dates needed for a given level of power was much higher. For example, the smaller effect sizes (estimated from sites far from the diffuser) resulted in more than half of the parameters in each of the three groups requiring >100 sampling dates to achieve 80% power. Only 26% of the individual-based parameters (all from transplants) required < 30 sampling dates, while none of the chemical-physical or population-based parameters achieved the same power with 30 dates.

Serial Correlation

Our analyses suggested that impacts on individual-based parameters are the most likely to be detected with a limited number of sampling dates. The analyses assumed that each sampling date provided an independent estimate of the true deltas (i.e., the underlying difference in parameter values between the Control and Impact sites). We examined patterns of serial correlation from the long-term study to gain insight into the frequency with which samples could be collected without grossly violating the assumption of temporal independence. This provided information on the time frame needed to collect series of independent samples.

There were no cases of significant ($P < 0.05$) negative serial correlation, and only 8% (4 of 50) of the parameters exhibited significant positive serial correlation (e.g., see Figure 6.7). Of the four parameters with positive serial correlation, two were chemical-physical parameters (seston sedimentation rate and seston percent organic matter), and two were population-based parameters (densities of sea pens and sea urchins: Figure 6.7c,d). None of the individual-based parameters exhibited significant serial correlation.

Serial correlation appeared to arise in the population-based parameters as a result of long-term trends in the deltas (Figures 6.7c,d). For example, the white sea urchin (*Lytechinus anamesus*) exhibited strong seasonal migrations, and was present during the winter and spring but absent during the summer and fall. The relative density at the two sites appeared to be set when urchins reappeared in winter; the ranking of the two sites were consistent within a year, but varied greatly among years (Figure 6.7c). This suggests that replicates should be collected only once per year, or a yearly average obtained from more frequent collections.

Density of sea pens (*Acanthoptilum* sp. and *Stylatula* sp.) exhibited an even longer term trend (Figure 6.7d). One site tended to have a greater density than the other site prior to October 1989, but the reverse was true for all samples collected after this date (Figure 6.7d). This could have arisen, for example, by a strong recruitment event in the Fall of 1989 at only one of the sites.

Despite these two examples, serial correlation was not a general problem for the various parameters estimated in our long-term study (e.g., Figure 6.7a,b). On average, the serial correlation for each of the three parameter groups was only

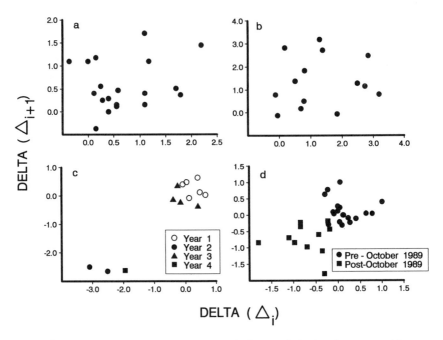

Figure 6.7. Patterns of serial correlation in deltas for four population-based parameters. These are the difference in density of: (a) cerianthid (burrowing) anemones (from band transect estimates); (b) copepods (from emergence traps); (c) white sea urchins (*Lytechinus anamesus*) (from quadrat samples); and (d) sea pen density (from band transects). There is significant serial correlation in (c) and (d), and data are separated into temporal groups to help distinguish the long-term patterns.

0.1–0.2 (Figure 6.8). Simulations suggest that serial correlation of this order introduce only small error into tests of impacts (Carpenter et al. 1989, Stewart-Oaten et al. 1992).

Based on these results, we assumed that sampling could occur every 60 days without yielding substantial amounts of serial correlation. Assuming that six samples are collected per year and the Before and After periods are of equal duration, the estimates of sample size (number of independent sampling events) can be translated into the number of years the assessment study must be conducted. Achieving 80% power would require 16 years for population-based parameters, 15 years for chemical–physical parameters, 4 years for individual-based parameters from field collections, and 1 year for individual-based parameters from transplants. To achieve 80% power for only a quarter of the parameters in each group, the required study duration is reduced to 11 years for population-based parameters, 7 years for chemical–physical parameters, 3 years for individual-based parameters from field collections, and <1 year for individual-based parameters from transplants.

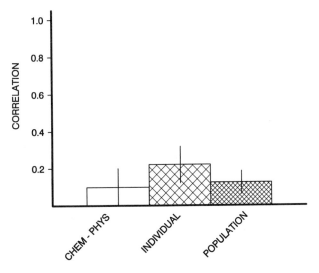

Figure 6.8. Degree of serial correlation in deltas for each parameter group. Shown are means (± SE), based on 18, 7, and 25 parameters for chemical-physical, individual and population groups respectively.

Discussion

Because relatively few well-designed studies of planned perturbations have been completed, there is a sparse empirical base to guide the design of future assessment programs (e.g., Carney 1987, Spies 1987, Underwood 1991, Stewart-Oaten, Chapter 7, Ambrose et al., Chapter 18). Recent discussions have highlighted general design considerations that should be incorporated in Before-After-Control-Impact approaches (e.g., Stewart-Oaten et al. 1986, 1992, Stewart-Oaten, Chapter 7, Underwood, Chapter 9), but these say little about specific considerations regarding sampling frequency and parameter selection. Often, a study must be planned in the absence of sufficient preliminary data to properly guide sampling decisions (Stewart-Oaten, Chapter 7). It is crucial to obtain good estimates of sampling variability and the size of impacts that might arise (or that are deemed ecologically important: Underwood and Peterson 1987, Yoccoz 1991), but this information typically is lacking. In the absence of a BACIPS (or analogous) study conducted previously on a similar perturbation in a similar habitat, it is vital that other existing data be used to guide specific design considerations.

Given limitation on time and funding, the selection of parameters and frequency of sampling are especially crucial features of the design process. One of the most acute constraints is the time available to collect data prior to the perturbation. In many situations, the Before period probably will be rather abbreviated for a variety of reasons beyond scientific control (Piltz, Chapter 16).

Therefore, parameter selection and sampling design should take into account the low numbers of temporal replicates that likely can be collected prior to the commencement of the disturbance (see Stewart-Oaten, Chapter 7 for discussion of model development based on these data). Key considerations in this regard are the likely variability in the parameter estimate (e.g., delta) and the probable magnitude of response to the disturbance, both of which influence statistical power to detect an effect. Constraints on the number of temporal replicates in the Before period are most likely to hamper detection of impacts on population density and chemical-physical characteristics, and least likely to affect detection of effects on individual performance. Unfortunately these results suggest that many field monitoring programs might be compromised because individual-based parameters rarely are examined (e.g., Carney 1987).

There are, however, compelling reasons to examine population and chemical-physical parameters despite the expected low power. First, chemical and physical properties describe the direct effect of many perturbations, and in many cases impacts could be ameliorated by subsequent intervention (e.g., source reduction, reduced discharge limits). Second, population attributes, such as density, reflect the ecological consequences of the disturbance, and are features of fundamental concern to resource managers and regulatory agencies. In addition, some species receive special regulatory consideration. Another reason is that, while the average power for population or chemical-physical parameters is low, some species or chemical-physical parameters will have greater power than others. The approach described here is equally useful in identifying promising candidates within a parameter group as it is in guiding allocation of effort among groups. Finally, the actual impacts of the new disturbance, of course, cannot be known *a priori*, and effects on populations and chemical-physical parameters certainly can be much larger (or variation much smaller) than anticipated based on extrapolations from other data sets.

It is useful to consider why the population and chemical-physical parameters had low and similar power, because low power arose for different reasons. Population parameters were highly responsive to produced water (i.e., larger impact), but exhibited much greater natural variability. In contrast, the chemical–physical parameters had much lower variability in deltas, but were not greatly altered by the discharge of produced water. It appears that these results generally will hold for other types of point source disturbances in the marine environment. Many chemical–physical parameters probably are influenced largely by large-scale oceanographic processes that similarly affect nearby sites. For example, certain chemical–physical attributes (e.g., sedimentation rate, water temperature, nutrient flux) are strongly associated with upwelling conditions, which is a region-wide phenomenon (e.g., Landry and Hickey 1989). In these situations, differences in these parameter values between Control and Impact sites (i.e., the deltas) will be similar through time (see also a related discussion in Magnuson et al. 1990, which discussed temporal coherence of chemical–physical and biological parameters in freshwater lakes).

The relatively small response of chemical–physical parameters we observed to the discharge of produced water is also consistent with recent analyses of the general effect of waste discharges on the distribution of trace elements in coastal waters. For example, massive discharges (10^9 L d^{-1}) of wastewaters in the Southern California Bight have had a negligible (< 1%) impact on concentration of cadmium in those waters (Sanudo-Wilhelmy and Flegal 1991). Similarly, Schmidt and Reimers (1991) found that, in the Santa Barbara Basin, the fraction of certain metals (Cd, Cu, Ni, Pb) from human sources that is deposited in sediments near municipal outfalls is quite small (< 1%) compared to the amount released. In both cases, natural inputs and physical mixing processes appeared to have reduced the contribution from human inputs to a small fraction of the background level. So for chemical–physical parameters, the large spatial scale of events that drive natural variation can lead to low variability in deltas, while other naturaľ processes can greatly diminish the signal provided by anthropogenic perturbations.

Population density, by comparison, is known to be highly responsive to local conditions, and can exhibit considerably different temporal patterns among neighboring sites (e.g., Holbrook et al. 1990, Magnuson et al. 1990, Schmitt and Holbrook 1990). The high sensitivity to local conditions potentially can translate into strong local responses to natural phenomena (thus increasing S_Δ) as well as anthropogenic perturbations such as wastewater discharges (thus increasing effect size). Within-site sampling error also can contribute to the high variability as benthic populations are notoriously difficult to sample (Vezina 1988, Thrush et al., Chapter 4).

It is important to note that the variability reported here (e.g., Figure 6.2) is a measure of the variability (over time) in *estimates* of the *differences* between sites. This variability includes both the true temporal variation in deltas and variation due to sampling error within a site (which adds error to the estimation of delta on any date). The contribution of sampling error will be a function of spatial variability within a site and sampling intensity, and therefore will vary with the within-site sampling design. This suggests that the variation in deltas (S_Δ) for population-based parameters could be reduced by more intensive sampling on each date, rather than increasing the number of dates. However, partitioning of observed variation for the long-term data set revealed that the deltas for population-based parameters were more variable due both to sampling error (i.e., high within-site spatial variation) and site-specific temporal variability (i.e., high variation in the actual deltas through time) (Osenberg, personal observation); increasing the sampling intensity within a date would reduce the observed variation (S_Δ) by only about 50%. Therefore, even if sampling error were removed (e.g., through more exhaustive sampling), population-based parameters still would be more variable than the chemical-physical or individual-based parameters (see Figure 6.2).

Our estimates of S_Δ probably are typical because the within-site sampling design of our long-term study is similar to that used in many assessment studies (see Thrush et al., Chapter 4). The costs and benefits of adjusting within-site sampling intensity to achieve greater power can be analyzed (e.g., the importance of within-site accuracy versus more sampling dates), although with limited resources, greater precision ultimately would be accomplished at the cost of fewer sampling dates (which is the unit of replication in a BACIPS design).

Difficulty in sampling populations or other parameters within sites not only can affect the variance of the estimate, it also might lead to overestimation of effect sizes from After-only studies (Figures 6.3, and 6.4). This would be true especially for a parameter that is not affected by the perturbation, and thus should have an effect size of 0. Our approach would overestimate this effect by confounding sampling error and any underlying spatial gradient as an effect of the perturbation. If so, the calculated number of surveys (sample size) needed for a given level of power would be underestimated. While this bias will exist for any parameter, our data suggest that, on average, it will be most acute for population-based parameters. Hence, limitations on detecting impacts at the population level may be even more difficult than our analyses suggest.

In contrast to population and chemical–physical parameters, individual-based parameters had relatively high power owing to relatively low levels of variability (Figure 6.2) and intermediate effect sizes (Figure 6.4). Although 80% power could be achieved for many of the parameters we examined with fewer than 10 sampling dates, it is unwise to reduce the sampling intensity below a level at which model development and testing can be performed (Stewart-Oaten, Chapter 7). Our data also indicate that variability in the deltas for individual-based parameters can be reduced by use of transplants (Figure 6.4), which results in increased power (Figure 6.5). This presumably occurs because, compared with estimates from field collections, transplants remove noise introduced by individual variation as well as variability between sites over time. For example, size-specific growth rates can be assessed accurately using marked individuals of known size; because size can influence growth and size-distributions can vary among sites (e.g., Osenberg et al. 1988), an analysis based on marked individuals is likely to be more powerful than one based on field collections.

It should be noted that several of the transplant-derived parameters for which we had relatively high power are closely related to population-based parameters, which had much lower power. For example, transplants of abalone larvae provided estimates of per capita settlement rates. In the field, natural rates of per capita settlement can be estimated from observed settlement rates and/or larval supply, both of which require estimation of density (e.g., Olson 1985; Keough 1986; Victor 1986; Raimondi 1990). Therefore, these field estimates would have considerable error for the same reasons that population parameters were highly variable. The use of transplants surmounted much of this problem by using cohorts of known size, thereby eliminating much of the variability that plagues the population parameters.

The observation that individual-based parameters may yield more powerful assessments is troubling given the rarity with which they are measured in field assessments. Care must be taken to guard against only considering parameters that yield low probabilities of demonstrable results (e.g., chemical–physical and population attributes); inclusion of individual-based parameters could greatly increase the sensitivity of assessment studies (Carney 1987, Osenberg et al. 1992a; see also Jones et al. 1991). However, the need to investigate individual-based parameters goes far beyond power considerations; it is the individual-based (and demographic) parameters that provide the mechanisms that underlie changes at the population (and therefore community) level. Furthermore, these individual-based parameters provide an explicit connection with detailed laboratory studies that focus on individuals and mechanisms of toxicity. What is needed are more realistic studies of individual-based effects under field conditions combined with both mechanistic laboratory studies and field assessments of population-level consequences. Recent advances with individual-based models (DeAngelis and Gross 1992) provide an explicit framework for making these fundamental linkages among environmental chemistry, physiology and population ecology (e.g., Hallam et al. 1990). Such models provide a powerful, mechanistic approach to assessing impacts on natural populations and complement the traditional approach of monitoring environmental impacts.

Acknowledgments

The assistance of D. Canestro is gratefully acknowledged. Also assisting in the field or laboratory were S. Anderson, M. Carr, M. Edwards, D. Forcucci, D. Heilprin, T. Herrlinger, B. Hoffman, T. Kaltenberg, P. Krause, S. Lin, A. Martinez, M. Perez, P. Raimondi, D. Reed, D. Steichen, K. Sydel, and V. Vrendenburg. G. Scelfo performed trace metal analysis of seawater samples, and Bonnie Williamson provided technical and logistical assistance. P. Raimondi provided the data on abalone larvae and A. Stewart-Oaten provided helpful discussions. We also appreciate the critical comments of A. Stewart-Oaten, J. Levinton and an anonymous reviewer on an earlier draft. This research was funded by the Minerals Management Service, U.S. Department of Interior under MMS Agreement No. 14-35-001-3071, and by the UC Coastal Toxicology Program. The views and conclusions in this chapter are those of the authors and should not be interpreted as necessarily representing the official policies, either expressed or implied, of the U.S. Government.

References

Alldredge, A. L., and J. M. King. 1980. Effects of moonlight on the temporal migration patterns of demersal zooplankton. Journal of Experimental Marine Biology and Ecology **44**:133–156.

Bernstein, B. B., and J. Zalinski. 1983. An optimum sampling design and power tests for environmental biologists. Journal of Environmental Management **16**:35–43.

Boesch, D. F., and N. N. Rabalais, editors. 1987. Long-term environmental effects of offshore oil and gas development. Elsevier Applied Science, New York, New York.

Bruland, K. W., K. H. Coale, and L. Mart. 1985. Analysis of seawater for dissolved cadmium, copper and lead: an intercomparison of voltametric and atomic absorption methods. Marine Chemistry **17**:285–300.

Campbell, D. T., and J. C. Stanley. 1966. Experimental and quasi-experimental designs for research. Rand McNally, Chicago, Illinois.

Carney, R. S. 1987. A review of study designs for the detection of long-term environmental effects of offshore petroleum activities. Pages 651–696 in D. F. Boesch and N. N. Rabalais, editors. Long-term environmental effects of offshore oil and gas development. Elsevier, New York, New York.

Carpenter, S. R., T. M. Frost, D. Heinsey, and T. K. Kratz. 1989. Randomized intervention analysis and the interpretation of whole-ecosystem experiments. Ecology 70:1142–1152.

Cohen, J. 1977. Statistical power analysis for the behavioral sciences. Academic Press, New York, New York.

Cooper, S. D., and L. A. Barmuta. 1993. Field experiments in biomonitoring. Pages 399–441 in D. M. Rosenberg and V. H. Resh, editors. Freshwater biomonitoring and benthic macroinvertebrates. Chapman and Hall, New York, New York.

DeAngelis, D. L., and J. Gross, editors. 1992. Individual-based models and approaches in ecology: populations, communities, and ecosystems. Chapman and Hall, New York, New York.

Eberhardt, L. L. 1976. Quantitative ecology and impact assessment. Journal of Environmental Management 4:27–70.

Eberhardt, L. L., and J. M. Thomas. 1991. Designing environmental field studies. Ecological Monographs 61:53–73.

Fairweather, P. G. 1991. Statistical power and design requirements for environmental monitoring. Australian Journal of Marine and Freshwater Research 42:555–567.

Faith, D. P., C. L. Humphrey, and P. L. Dostine. 1991. Statistical power and BACI designs in biological monitoring: comparative evaluation of measures of community dissimilarity based on benthic macroinvertebrate communities in Rockhole Mine Creek, Northern Territory, Australia. Australian Journal of Marine and Freshwater Research 42:589–602.

Ferris, V. R., and J. M. Ferris. 1979. Thread worms (Nematoda). Pages 1–33 in C. W. J. Hart and S. L. H. Fuller, editors. Pollution ecology of estuarine invertebrates. Academic Press, New York, New York.

Flegal, A. R., G. J. Smith, G. A. Gill, S. Sanudo-Wilhelmy, and L. C. D. Anderson. 1991. Dissolved trace element cycles in the San Francisco Bay estuary. Marine Chemistry 36:329–363.

Frost, T. M., S. R. Carpenter, and T. K. Kratz. 1992. Choosing ecological indicators: effects of taxonomic aggregation on sensitivity to stress and natural variability. Pages 215–227 in D. H. McKenzie, D. E. Hyatt and V. J. McDonald, editors. Ecological indicators. Volume 1. Elsevier Applied Science, New York, New York.

Gill, J. L. 1978. Design and analysis of experiments in the animal and medical sciences. Volume 1–3. Iowa State University Press, Ames, Iowa.

Green, R. H. 1979. Sampling design and statistical methods for environmental biologists. Wiley and Sons, New York, New York.

Green, R. H. 1989. Power analysis and practical strategies for environmental monitoring. Environmental Research 50:195–205.

Hallam, T. G., R. R. Lassiter, J. Li, and W. McKinney. 1990. Toxicant-induced mortality in models of Daphnia populations. Environmental Toxicology and Chemistry 9:597–621.

Herman, P. M. J., and C. Heip. 1988. On the use of meiofauna in ecological monitoring: who needs taxonomy? Marine Pollution Bulletin 19:665–668.

Higashi, R. M., G. N. Cherr, C. A. Bergens, T. W.-M. Fan, and D. Crosby. 1992. Toxicant isolation from a produced water source in the Santa Barbara Channel. Pages 223–233 in J. P. Ray and F. R. Englehardt, editors. Produced water: technological/environmental issues and solutions. Plenum Press, New York, New York.

Holbrook, S. J., R. J. Schmitt, and R. F. Ambrose. 1990. Biogenic habitat structure and characteristics of temperate reef fish assemblages. Australian Journal of Ecology 15:489–503.

Jones, M., C. Folt, and S. Guarda. 1991. Characterizing individual, population and community effects of sublethal levels of aquatic toxicants: an experimental case study using Daphnia. Freshwater Biology 26:35–44.

Keough, M. J. 1986. The distribution of the bryozoan Bugula neritina on seagrass blades: settlement growth and mortality. Ecology 67:846–857.

Krause, P. R., C. W. Osenberg, and R. J. Schmitt. 1992. Effects of produced water on early life stages of a sea urchin: state-specific responses and related expression. Pages 431–444 in J. P. Ray and F. R. Englehardt, editors. Produced water: technological/environmental issues and solutions. Plenum Press, New York, New York.

Landry, M. R., and B. M. Hickey, editors. 1989. Coastal oceanography of Washington amd Oregon. Elsevier Science Publishers, Amsterdam, The Netherlands.

Luoma, S. N., D. J. Cain, C. Brown, and E. V. Axtman. 1991. Trace metals in clams (*Macoma balthica*) and sediments at the Palo Alto mudflat in South San Francisco Bay: April, 1990 - April, 1991. U.S. Geological Survey Report 91-460. 47 pp.

Magnuson, J. J., B. J. Benson, and T. K. Kratz. 1990. Temporal coherence in the limnology of a suite of lakes in Wisconsin, U.S.A. Freshwater Biology **23**:145–159.

Middleditch, B. S. 1984. Ecological effects of produced water effluents from offshore oil and gas production platforms. Ocean Management **9**:191–316.

Neff, J. M. 1987. Biological effects of drilling fluids, drill cuttings and produced waters. Pages 469–538 in D. F. Boesch and N. N. Rabalais, editors. Long-term environmental effects of offshore oil and gas development. Elsevier Applied Science, New York, New York.

Neff, J. M., N. N. Rabalais, and D. F. Boesch. 1987. Offshore oil and gas development activities potentially causing long-term environmental effects. Pages 149–173 in D. F. Boesch and N. N. Rabalais, editors. Long-term environmental effects of offshore oil and gas development. Elsevier Applied Science, New York, New York.

Oakden, J. M., J. S. Oliver, and A. R. Flegal. 1984. Behavioral responses of a phoxocephalid amphipod to organic enrichment and trace metals in sediments. Marine Ecology Progress Series **14**:253–257.

Olson, R. R. 1985. The consequences of short-distance larval dispersal in a sessile marine invertebrate. Ecology **66**:30–39.

Osenberg, C. W., E. E. Werner, G. G. Mittelbach, and D. J. Hall. 1988. Growth patterns in bluegill(*Lepomis macrochirus*) and pumpkinseed(*L. gibbosus*) sunfish: environmental variation and the importance of ontogenetic niche shifts. Canadian Journal of Fisheries and Aquatic Sciences **45**:17–26.

Osenberg, C. W., S. J. Holbrook, and R. J. Schmitt. 1992a. Implications for the design of environmental impact studies. Pages 75–89 in P. M. Griffman and S. E. Yoder, editors. Perspectives on the marine environment of southern California. USC Sea Grant Program, Los Angeles, California.

Osenberg, C. W., R. J. Schmitt, S. J. Holbrook, and D. Canestro. 1992b. Spatial scale of ecological effects associated with an open coast discharge of produced water. Pages 387–402 in J. P. Ray and F. R. Englehardt, editors. Produced water: technological/environmental issues and solutions. Plenum Press, New York, New York.

Pearson, T. H., and R. Rosenberg. 1978. Macrobenthic succession in relation to organic enrichment and pollution of the marine environment. Oceanography and Marine Biology Annual Review **16**:229–311.

Peterman, R. M. 1990. Statistical power analysis can improve fisheries research and management. Canadian Journal of Fisheries and Aquatic Sciences **47**:2–15.

Raimondi, P. T. 1990. Patterns, mechanisms, consequences of variability in settlement and recruitment of an intertidal barnacle. Ecological Monographs **60**:283–309.

Raimondi, P. T., and R. J. Schmitt. 1992. Effects of produced water on settlement of larvae: field tests using red abalone. Pages 415–430 in J. P. Ray and F. R. Englehardt, editors. Produced water: technological/environmental issues and solutions. Plenum Press, New York, New York.

Sanudo-Wilhemy, S., and A. R. Flegal. 1991. Trace element distributions in coastal waters along the U.S.-Mexican boundary: relative contributions of natural processes vs. anthropogenic inputs. Marine Chemistry **33**:371–392.

Schmidt, H., and C. E. Reimers. 1991. The recent history of trace metal accumulation in the Santa Barbara basin, Southern California borderland. Estuarine, Coastal and Shelf Science **33**:485–500.

Schmitt, R. J., and S. J. Holbrook. 1990. Contrasting effects of giant kelp on dynamics of surfperch populations. Oecologia **84**:419–429.

Schroeter, S. C., J. D. Dixon, J. Kastendiek, R. O. Smith, and J. R. Bence. 1993. Detecting the ecological effects of environmental impacts: a case study of kelp forest invertebrates. Ecological Applications **3**:331–350.

Skalski, J. R., and D. H. McKenzie. 1982. A design for aquatic monitoring systems. Journal of Environmental Management **14**:237–251.

Spies, R. B. 1987. The biological effects of petroleum hydrocarbons in the sea: assessments from the field and microcosms. Pages 411–467 in D. F. Boesch and N. N. Rabalais, editors. Long-term enviromental effects of offshore oil and gas development. Elsevier Applied Science, New York, New York.

Spies, R. B., and D. J. DesMarais. 1983. Natural isotope study of trophic enrichment of marine benthic communities by petroleum seepage. Marine Biology **73**:67–71.

Steichen, D. S. 1994. The response of benthic and demersal macrofauna to organic enrichment at a natural oil seep. M.A. Thesis. University of California, Santa Barbara, California.

Stetch, J. J. 1983. Habitat selection and vertical migration of sand-dwelling demersal gammarid amphipods. Ph.D. Dissertation. University of California, Santa Barbara, California.

Stewart-Oaten, A., W. W. Murdoch, and K. R. Parker. 1986. Environmental impact assessment: "psuedoreplication" in time? Ecology **67**:929–940.

Stewart-Oaten, A., J. R. Bence, and C. W. Osenberg. 1992. Assessing effects of unreplicated perturbations: no simple solutions. Ecology **73**:1396–1404.

Underwood, A. J. 1991. Beyond BACI: experimental designs for detecting human environmental impacts on temporal variations in natural populations. Australian Journal of Marine and Freshwater Research **42**:569–587.

Underwood, A. J., and C. H. Peterson. 1988. Towards an ecological framework for investigating pollution. Marine Ecology Progress Series **46**:227–234.

Vezina, A. F. 1988. Sampling variance and the design of quantitative surveys of the marine benthos. Marine Biology **97**:151–156.

Victor, V. C. 1986. Larval settlement and juvenile mortality in a recruitment-limited coral reef fish population. Ecological Monographs **56**:145–160.

Warwick, R. M. 1988. The level of taxonomic discrimination required to detect pollution effects on marine benthic communities. Marine Pollution Bulletin **19**:259–268.

Yoccoz, N. G. 1991. Use, overuse and misuse of significance tests in evolutionary biology and ecology. Bulletin of the Ecological Society of America **72**(2):106–111.

CHAPTER 7

PROBLEMS IN THE ANALYSIS OF ENVIRONMENTAL MONITORING DATA

Allan Stewart-Oaten

This chapter discusses problems in the statistical analysis of data which monitor the environmental effects of planned human alterations. "Alteration" indicates a long-term (i.e., press) change, like the installation of a power plant, sewage outfall or oil platform, rather than a short-term (i.e., pulse) change, like an accident or the temporary effects of building the power plant, etc. "Planned" indicates that data are available from both before and after the alteration. A common goal is to compare the value of some biological parameter at the affected site before the alteration to the value after.

Many biological parameters, such as abundance, average size, age distribution, various measures of diversity, etc., fluctuate over time. Much of this fluctuation is currently unpredictable, and must be regarded as random. It must be allowed for as part of the "error" in formal statistical inference. This requires sampling at several different times both before and after the alteration. Since the times cannot be randomly assigned to "treatments" (Before or After), a monitoring study cannot be analyzed as an experiment. The generic model "observation = treatment mean + random error" is not automatically justified. Instead, the models underlying the analyses will be guesses, needing justification by plausibility and fit to the data, and often including complications like deterministic functions with unknown parameters (e.g., to deal with seasons) and heteroscedastic or correlated errors.

The first section discusses a Before-After design, in which samples are taken at several times before and after the alteration. This design has been used to assess impacts on temporally varying phenomena in many contexts following Box and Tiao's (1975) analysis of the effect of "interventions" (new laws and a new freeway) on Los Angeles air pollution. In cases where there is no feasible comparison site (e.g., global warming), some variant of this design seems the only possibility. It is introduced here mainly for illustration, since its problems are not qualitatively different from those of other designs, but stand out more

Detecting Ecological Impacts: Concepts and Applications in Coastal Habitats, edited by R. J. Schmitt and C. W. Osenberg

clearly. Even to define the parameters describing biological or ecological change between one period and another requires a model of a stochastic process. For some variables of interest, e.g., abundance, such models may involve strong temporal fluctuations. Deterministic fluctuations will lead to bias unless model forms can be guessed approximately correctly. Random fluctuations are likely to have significant long-term serial correlation of unknown structure, which can cause estimates of effects to have large variances and these variances to be badly underestimated from the data.

The second section outlines how the use of a Control site as a covariate may make acceptable effect estimates, and variance estimates, possible. In this "BACIPS" (Before-After-Control-Impact Paired Series) design, data are taken "simultaneously" at one or more Impact sites near the alteration and at one or more Control areas, nearby and similar but far enough from the alteration to be little affected by it, on sequences of sampling occasions Before and After the alteration. This general design has also been called a "multiple time series quasi-experiment" (Campbell and Stanley 1966), "pseudo-experiment" or "pseudodesign" (Eberhardt 1976), "Control-Treatment Pairs (CTP)" (Skalski and McKenzie 1982), and "BACI" (Stewart-Oaten et al. 1986). The idea is that suitable Controls will "track" the Impact sites in some sense. A change in this tracking relationship following the alteration will be evidence for an effect.

However, this analysis depends on how the tracking relationship is modeled. Some alternative models are introduced and briefly discussed. It is argued that analysis using more than one model may be needed, with necessarily rough ways of checking the compatibility of their conclusions. Some additional problems of basing assessment on parameters other than the mean (or median), such as variances, are discussed.

The third section discusses some approaches to causal assessment.

This chapter is guided by Tukey's (1962) dictum: "Far better an approximate answer to the right question, which is often vague, than an exact answer to the wrong question, which can always be made precise". Its purpose is not to disparage formal methods, but to argue that the right questions in impact assessment are likely to have only approximate, messy, and possibly multiple answers, requiring informal combinations of formal results.

Before-After Studies

Defining Parameters

Suppose we are interested in how the abundance of a given species has changed following the alteration. Let $N_B(t)$ be the abundance at the Impact site at time t during the period Before the alteration was installed (or began operating), and $N_A(t)$ the abundance at time t during the period After installation. These are *true* abundances, assumed known exactly over a Before period $T_s < t < T_0$ and an After period $T_0 < T_E$, respectively. Sampling error is an additional complication, but is distracting at this point.

We might say that "the" abundance has decreased if $N_A(t)$ is in some sense smaller than $N_B(t)$. But neither of these is a simple number: both are functions of t. Nor can these two functions be compared in the usual way, by asking whether $N_B(t) > N_A(t)$ for each time, t, or by comparing their averages over some period. There is no time for which both functions exist: $N_B(t)$ exists only for $t < T_0$, the installation time, and $N_A(t)$ only for $t > T_0$.

We could define the function $N_A(t)$ to be smaller than $N_B(t)$ if the average of $N_A(t)$ over the After period ($T_0 < t < T_E$) is smaller than the average of $N_B(t)$ over the Before period. One objection to this is that it depends on the periods chosen. If the After period's "Winter" fraction is larger than the Before period's, then we might have $N_A(t) < N_B(t)$ by this definition, even though $N_A(t)$ may have a greater "Winter" average *and* a greater "Summer" average than $N_B(t)$. Matching seasons or using a weighted average might solve this problem (Cochran and Rubin 1974).

A second objection is that both $N_B(t)$ and $N_A(t)$ are determined partly by "random" factors such as births, deaths, and movements by individuals, invasions by predators, competitors or disease, short-term events like storms and upwellings, etc. It may be that $N_A(t) < N_B(t)$ by this definition, only because of random factors which had nothing to do with the alteration. This suggests that judgment should be based on some kind of average or distribution of what *could have* happened, rather than directly on what did happen.

This conclusion especially applies if the consequences of the decision to be made will depend on future abundances, rather than past ones, as in decisions about ceasing operations, modifying designs, or compensatory mitigation. In these cases, the abundances up to the time of the decision are useful mainly as guides to future abundances —i.e., to their probability distributions. Even decisions concerning punishment or reparations for damage already done require a comparison between what has happened following the alteration, i.e., $N_A(t)$ for $T_0 < t < T_E$, and what *would have happened* had the alteration not occurred, i.e., the distribution of possibilities for $N_B(t)$ for the same period.

Time-Series Modeling

To define distributions of what the abundances could have been, or could be in future, we need to regard $N_B(t)$ and $N_A(t)$ as stochastic processes. Such processes can be modeled by giving a formula from which, given past values of a function and also the values of a collection of random variables generated independently in a specified way, all future values of the function could be determined. For example, the abundance of an annual population in a constant environment might be described by $N(t) = r_t N(t-1)/(1 + cN(t-1))$, where the r_t's are independent draws from some specified distribution. Given $N(0)$ and the set of random values $r_1, r_2, ...$, the entire process could be calculated.

Given the past values of a process, we can simulate a possible future on a computer, by using a random number generator and applying the formula. This future will depend in part on the chance values generated (e.g., the r_t's). We could then simulate another future, using the same past and formula, but a new set of numbers from the same generator. Each future constitutes a "realization" of the process: each is a function of time. For any time, t, in the future, and any number x, we can determine the probability that a realization generated in this way will have a value $\leq x$ at time t, i.e., $P\{N(t) \leq x\}$. For instance, we could simulate a large number of futures and count the fraction with this property. These probabilities give the distribution of the process at time t. Similarly, for any two times, t_1 and t_2, we can determine $P\{N(t_1) \leq x_1 \text{ and } N(t_2) \leq x_2\}$ for any x_1 and x_2, and thus the joint distribution for these two times. From these distributions, means, variances and covariances can be obtained, all functions of time (or of two times). Similar probabilities can be computed for any finite set of future times; the collection of all such probabilities gives the distribution of the entire process.

Thus we wish to compare the distributions of the Before and After processes, N_B and N_A, generated by possibly different formulae and random number generators, using a single partial realization of each, $N_B(t)$ $(T_S < t < T_0)$ and $N_A(t)$ $(T_0 < t < T_E)$.

Focus on the distribution circumvents the problem that the realizations, $N_B(t)$ and $N_A(t)$, are never observed at the same time, t. Even though the alteration prevents any actual realizations from occurring, there is a distribution of possible realizations of $N_B(t)$ for $t > T_0$, because the distribution depends only on the past, the appropriate formula, and the distributions of the random variables. Thus the distributions, or key parameters like means and variances, could be compared at any time t, even though realizations cannot be. The effect of the alteration (or the change coincident with the alteration) could be defined as the difference between the means of $N_A(t)$ and $N_B(t)$. If no other causes are operating, this is the difference between the mean abundance obtained with the alteration and the mean that *would have been* obtained had the alteration not occurred.

Estimating a Varying Mean

Unfortunately, the distributions, or their parameters, are still functions of time. For example, given a "history" of past values, H, at time t_0 say, the mean of $N_B(t)$ at time t is

$$M_B(t,H,t_0) = E\{N_B(t) \mid H,t_0\}, \tag{1}$$

the mean of the values at time t of all possible realizations beginning from a history H at time t_0. This is *not* an average over time: t, H and t_0 are all fixed. We are averaging over the possible values of $N_B(t)$, each possible value corresponding to a set of possible values of the random variables involved in the algorithm (e.g., r_1, r_2, ..., in the example above). Similarly we have $M_A(t,H,t_0)$, the variance functions $V_B(t,H,t_0)$ and $V_A(t,H,t_0)$, and the covariance functions $C_B(t_1,t_2,H,t_0) = \text{Cov}\{N_B(t_1),N_B(t_2) \mid H,t_0\}$ and $C_A(t_1,t_2,H,t_0)$. These functions are not known because the algorithms (the formulae and the distributions of the random variables) for generating realizations are not known. They must be partly guessed, guided by biological knowledge and intuition, tractability and flexibility, and partly estimated from the data (i.e., from the observed realizations).

Some judgment is unavoidable. To make any inferences at all, or even to produce meaningful descriptive summaries of the data, some assumptions are necessary so that observations taken at different times can be combined to estimate parameters relevant to all times.

Modeling the Mean Function.   One possibility is to model the mean functions, M_B and M_A, as explicit functions of time, which are known except for a small number of time-independent parameters, to be estimated from the data and other information. If time is measured in years, we might assume

$$M_B(t,H,t_0) = \mu_B + \alpha_B \sin 2\pi t + \beta_B \cos 2\pi t + h(t,H,t_0), \qquad (2)$$

where $h \to 0$ as $t - t_0$ increases (the effects of past history die away). Assuming the process began far in the past, M_B oscillates sinusoidally (seasonally) about a fixed value. The effect of the alteration could be defined in terms of μ_B, α_B, and β_B and the corresponding After parameters, with all six estimated from the data.

An immediate problem is that this functional form may be wrong. The mean of the process may not be smoothly sinusoidal. Our estimates of the alteration's effects will then be biased. Alternative forms are available, e.g., replacing $\alpha_B \sin 2\pi t + \beta_B \cos 2\pi t$ by a polynomial or by separate fixed means for each "season" (defined by the biology, not the calendar), but these may be wrong too.

However, the greater problems are likely to be estimators with very high variances, and underestimation of these variances. These arise because of large, long-lasting fluctuations that are not allowed for in the mean functions, so must be treated as random deviations from them. These fluctuations can be caused by major environmental events, like El Niño or a large storm, or by biological events unrelated to the alteration, like an epidemic or a predator–prey cycle. Such a fluctuation may last through a significant part of the sampling program; if it does, then the observed abundances, $N(t_i)$, will be "serially correlated" (the series is "autocorrelated"): several of them, especially those close in time, will have essentially the same random deviation from the mean, so (i) this deviation does not "cancel out" in the average, and (ii) these observations will not vary much, so the true variance will be underestimated. Thus, effect estimates may be very unreliable (containing a large chance component), but may seem to be reliable.

One way to deal with this problem is to try to model major periodic phenomena deterministically as part of the mean function. Calling something "random" rather than "deterministic" is often rather arbitrary, based on what we are unable to predict in our present state of knowledge more than on what is inherently unpredictable. However, unless the phenomenon is well understood, this seems likely to complicate interpretation without reducing variance, by introducing additional parameters which explain little of the variation but whose estimates are correlated with the estimates of interest, i.e., of the alteration's effects, such as $\mu_A - \mu_B$ in Equation 2.

Modeling the Errors. An alternative is to write a model for the abundances, e.g.,

$$N(t_i) = M(t_i) + \varepsilon_i, \tag{3}$$

where M is the parametric mean function, and the errors, ε_i, may be correlated. The past history, H, being unknown, can be taken as being far in the past: it is usually plausible that, if H is information about $N(t)$ for $t < t_0$, then H will be irrelevant if t_0 is long enough before the first observed time, t_1. The errors could be modeled in a way that takes specific account of their likely mechanisms, but this is hard to do when there are several sources of error whose mechanisms are little known. Generic models, especially linear models (ARMA models, Box and Jenkins 1976) of the form

$$\varepsilon_i = \Sigma_j b_j \varepsilon_{i-j} + a_i + \Sigma c_j a_{i-j}, \tag{4}$$

where the a_i's are uncorrelated random values and the b's and c's are unknown constants, are usually preferred. This class of models can allow for trends or seasonal patterns (so that "$\alpha_B \sin 2\pi t + \beta_B \cos 2\pi t$" can be omitted from Equation 2), by focusing on differences of successive observations, or observations a year apart, and allowing the current error to depend directly on the error a year earlier.

However, these linear models may not fit long-term phenomena like El Niño well: e.g., a recent pattern of declining abundances may indicate the beginning of an El Niño, and thus foretell a continuing decline followed by a long period of low abundance, but other patterns of recent abundances may be uninformative. Patterns of this kind may be better described by models with long-range dependence, e.g., with correlations that decay slower than exponentially over time (Beran 1992).

Attributing seasonal patterns to errors is mainly for series with no clear physical mechanism for seasonality, and better suited to forecasting than to estimation. Without differencing, the forecasts have exponentially declining seasonal patterns. The use of differences implies "homogeneous" behavior, independent of the current level of the process: there is no tendency to return to a mean that is a periodic function of time, like Equation 2—i.e., no "density-dependent" regulation, so a time series made up of yearly averages would be "nonstationary". This seems unlikely for abundances.

Finally, these models imply constant variances and correlations that depend only on the number of intervening samples; this seems doubtful with unevenly spaced observations, or if disturbance is greater at some times of the year, but the analytical consequences may be mild (cf. Stigler 1976).

Using Covariates. The typically small numbers and span of sampling times in impact assessment data will make it difficult to carry out either the mean function or the ARMA approach. Both introduce new parameters to estimate. Even if a simple assumed linear form is correct, variances and covariances may still be underestimated, because the series is so short that the variance of the average of the errors, i.e., $V\{\sum e_i/n\}$, is not negligible compared to the variance of a single observation, i.e., $V\{e_i\}$ (see Priestley 1981, Equation 5.3.12). Above all, even when these problems are minor or resolved, the main achievement of such models will be realistic estimates of the variances of our estimates of the alteration's effects. The models do relatively little to reduce the variances. Even if the correct error model were given to us, the estimated effects would often still have variances too large for practical use.

A third approach can potentially both reduce the variances of estimated effects and estimate these variances accurately. This is to include in the model other observable variables which are affected by the natural fluctuations but not affected by the alteration. These "covariates" can be used to estimate the contribution of the natural fluctuations to the abundance. By removing this contribution, we obtain a "corrected" or "adjusted" abundance which has smaller temporal variance and smaller serial correlation than the raw abundance but is equally affected by the alteration. We can estimate what the abundance would have been under "standard" conditions, and estimate the effect of the alteration by the change it would cause under these conditions. This approach can also reduce deterministic bias: e.g., a covariate like water temperature may be a better indicator of seasonal variation than the time of year itself.

This approach also has difficulties. It requires a model for estimating the natural fluctuations on the basis of the covariates. Most commonly, some form of regression of the observed values (the abundances) against the covariates is used, possibly with either or both being transformed first. If the model form is wrong, estimated alteration effects are likely to be biased. It also requires that the covariates be reasonably good indicators of the natural fluctuations. The covariate model will include some additional parameters to be estimated, thus reducing the information available for estimating effects. If these parameters have little explanatory value, the variances of our effect estimates may actually increase, and our estimates of these variances will become more complicated but no more accurate. This can occur either if the covariate is not strongly correlated with the natural fluctuation or if it is observed with substantial error.

Before-After-Control-Impact Paired Series Designs

Impact-Control Differences: Model

In many cases, the most effective covariate for the abundance at the Impact site is likely to be the simultaneous abundance at a Control site. This is not an experimental control, since treatments are not assigned, randomly or otherwise, by the investigator. In this discussion, a Control is an area which is similar to the Impact area in features judged to be important (e.g., depth, topography, current patterns, suites of species) and near enough to experience similar environmental fluctuations (storms, upwellings, etc.) but far enough away to be unaffected or little affected by the alteration. The discussion will also apply to a set of Control areas represented by a single value at each time.

We now write $N_{IB}(t)$, $N_{CB}(t)$, etc., to indicate the site (I or C) as well as the period. If several Control sites are used, $N_{CB}(t)$ can be thought of as the average or some other suitable summary (e.g., the median) of their abundances. We also write $M_{IB}(t,H,t_0)$ for the mean at time t of all possible realizations of N_{IB} given a history H at a starting point t_0. Other means are defined similarly. $M_{IB}(t,H,t_0)$ is defined for all t, including $t > T_0$, the time of the alteration, even though $N_{IB}(t)$ cannot be observed at these times.

Suppose both the Impact and the Control areas are contained in a larger region R, all of whose sub-areas experience similar environmental variation, such as seasons, major storms, climatic disturbances like El Niño, etc. It might then be reasonable to assume that the mean (over realizations) abundance per unit area or volume at any one location (sub-area) differs from the mean for the region only because of (i) particular features of the location itself, which are constant, and (ii) lingering effects of past abundances, expected to shrink rapidly as a result of births, deaths, and movements. One model for this is

$$M_{LP}(t,H,t_0) = M_{RP}(t) + \alpha_{LP} + h_{LP}(t,H,t_0) \tag{5}$$

for the mean of the abundance process at location L (Impact or Control) in period P (Before or After). Here, $M_{RP}(t)$ is the mean abundance for the region as a whole, and the other terms are the two types of deviation.

Random environmental variation would cause the actual abundances to differ from their means. If $E_{LP}(t)$ represents this deviation at location L, then the deviation for the region, $E_{RP}(t)$, is the average of $E_{LP}(t)$ over locations, L, in the region, and we can write $\eta_{LP}(t) = E_{LP}(t) - E_{RP}(t)$ for the difference between the deviation at L and the average deviation. The model then describes the abundance at L during period P by

$$N_{LP}(t) = M_{RP}(t) + E_{RP}(t) + \alpha_{LP} + \eta_{LP}(t), \tag{6}$$

where $M_{RP}(t)$, and α_{LP} are deterministic, and $E_{RP}(t)$ and $\eta_{LP}(t)$ are stochastic processes with mean 0 for each t (since they describe deviation from the mean). As for the "Before-After" model (Equation 3), the history, H, is taken to be far in the past and irrelevant.

Impact-Control Differences: Estimating the Effect

If Equation 6 is accepted, then the difference between the Impact and Control abundances, $D_P(t) + N_{IP}(t) - N_{CP}(t)$ is

$$D_P(t) = \alpha_{IP} - \alpha_{CP} + \varepsilon_P(t) \tag{7}$$

where $\varepsilon_P(t) = \eta_{IP}(t) - \eta_{CP}(t)$. Given observations $N_{IP}(t)$ and $N_{CP}(t)$ at times t_{P1}, ..., $t_{Pn(P)}$ in period P, a natural unbiased estimate of $\alpha_{IP} - \alpha_{CP}$ is $D_{P\bullet}$, the average of the $D_P(t_{Pi})$'s.

Assuming Equation 6 holds for both periods, we can regard

$$\delta = (\alpha_{IA} - \alpha_{CA}) - (\alpha_{IB} - \alpha_{CB}) \tag{8}$$

as the change in the mean at the Impact area relative to that at the Control area, between the Before and After periods. If we can assume that this change would have been zero without the alteration, i.e., that the mean of $N_{IA}(t) - N_{CA}(t)$ would have continued to be $(\alpha_{IB} - \alpha_{CB})$, then $(\alpha_{IA} - \alpha_{CA}) - (\alpha_{IB} - \alpha_{CB})$ gives the change in mean at the Impact area due to the alteration. Thus $D_{A\bullet} - D_{B\bullet}$ is an unbiased estimate of the effect of the alteration on mean abundance at the Impact site—if the model is correct and the alteration caused the change.

The Variance of the Effect Estimate. Under Equation 6, the use of the Control as a covariate has allowed us to define a parameter representing the effect without further assumptions about $M_{RP}(t)$, the temporally fluctuating component of the mean Impact site abundance. Equation 6 also implies that the difference, $N_{IP}(t) - N_{CP}(t)$, removes the "regional" random term, $E_{RP}(t)$, as well as $M_{RP}(t)$, thus potentially removing much of the variance and much of the serial correlation which the Before-After design must contend with.

To describe this, we write $V_{\varepsilon P}(t)$ and $C_{\varepsilon P}(t_1, t_2)$ for the variance and covariance functions of the error, $\varepsilon_P(t)$, and $C_{BA}(t_1, t_2)$ for $\mathrm{Cov}(\varepsilon_B(t_1), \varepsilon_A(t_2))$, the covariance between a Before and an After difference. From Equation 7, the variance of $D_{A\bullet} - D_{B\bullet}$ is:

$$V\{D_{A\bullet} - D_{B\bullet}\} = \Sigma_P V_{\varepsilon P\bullet}/n(P) + \Sigma_P[1 - 1/n(P)]C_{\varepsilon P\bullet} - 2C_{BA\bullet}, \tag{9}$$

where $n(P)$ is the number of observations in period P, $V_{\varepsilon P\bullet} = \Sigma_j V_{\varepsilon P}(t_{Pj})/n(P)$ and $C_{\varepsilon P} = 2\Sigma_k\Sigma_{j<k}C_{\varepsilon P}(t_{Pj}, t_{Pk})/n(P)[n(P) - 1]$, the averages of the variances and covariances of the differences in period P, and $C_{BA\bullet} = \Sigma_k\Sigma_j C_{BA}(t_{Bj}, t_{Ak})$ $/n(B)n(A)$, the average covariance between a Before and an After difference.

The standard estimate of $V\{D_{A\bullet} - D_{B\bullet}\}$ is

$$s^2 = \Sigma_P\Sigma_j(D_P(t_{Pj}) - D_{P\bullet})^2/n(P)[n(P) - 1] \tag{10}$$

assuming possibly unequal Before and After variances. It is biased low by

$$b(s^2) = V\{D_{A\bullet} - D_{B\bullet}\} - E\{s^2\} = \Sigma_P C_{\varepsilon P\bullet} - 2C_{BA\bullet}. \tag{11}$$

This result holds even if the variance function, $V_{\varepsilon P}(t)$, varies over time (see also Stigler 1976, Cressie and Whitford 1986).

Thus, Equation 9 gives the variance of the effect estimate, and Equation 11 gives the amount by which this variance will be underestimated (on average) if serial correlation is ignored. Previously it was argued that both will often be unacceptably large when the Before-After design is used. This may not be so when the differences are used.

Large fluctuations will often be the result of large-scale disturbances affecting the whole region, so that most of the variance at a site will be due to the "common" regional variation, $E_{RP}(t)$ in Equation 6, which is removed when we take differences. If the Control and Impact sites are not far apart, $\eta_{IP}(t)$ and $\eta_{CP}(t)$, the local deviations from the average regional fluctuation, may be highly correlated, further reducing $V_{\varepsilon P}(t) = V\{\eta_{IP}(t) - \eta_{CP}(t)\}$. These local deviations should be more quickly removed than regional deviations, not only because they are smaller but also because of mixing (of nutrients or planktonic stages) and movement within the region: e.g., a chance increase at a site, unrelated to changes in long-term physical or chemical conditions at the site, should be quickly dissipated to neighboring sites. If so, serial correlation of the differences, $D_P(t)$, will decrease rapidly with time.

Reducing Variance and Bias. Smaller variances and correlations will reduce $V = V\{D_A\bullet - D_B\bullet\}$. The latter will also reduce the bias, $b = b(s^2)$ in Equation 11, both absolutely and relative to the variance. Widely spaced sampling times will reduce b, but also reduce the number of observations, thus increasing V, unless the sampling period is lengthened.

Sampling error will increase $V_{\varepsilon P}(t)$, the variance of the errors, but would not usually affect the covariances. Thus, reducing the sampling error will reduce V, but not b. A confidence interval for δ (Equation 8) should have length about $2t\sqrt{V}$ (where t is from the t distribution). The standard interval, using s^2, has length about $2t\sqrt{[V - b]}$, so is too short by about $2tb/\{\sqrt{V} + \sqrt{[V - b]}\}$. Both the absolute and relative error increase if V decreases but covariances do not.

If the underestimate of variance seems likely to be serious, the autocorrelation of the errors can be allowed for, e.g., by writing an explicit model. If variances within periods are equal, differences at t_j and t_{j+m} in the same period have correlation ρ^m, and differences in different periods are uncorrelated, then $b(s^2) = 2\rho V/(1 - \rho)$. Thus multiplying s^2 by $1 + 2r/(1 - r)$, where r is the first order serial correlation (estimating ρ), may approximate the right adjustment, though only roughly.

Alternative Models

The model of Equation 6 assumes that spatial and temporal variation are additive: i.e., systematic and random large-scale fluctuations, like seasons and storms, are assumed to affect all sites in the region approximately equally, so that they largely cancel in the differences, $N_I(t) - N_C(t)$. We now consider some alternatives. For applications of some of these, see Bence et al. (Chapter 8).

Additivity after Transformation. Spatial and temporal variation may not be additive on the abundance: e.g., they could be additive on the log of the abundance (i.e., multiplicative on the abundance) so Equations 6 and 7 hold for

$\log[N_{LP}(t \mid H, t_0)]$, or additivity may apply to the reciprocal (the area or volume needed to support one individual) or to some other transformation.

One way to approach this problem is to seek the right transformation, e.g., using a power transformation with power chosen by the Tukey test (Tukey 1949, Snedecor and Cochran 1980, p. 283) or a similar method (e.g., Andrews 1971, Berry 1987). While useful, this approach is not trouble free (Smith et al. 1991, 1993). Some transformations require rather arbitrary adjustments when the observed abundance is zero (e.g., due to sampling error). The change in the difference of the means of the transformed variables may not be easy to interpret in terms of the original abundances. There may not be a "right" transformation, e.g., if abundances at Impact are higher than at Control under some conditions or seasons, but lower under others. Even if there is a "right" transformation for the Before period, it may not be right for the After period if the alteration does not act in the same way as other effects.

Ratio Models. Suppose the Impact and Control sites are sub-areas of a larger region subject to mixing and experiencing similar environmental variation, both systematic and random. The total population of the region, $N_R(t)$, fluctuates in response to this variation, and also to local variation at the sub-areas, but is then redistributed by movement, births and deaths. The abundance in a sub-area might then tend to be a roughly fixed proportion of the abundance of the region, the proportion being determined by such factors as water movement (bringing in recruits), usable space, and local survivorship.

If, given $N_R(t)$, Impact and Control abundances are given by independent Poisson variables, $N_I(t)$ and $N_C(t)$, with means $\alpha_I N_R(t)$ and $\alpha_C N_R(t)$, then standard tests and confidence intervals for the ratio, $r_{IC} = \alpha_I/\alpha_C$, are derived from those for the parameter $p_I = \alpha_I/(\alpha_I + \alpha_C)$, the probability of "success" in the Binomial distribution for $N_I(t)$ successes, from $N_C(t) + N_I(t)$ trials (Lehmann 1959, p. 180). If, given the values of the sequence $\{N_R(t_i)\}$, the pairs $\{(N_C(t_i), N_I(t_i))\}$ are independent (over time), then the combined estimate of p_I is $\hat{p}_I = 1/[1 + \Sigma N_C(t_i)/\Sigma N_I(t_i)]$. If the totals, $\Sigma N_C(t_i)$ and $\Sigma N_I(t_i)$, are not small, then $\hat{r}_{IC} = \hat{p}_I/(1 - \hat{p}_I) = \Sigma N_I(t_i)/\Sigma N_C(t_i)$ is approximately Normal with mean r_{IC} and variance $r_{IC}^2/\Sigma[N_C(t_i) + N_I(t_i)]$. We might measure the change at Impact relative to Control by the difference between the Before and After r_{IC}'s, substituting $\hat{r}_{IC}$ for r_{IC} in the variance formulae for a confidence interval. Eberhart (1976) suggests a similar approach.

Three uncertain assumptions in this model are (i) that α_I/α_C is the same at each time, (ii) that $N_C(t)$ and $N_I(t)$ are Poisson, and (iii) that the $\{(N_C(t_i), N_I(t_i))\}$ pairs are independent given the $N_R(t)$'s. These have the unlikely corollary that, given the sample size, $\Sigma[N_C(t_i) + N_I(t_i)]$, the number of sampling times is irrelevant: a single Before time and a single After time would suffice. More realistic approaches include: (i) treating $p_I = p_I(t)$ as variable in time, e.g., as a stochastic process with mean p_I; (ii) assuming the mean of $N_L(t)$ ($L = C$ or I) to be θ_L which is itself gamma distributed with mean $\alpha_L N_R(t)$, so that $N_C(t)$ and $N_I(t)$ have negative binomial distributions (this leads to a special case of (i), with $p_I(t_i)$ having a Beta distribution, if the two gamma distributions have the same scale parameter); (iii) treating the values $\hat{p}_I(t_i) = N_I(t_i)/[N_C(t_i) + N_I(t_i)]$ as a possibly correlated series, with variances, $V\{\hat{p}_I(t_i)\}$, not necessarily proportional to $1/[N_C(t_i) + N_I(t_i)]$.

A Predictive Model.   Perhaps the most common covariate model is

$$N_I(t_i) = \gamma + \beta N_C(t_i) + \varepsilon_i, \tag{12}$$

where the error, ε_i, is uncorrelated with $N_C(t_i)$. With this model, we could estimate an alteration effect as a change in γ or in β between Before and After. But it seems preferable to estimate or describe the change (i) by presenting both regression lines, thus showing that the effect varies with "environmental conditions", as represented by $N_C(t)$ (e.g., Mathur et al., 1980; Bence et al., Chapter 8), and (ii) presenting as the overall estimate the change in $\gamma + \beta N_C$, where N_C is a "typical" Control value, possibly the average of all observed $N_C(t)$ values, both Before and After. The hope is that most of the variation in $N_I(t)$ is "explained by" variation in $N_C(t)$. Large, low-frequency variation, like El Niño, which can be difficult to model but affects both N_C and N_I, might then not play a significant role, so that the ε_i's can be treated as independent or as obeying a simple generic model, e.g., autoregressive of order 1 ($\varepsilon_i = b\varepsilon_{i-1} + a_i$, where $b < 1$: see Equation 4).

This model seems a potentially useful combination of the ideas of additive and multiplicative differences between the Impact and Control sites, but an attempt to derive it from a rough mechanistic model shows that nuisance variation may not be completely removed this way. Suppose $N_R(t)$, the regional average abundance at time t, has mean $\mu(t)$ and variance $\sigma^2(t)$; that $N_R(t)$, $N_C(t)$ and $N_I(t)$ are jointly Normal; and that, given $N_R(t)$, $N_C(t)$ and $N_I(t)$ have means $\alpha_C N_R(t)$ and $\alpha_I N_R(t)$, variances ϕ_{CC} and ϕ_{II}, and covariance ϕ_{CI}. Then the unconditional distribution of $N_I(t)$ and $N_C(t)$ is Normal with means $\alpha_I \mu(t)$ and $\alpha_C \mu(t)$, variances $\tau_{II} = \phi_{II} + \alpha_I^2 \sigma^2(t)$ and $\tau_{CC} = \phi_{CC} + \alpha_C^2 \sigma^2(t)$, and covariance $\tau_{CI} = \phi_{CI} + \alpha_C \alpha_I \sigma^2(t)$. Standard manipulations show that, given $N_C(t)$, the distribution of $N_I(t)$ is Normal with mean $E\{N_I(t) \mid N_C(t)\} = (\alpha_I - b\alpha_C)\mu(t) + bN_C(t)$ and variance $V = \tau_{II} - \tau_{CI}^2/\tau_{CC}$, where $b = \tau_{CI}/\tau_{CC}$.

Thus neither $\mu(t)$ nor $\sigma^2(t)$, the "regional" mean and variance, drop out in this version: both the slope and the intercept in Equation 12 are functions of time. This is a form of the "errors in variables" problem (Fuller 1987, Snedecor and Cochran 1980, p. 171): Equation 12 holds with "$\alpha_C N_R(t_i)$" instead of "$N_C(t_i)$"; the latter is an estimate, with error, of the former; when it is substituted, its error becomes part of the "ε_i", which is thus correlated with the "independent" variable, $N_C(t_i)$.

Thus, the "predictive" model of Equation 12 does not follow from this argument. It may not follow from any simple mechanistic argument, though more careful attention to mechanisms and distribution choices may do better. This does not mean it should not be used: statistical analyses are frequently based on generic models chosen more because they are simple, well understood, and have about the right behavior, than because of a mechanistic derivation. If σ^2 is large compared to the ϕs, or if it does not vary much over time, and if $\mu(t)$ can be modeled as a seasonal function, then the modified Equation 12,

$$N_I(t_i) = \gamma + \alpha_1 \sin 2\pi t + \alpha_2 \cos 2\pi t + \beta N_C(t_i) + \varepsilon_i, \qquad (13)$$

might be satisfactory. This model would allow the relative advantages of the sites, or the effect of the alteration, to vary with seasons.

Matching. In some cases, matching could be used to remove deterministic time effects, including cases where the alteration itself has different effects under different conditions. For example, data could be analyzed separately for winter and summer or for periods of upcoast and downcoast currents (Reitzel et al. 1994).

Model Uncertainty

The previous section suggests that there may be many plausible models on which assessment could be based. It may be possible to rule some of these out by goodness-of-fit tests, diagnostic plots, or arguments based on mechanisms or auxiliary variables. These methods are informal (e.g., there is no clear criterion for choosing the level of a goodness-of-fit test), but honest use would usually retain several models for assessment. Thus model uncertainty is a part of the uncertainty in estimates of change.

It is likely that none of the models remaining is "correct". A strategy is to begin with a broad enough range of realistic models to have a high likelihood that at least one of them is close enough to the truth for effective decision making. Impact assessment could benefit from a "kit" of generic stochastic spatio-temporal models which can allow for major systematic and random effects and reflect the physiology and behavior of groups of organisms, but allow comparison of neighboring sites over time without an excess of unknown parameters. A possible starting set might consist of the three types of models—additive (perhaps after transformation), ratio, and predictive—discussed in the previous section.

If all the unelimated models give similar answers, model uncertainty could be displayed by giving the results from the simplest or most plausible model, with bounds showing the range of variation due to model differences. But "similar answers" may not be easy to define: e.g., models assuming multiplicative effects must give different answers from models assuming additive effects. This is a case where estimates and confidence intervals for effect sizes are messier than P-values for a test of "no effect" (see Stewart-Oaten, Chapter 2)—although the tidiness of the latter is misleading, since tests using different models are testing different things (Sampson and Guttorp 1991). Many of the problems in this area are discussed by Cochran and Rubin (1974); there are few tidy answers.

Some Suggestions. Models giving $E\{N_I(t) \mid N_C(t)\}$, like Equations 12 or 13, are the easiest to deal with. For any given N_C, we can calculate confidence intervals for $E_B\{N_I(t) \mid N_C\} - E_A\{N_I(t) \mid N_C\}$, the difference between the Before and After mean Impact values when the Control has the value N_C. Thus we could compute a "typical" loss, e.g., with N_C = the average of all $N_C(t)$ values, both Before and After. Bence et al (Chapter 8) suggest constructing a confidence interval for the "average" percent loss by a jackknife method, using the values L_i = estimated average percent loss when the ith sampling time is omitted from the data set. The result compares well with the estimate based on Equations 6 to 8, using $N(t) = \log(\text{abundance})$.

It is harder to deal with models which give Before and After estimates of $E\{F(N_I(t), N_C(t))\}$, for some function F: e.g., $F(N_I(t), N_C(t)) = 1/N_I(t) - 1/N_C(t)$ for the "difference" model, Equation 7, with the reciprocal transformation, or $F(N_I(t), N_C(t)) = N_I(t)/[N_C(t) + N_I(t)]$ for the ratio model. For these, one approach might be to choose a "typical" Control value, N_C^* (e.g., the average of $N_C(t)$ for the entire study), equate $F(N, N_C^*)$ to its Before and After means, and solve to find Before and After values of N, interpreted as "typical" Before and After Impact values when Control is at N_C^*. Thus if $Dp_\bullet$ is the average of $F(N_I(t_i), N_C(t_i))$ for period P, and the equation $F(N, N_C^*) = D_{P\bullet}$ has the solution $N = G(N_C^*, D_{P\bullet})$, we could estimate the change in the typical Impact value as $G(N_C^*, D_{B\bullet}) - G(N_C^*, D_{A\bullet})$. If G is expanded in Taylor series, with $G_1 = \partial G/\partial D$, we obtain the approximation $V\{G(N_C^*, D_{P\bullet})\} \approx V\{D_{P\bullet}\}G_1^2(N_C^*, D_{P\bullet})/2$. This could be used for an approximate confidence interval for the change. E.g., if $F(N_I, N_C) = 1/N_I - 1/N_C$, so $D_{P\bullet}$ = the average of $1/N_I(t_i) - 1/N_C(t_i)$ in period P, then $G(N_C^*, D_{P\bullet}) = [1/N_C^* + D_{P\bullet}]^{-1} = \hat{N}_{IP}$, say; so an approximate confidence interval is $\hat{N}_{IB} - \hat{N}_{IA} \pm t\sqrt{\{s_B^2 \hat{N}_{IB}^4/n(B) + s_A^2\hat{N}_{IA}^4/n(A)\}}$ where t is from the t distribution and $n(P)$ and s_P^2 are the number of observations and the sample variance of $1/N_I(t_i) - 1/N_C(t_i)$, in period P.

A similar approach is to use the estimate of δ = the change in the mean of $F(N_I(t_i), N_C(t_i))$ (i.e., the Before mean minus the After mean, as in Equation 8) directly. If $D = D_{B\bullet} - D_{A\bullet}$ is the estimate of δ, we construct a "no alteration" sample consisting of the Before values of $N_I(t_{Bi})$ and the estimated After values $\hat{N}_I(t_{Ai})$ which solve $F(\hat{N}_I(t_{Ai}), N_C(t_{Ai})) = F(N_I(t_{Ai}), N_C(t_{Ai})) + D$. Thus, $\hat{N}_I(t_{Ai})$ estimates the value we would have got at time t_{Ai} had the alteration not occurred. We also construct an "alteration" sample consisting of the After values $N_I(t_{Ai})$ and the estimated Before values $\hat{N}_I(t_{Bi})$ which solve $F(\hat{N}_I(t_{Bi}), N_C(t_{Bi})) = F(N_I(t_{Bi}), N_C(t_{Bi})) - D$. Thus, $\hat{N}_I(t_{Bi})$ estimates the value we would have got at time t_{Bi} had the alteration existed then. We then estimate the "typical" effect by the difference between the averages of these samples. A more elaborate scheme would be to use the upper and lower boundaries of the confidence interval for δ, instead of using D. Thus if the confidence interval for the change in $E\{1/N_I(t) - 1/N_C(t)\}$ is (D_L, D_U), then the upper boundary for an approximate confidence interval for the change in N_I is found by (i) construct the "no alteration" sample $N_I(t_{B1}), ..., N_I(t_{Bn(B)}), \hat{N}_I(t_{A1}), ... ,\hat{N}_I(t_{An(A)})$, where $\hat{N}_I(t_{Ak}) = [1/N_I(t_{Ak}) + D_L]^{-1}$, and calculate its mean, $M_{U,NA}$; (ii) construct the "alteration" sample, $\hat{N}_I(t_{B1}), ..., \hat{N}_I(t_{Bn(B)}), N_I(t_{A1}), ... , N_I(t_{An(A)})$ where $\hat{N}_I(t_{Bk}) = [1/N_I(t_{Bk}) - D_L]^{-1}$, and calculate its mean, $M_{U,A}$. The approximate upper confidence limit is $M_{U,NA} - M_{U,A}$. (D_L is used for the upper limit because $F(N_I, N_C) = 1/N_I - 1/N_C$ is a decreasing function of N_I. If the alteration reduces abundance, D_L and D_U should be negative.)

Approximate Answers to the Right Questions. Such comparisons are very rough. Presenting confidence intervals from several different models is an attempt to combine the ranges of their effect estimates and their error (standard deviation) estimates. The approach ignores both the "errors in variables" problem and that means are not preserved by nonlinear transformations, e.g., $E\{1/N_I(t)\} \neq 1/E\{N_I(t)\}$. The last problem may not be severe for differences, since the errors may approximately cancel. A Taylor series approach to it is described by Sampson and Guttorp (1991), but seems hard to apply here: the pairs $\{(N_I(t_i), N_C(t_i))\}$ are assumed to be independent (for different times). Medians are preserved, so it might be possible to improve model comparisons by using confidence intervals for medians (e.g., based on the sign test) rather than for means.

But these, or similar, comparisons could be useful. Although the models may have very different forms, e.g., additive versus multiplicative, they may give similar results, especially if the Before and After series of Control values are similar. When the series are dissimilar, we need to distinguish a change at one site that does not occur at the other from a change in a comparison measure (the difference or the ratio) that is due solely to a natural change over the entire region. Even when a single model seems clearly "best", we may want to present the results in a different measure or "scale": e.g., the ratio model may be the most plausible, and fit the data best, but a decision maker might want to know "about how many individuals" of a given species will be "lost", i.e., the arithmetic difference between N_I and what it would have been. When several dissimilar models remain in the running, P-values for a test of "no effect" might be useful, not as a measure of the strength of the effect but to help indicate the compatibility of the models. In some cases it may be necessary to report more than one set of results, with arguments for preferring some models to others.

There are ways to avoid (or evade) model uncertainty. One is to choose a standard model (e.g., Equation 7 with independent errors, $\varepsilon_P(t)$), or a standard analysis (e.g., a t-test or ANOVA), and report the results of this alone. By implying that the model used (often an implausibly simple one) is known with certainty to be true, this approach seems misleading. It is sometimes supported by subjecting the model to a goodness-of-fit test, but other plausible models might also pass this test while giving different assessment results, especially if data are sparse.

Estimating Other Parameters

Underwood (1991, Chapter 9) has suggested that effects on other parameters, notably the variance, should also be assessed. This is attractive, but estimating the variance functions, $V_B(t,H,t_0)$ and $V_A(t,H,t_0)$, or the covariance functions, $C_B(t_1,t_2,H,t_0)$ and $C_A(t_1,t_2,H,t_0)$, would be harder than estimating the means. If we observe $N(t)$ at times t_1, t_2, ..., t_n, the mean (over all possible realizations) of the usual estimate of variance,

$$s_N^2 = \Sigma[N(t_i) - N.]^2/(n - 1), \tag{14}$$

is

$$\phi = \Sigma[M(t_i,H,t_0) - M(\textbf{.},H,t_0)]^2/(n - 1) + V(\textbf{.},H,t_0) - 2\Sigma_j\Sigma_{j<k}C\,(t_j,t_k,H,t_0)/n(n - 1), \tag{15}$$

where $M(\textbf{.},H,t_0) = \Sigma M(t_j,H,t_0)/n$ and $V(\textbf{.},H,t_0) = \Sigma V(t_j,H,t_0)/n$. If we knew how to adjust to eliminate the two sums in Equation 15, most of the difficulty in inference concerning the mean function would be removed. But inference concerning the variance function would still face all the problems discussed so far.

Inference (estimation and uncertainty measurement) concerning mean functions is difficult because their functional form is unknown, and temporal variation and serial correlation cause (i) effect estimates to have large variances and (ii) these variances to be hard to estimate. Inference concerning variance functions has all three difficulties in more severe form. The estimate of the variance function can also be biased by variation in the mean function, and the variance of this estimate is affected by higher temporal moments, e.g., by the kurtosis at time t, or $E\{N(t_1)N(t_2)N(t_3)N(t_4)\}$ for four distinct times.

The standard "fixes" of deterministic modeling, time-series modeling and the use of covariates are all harder to achieve. Deterministic modeling requires intuition or knowledge about the behavior of these functions. Variances seem as likely to vary over time as means; e.g., $V(t,H,t_0)$ seems likely to be higher if t is in a period with high levels of disturbance (more storms, upwellings, or migrations), and the part due to sampling error may depend on population size. Covariances, $C(t_1,t_2,H,t_0)$, would be expected to be higher if t_1 and t_2 are close, but lower if the period between t_1 and t_2 is one of high disturbance. It seems harder to base plausible deterministic models for variances and covariances on these mechanisms than it is for means. Deterministic models for higher moments seem even more remote. Cox (1981) briefly discusses approaches to monotone and cyclical variation in variances.

If $V(t,H,t_0)$ varies in time, we never observe an estimate of it. In contrast, the observed $N(t)$ is itself an unbiased estimate of $M(t,H,t_0)$, so plotting the path traced out by the observations can give us some indication of functional form. If

$M(t,H,t_0)$ is known or accurately estimated (i.e., if the problem for means is largely solved!), plots of squared residuals, and of products of residuals, give similar, but weaker, guidance for variances or covariances. A plot could be made by calculating s_N^2, in Equation 14, for restricted values of i, e.g., s_{N1}^2 uses only $N(t_1)$, ..., $N(t_k)$, s_{N2}^2 uses only $N(t_2)$, ..., $N(t_{k+1})$, etc.; or nonoverlapping blocks, e.g., the first k times, the next k, etc., could be used if the series is not short. But these would be guides to $V(t,H,t_0)$ only if the two sums in Equation 15 were missing.

It is possible that some of the methods for dealing with ARMA models having missing values would be useful, but the difficulties seem much greater than for estimating means. More directly, generic, ARMA-like error models designed to cope with varying temporal variances and covariances has been a busy research area in financial time-series analysis since Engle (1982): see Engle and Rothschild (1992) and Bollerslev et al. (1992). These ARCH (AutoRegressive Conditional Heteroskedasticity) and GARCH (Generalized ARCH) models are, like ARMA models, mainly concerned with forecasting and with conditional behavior, rather than parameter estimation: e.g., the variance of future observations is usually assumed to vary in response to past values, although systematic influences like seasons or day-of-the-week can be included. They also seem to need large, high frequency data sets for effective analysis.

Covariate adjustment, including the use of a Control site, also seems difficult. For example, we can use regression of $N(t)$ against a covariate, $X(t)$, to estimate the conditional mean of $N(t)$ for a given value of $X(t)$, i.e., $E\{N(t) \mid X(t) = x\}$, for any x. This is because the observed $N(t)$ is itself an unbiased estimate of $E\{N(t) \mid X(t)\}$. But we do not observe an unbiased estimate of the conditional variance, i.e., of $V\{N(t) \mid X(t)\}$. (This is variance among realizations so it cannot be estimated from repeated estimates of $N(t)$ at one time: these vary only because of sampling error.) Thus the covariate adjustment may require strong assumptions to (i) determine the form of the relationship between the variances at the two sites (there seems no reason to expect it to be simpler than the relationship between the means), and (ii) estimate the parameters of this relationship.

Finally, a variance change seems hard to interpret. A decrease in the mean would indicate that conditions have deteriorated for the species. The amount of the decrease is also significant for decision making. Although individual assessments would differ, there are reasonable bases for a decision maker to compare a 30% loss of species A to a 60% loss of Species B, and perhaps even to a 5% increase in local unemployment. But it is not clear what a change in "the average variance" (or some parameter of a deterministic variance function) would signify, let alone how one would weigh a 30% increase in it against an economic effect.

Causal Uncertainty

Both statistical uncertainty (as measured in confidence intervals) and model uncertainty apply initially to estimates of change. There is additional uncertainty as to whether the alteration caused the change (i.e., whether the change is an "effect"), since assignment of sampling times and sites to "unaffected" or "potentially affected" is not under the investigator's control, and in particular is not random.

Experiments with randomized assignments seem clearly the best way to establish causes as opposed to associations (Barnard 1982), but these were invented relatively recently. (Fisher 1925). Much accepted scientific "truth" is still probably based on nonrandomized studies. Problems of causality in observational and quasi-experimental studies have attracted increased attention from statisticians recently (e.g., Cochran 1972, Rubin 1974, Pratt and Schlaifer 1984, Rosenbaum 1984, Cox 1992).

Hill (1965, see also 1971, Chapter 24) lists characteristics favoring causal interpretation of results from observational studies: (i) strength of effect, (ii) consistency (among studies), (iii) specificity, (iv) temporality (does the cause precede the effect), (v) biological gradient (monotone dose-response curve), (vi) plausibility, (vii) coherence, (viii) experimental or "semi-experimental" evidence, and (ix) analogy (with similar causes which led to similar effects). In impact assessment, (iv) seems covered by the Before data.

Hill (1965) stresses (i), which seems to favor confidence intervals over hypothesis tests. The larger the effect, the less likely it is to be the result of some overlooked factor. He points out that the measure of "strength" does not need to be the same as the measure of "importance": e.g., in medical studies, the relative difference between groups in death rates from a specific illness may be convincing, even though the absolute rates are both small. Thus an impact analysis in terms of the ratio model can help suggest cause, even though importance may be judged by an estimate of absolute change.

Schroeter et al. (1993) stress (iii), (v), (vi) and (vii): "In the absence of a demonstrated causal chain, a convincing case requires that the results for a number of different species tie together and be consistent, that plausible mechanisms for an ecological impact be identified, and that reasonable alternative mechanisms be explored and ruled out." In their study of kelp bed invertebrates affected by the plume of a power plant cooling system, they (vii) demonstrate similar declines for similar species of snails, at two Impact sites, with (v) the site nearer the plant showing greater effects; (vi) suggest two plausible mechanisms (reduced supplies of drift kelp, and increased abrasion due to the flux of fine particles); and (iii) reject several alternative explanations (e.g., by carrying out separate analyses omitting samples possibly affected by an urchin feeding front). Another version of (iii) might be that species which should *not* be affected should not change. Also, a time trend in the estimated change might indicate the temporary effects of installation, as opposed to the long-term effects of existence or operation.

A version of (ii) may be comparison of results with other assessment studies of similar types. Note, however, that impact studies are usually concerned with effects at a particular place and time, not with generalizations: they are analogous to asking not whether smoking causes cancer but whether it caused a particular smoker's cancer. Other impact studies are perhaps better seen as a version of (ix), e.g., assessments in temperate coastal waters of different continents may involve different species but similar feeding and motility groups, etc. These other studies would be subsidiary in an assessment of a particular alteration, rather than "equivalent" as in studies aimed at generalizations.

Item (viii) is what the BACIPS design is intended to achieve (see Campbell and Stanley 1966). It seems to lie between an experiment and an observational study. In each case, we compare a "treated" and an "untreated" population on the basis of a sample, but causal inferences from observational studies are less reliable for two reasons. First, the allowance for error in inferences from sample to population are more likely to be wrong, since they depend on detailed models, rather than on randomized assignment and a treatment-unit additivity assumption (i.e., that a treatment has about the same effect on each unit). Second, the populations may be different not because of the treatment but because of other factors which are correlated with the treatment assignment. For example, former smokers seem to be less healthy than current smokers (Freedman et al. 1991, p. 23), but the "outcome" variable, health, may be a cause, rather than a result, of the assignment (the decision to give up smoking).

An Impact-Control comparison using only After data risks both problems. The error problem arises because the analysis requires a time series model; the assignment problem arises because features which cause a site to be chosen for an alteration may be important in determining abundances, e.g., a well-intentioned developer might choose a site where abundances are already low.

A Before-After study risks mainly the error problem. A factor affecting abundance may vary more between periods than within periods, and thus mask or mimic an effect of the alteration. The assignment problem is unlikely: such factors rarely determine the startup time, i.e., which times are Before and which After.

The BACIPS setup avoids the assignment problem for the same reason. The chance of the error problem is reduced since the source of additional variation must affect one site differently from the other in a way not anticipated by the model. This can happen in two ways: broadscale changes which are incorrectly modeled (e.g., multiplicative effects represented as additive) and which differ more between than within periods; and large, long-lasting, local changes at either site, occurring at about the same time as the alteration but not related to it.

The use of several models is the main check on differential effects of broadscale changes. Environmental variables might also be used for this. For example, as a check on El Niño effects, Schroeter et al. (1993) found the same temperature changes at all sites, and greater bottom disturbance at the Control, suggesting that El Niño effects could not account for the greater decline at the Impact. However, this check cannot eliminate the possibility of identical environmental changes having different biological effects at different sites. Using environmental variables for blocking, or as covariates, to show that estimated biological effects of the alteration are similar under different environmental conditions, or (since some effects may be expected to vary with conditions) that they agree with expectations, may help, but this may be difficult if data are sparse.

Additional checks are possible with multiple Control sites. Results using different Controls should be similar, or the differences explainable. A single model incorporating all Control sites, and allowing for systematic location effects and for both temporal and spatial correlation, may allow more realism and flexibility, and reduce the "errors in variables" problem (since the "unaffected" abundance will be more accurately estimated), and still retain more degrees of freedom (number of observations − number of estimated parameters) for estimating variance. In particular, multiple Controls may give a much better idea of the likely variability due to naturally-arising, large, long-lasting local perturbations which might mimic or mask an alteration effect.

However, multiple Controls offer no guarantees. The sites are not random, since Impact, at least, will have been deliberately chosen. It may well be suggestive (in a study with After data only) that the mean of the Impact site over the After period is (say) smaller than all the Control means, or (in a Before-After study) that the difference between the Before and After means was greater at the Impact site than at any of the Controls, but it is not possible to attach a standard error, confidence or P-value to this without several dubious assumptions: selection of the Impact site was "effectively" random (e.g., the reasons for its selection are unrelated, directly or indirectly, to the abundance of the species under study), and the means or changes in means at the different sites are independent (e.g., neighboring sites are not expected to have more similar changes than distant sites). A "single control" study could get misleading results if an extraneous factor affects the Control. Multiple controls protect against this, but not against an extraneous factor affecting the Impact site, or affecting some Controls but not others.

Multiple controls might even be less reliable than a single control if an extraneous factor affects the Impact site and nearby, but not distant, Controls, or if the more distant Controls track the Impact site poorly. Thus, as noted by Underwood (1992, Chapter 9) and others, multiple Controls (and multiple Impacts, e.g., sites which let us check that the putative effect decreases with distance from the alteration) offer the possibility of more convincing causal arguments and of reductions in the effects of both natural temporal variation and model uncertainty, but these gains require careful modeling and analysis.

Stronger semi-experimental evidence arises if the potential cause can be applied, removed and then reinstated (Cox 1992). Effects on individuals during operation can be compared to effects during brief shutdowns (Raimondi and Schmitt 1992). Effects on populations may require long shutdowns, but an example is given by Granelli et al. (1990): the operation of a sewage plant was suspended for several months to allow comparison with "plant on" conditions. This seems to provide good evidence of the effect of an installation's operation; its existence may have other effects (e.g., provision of substrate or alteration of water movement) that cannot be checked this way.

Discussion

Impact assessment, like other observational studies, is likely to be messy, even after a conscientious effort to apply the formal techniques of mathematical statistics. Decision makers need to combine quantitative information of disparate kinds. The biological aspects alone may involve more than one general parameter (e.g., mean abundance, mean size), usually for more than one species.

Even for a single general parameter, like mean abundance, it may not be possible to describe the formal results succinctly and unambiguously. The main reason for this is model uncertainty. Formal results focus on estimates of the parameters of a stochastic model, but several models are likely to be both plausible and compatible with the data. It seems misleading to ignore this by considering only one simple model. Rather, estimates should be made from two or more models that seem to fit the data, and the variation in the results described as part of the measurement of uncertainty. However, this can be done only roughly if, as is likely, the parameters of these models mean different things.

These problems may not apply to all variables of interest. Appropriate models may be clearer for physical, chemical or physiological variables. Osenberg et al. (Chapter 6) give examples where the ratio of effect size to temporal standard error seems lowest for physiological variables. Multivariate analysis may reveal composite variables with low temporal variation and potentially high sensitivity to the alteration, such as linear combinations of (possibly transformed) abundances of different species or groups (Carney 1987), though a variable's suitability depends on its importance, and these composites may be hard to interpret.

Some of the difficulties described here may decline as our understanding grows. It would be useful to develop a kit of tractable models based on plausible assumptions, known mechanisms and empirical experience. Well-documented, archived data sets of assessment studies (even the Before or After data alone), or multisite studies that were not intended for monitoring, could give clearer ideas of temporal "tracking" between neighboring sites: the role of site features (depth, substrate, etc.) or of species life-history characteristics, the likelihood of periodic cycles and long-term serial correlations, and the most useful auxiliary environmental parameters. Impact assessment studies could also help classify the types of effects to be expected for given types of alteration and of impact site. They may also help weed out approaches (e.g., choices of parameters) that do not work well and identify others with promise: see Carney (1987) for examples.

However, given the variety of ways in which regions can differ, it is unlikely that model uncertainties will disappear. Indeed it is unlikely that we will ever have an exactly correct model. Thus formal inference will need to include both diagnostic checks to exclude plausible models that do not fit the data, and rough measures of model uncertainty from those not excluded.

Acknowledgments

This chapter owes a great deal to numerous suggestions from R. Schmitt, J. Bence, P. Switzer, and especially C. Osenberg. This work was supported in part by the Minerals Management Service, U.S. Department of the Interior, under MMS Agreement No. 14-35-0001-30471 (The Southern California Educational Initiative). The views and interpretation contained in this document are those of the author and should not be interpreted as necessarily representing the official policies, either express or implied, of the U.S. Government.

References

Andrews, D. F. 1971. A note on the selection of data transformations. Biometrika **58**:249–254.

Barnard, G. A. 1982. Causation. Pages 387–389 *in* S. Kotz, N. Johnson and C. Read, editors. Encyclopedia of statistical sciences, Vol. 1. Wiley and Sons, New York, New York.

Beran, J. 1992. Data with long-range dependence. Statistical Science **7**:404–427.

Berry, D. A. 1987. Logarithmic transformations in ANOVA. Biometrics **43**:439–456.

Bollerslev, T., R. Y. Chou, and K. F. Kroner. 1992. ARCH modeling in finance: a review of the theory and empirical evidence. Journal of Econometrics **52**:5–59.

Box, G. E. P., and G. M. Jenkins. 1976. Time series analysis: forecasting and control. Holden-Day, San Francisco, California.

Box, G. E. P., and G. C. Tiao. 1975. Intervention analysis with applications to economic and environmental problems. Journal of the American Statistical Association **70**:70–79.

Campbell, D. T., and J. C. Stanley. 1966. Experimental and quasi-experimental designs for research. Rand McNally, Chicago, Illinois.

Carney, R. S. 1987. A review of study designs for the detection of long-term environmental effects of offshore petroleum activities. Pages 651–696 *in* D. F. Boesch and N. N. Rabalais, editors. Long-term environmental effects of offshore oil and gas development. Elsevier, New York, New York.

Cochran, W. G. 1972. Observational studies. *in* T. A. Bancroft, editor. Statistical papers in honor of George W. Snedecor. Iowa State University Press, Ames, Iowa.

Cochran, W. G., and D. B. Rubin. 1974. Controlling bias in observational studies: a review. Sankhya, Series A **35**:417–446.

Cox, D. R. 1981. Statistical analysis of time series: some recent developments. Scandinavian Journal of Statistics **8**:93–115.

Cox, D. R. 1992. Causality: some statistical aspects. Journal of the Royal Statistical Society, A **155**:291–301.

Cressie, N. A. C., and H. J. Whitford. 1986. How to use the two sample t test. Biometrical Journal **28**:131–148.

Eberhardt, L. L. 1976. Quantitative ecology and impact assessment. Journal of Environmental Management **4**:27–70.

Engle, R. F. 1982. Autoregressive conditional heteroskedasticity with estimates of the variance of UK inflation. Econometrica **50**:987–1008.

Engle, R. F., and M. Rothschild. 1992. Editor's Introduction to special issue on ARCH models. Journal of Econometrics **52**:1–4.

Fisher, R. A. 1925. Statistical methods for research workers, 1st Edition. Hafner, New York, New York.

Freedman, D., R. Pisani, R. Purves, and A. Adhikari. 1991. Statistics. W.W. Norton, New York, New York.

Fuller, W. A. 1987. Measurement error models. Wiley and Sons, New York, New York.

Granelli, E., K. Wallstrom, U. Larsson, W. Granelli, and R. Elmgren. 1990. Nutrient limitation of primary production in the Baltic area. Ambio **19**:142–151.

Hill, A. B. 1965. The environment and disease: association or causation. Proceedings of the Royal Society of Medicine **58**:295–300.

Hill, A. B. 1971. Principles of medical statistics. Oxford University Press, New York, New York.

Lehmann, E. L. 1959. Testing statistical hypotheses. Wiley and Sons, New York, New York.

Mathur, D., T. W. Robbins, and E. J. Purdy. 1980. Assessment of thermal discharges on zooplankton in Conowingo Pond, Pennsylvania. Canadian Journal of Fisheries and Aquatic Sciences 37:937–944.

Pratt, J. W., and R. Schlaifer. 1984. On the nature and discovery of structure. Journal of the American Statistical Association 79:9–33.

Priestley, M. B. 1981. Spectral analysis and time series. Academic Press, London, England.

Raimondi, P. T., and R. J. Schmitt. 1992. Effects of produced water on settlement of larvae: field tests using red abalone. Pages 415–430 in J. P. Ray and F. R. Englehardt, editors. Produced water: technological/environmental issues and solutions. Plenum Press, New York, New York.

Reitzel, J., M. H. S. Elwany, and J. D. Callahan. 1994. Statistical analyses of the effects of a coastal power plant cooling system on underwater irradiance. Applied Ocean Research 16:373–379.

Rosenbaum, P. R. 1984. From association to causation in observational studies: the role of tests of strongly ignorable treatment assignment. Journal of the American Statistical Association 79:41–48.

Rubin, D. B. 1974. Estimating causal effects of treatments in randomized and non-randomized studies. Journal of Educational Psychology 66:688–701.

Sampson, P. D., and P. Guttorp. 1991. Power transformations and tests of environmental impact as interaction effects. The American Statistician 45:83–89.

Schroeter, S. C., J. D. Dixon, J. Kastendiek, R. O. Smith, and J. R. Bence. 1993. Detecting the ecological effects of environmental impacts: a case study of kelp forest invertebrates. Ecological Applications 3:331–350.

Skalski, J. R., and D. H. McKenzie. 1982. A design for aquatic monitoring systems. Journal of Environmental Management 14:237–251.

Smith, E. P., D. R. Orvos, and J. J. Cairns. 1991. Comments and concerns in using the BACI model for impact assessment. Pages 153–157 in American Statistical Association, 1991 Proceedings of the Section on Statistics and the Environment, Alexandria, Virginia.

Smith, E. P., D. R. Orvos, and J. J. Cairns. 1993. Impact assessment using the Before-After-Control-Impact (BACI) model: concerns and comments. Canadian Journal of Fisheries and Aquatic Sciences 50:627–637.

Snedecor, G. W., and W. G. Cochran. 1980. Statistical methods, 7th Edition. Iowa State University Press, Ames, Iowa.

Stewart-Oaten, A., W. W. Murdoch, and K. R. Parker. 1986. Environmental impact assessment: "psuedoreplication" in time? Ecology 67:929–940.

Stigler, S. M. 1976. The effect of sample heterogeneity on linear functions of order statistics, with applications to robust estimation. Journal of the American Statistical Association 71:956–960.

Tukey, J. W. 1949. One degree of freedom for non-additivity. Biometrics 5:232–242.

Tukey, J. W. 1962. The future of data analysis. Annals of Mathematical Statistics 33:1–67.

Underwood, A. J. 1991. Beyond BACI: experimental designs for detecting human environmental impacts on temporal variations in natural populations. Australian Journal of Marine and Freshwater Research 42:569–587.

Underwood, A. J. 1992. Beyond BACI: the detection of environmental impacts on populations in the real, but variable, world. Journal of Experimental Marine Biology and Ecology 161:145–178.

CHAPTER 8

ESTIMATING THE SIZE OF AN EFFECT FROM A BEFORE-AFTER-CONTROL-IMPACT PAIRED SERIES DESIGN

The Predictive Approach Applied to a Power Plant Study

James R. Bence, Allan Stewart-Oaten, and Stephen C. Schroeter

Study of unreplicated perturbations to ecological systems is of practical importance in both applied and basic research. Obviously, after the perturbation has occurred we cannot observe the state of the Impact site in the absence of the perturbation. Nevertheless, our basic goal is to estimate what this condition would have been, and compare this estimate with the observed (perturbed) condition. Here we consider this goal in the context of the Before-After-Control-Impact Paired Series (BACIPS) design (Stewart-Oaten et al. 1986, Chapter 7). In this design, paired samples are collected a number of times, both Before and After the perturbation, simultaneously (or nearly so) at both a Control and Impact location. In what follows we assume that the effect of the perturbation lasts through the After monitoring period, and to streamline the presentation we consider only the simplest case where its magnitude does not show a systematic trend with time, neither growing in size nor dying away.

The basic idea behind the BACIPS design is that there can be natural differences between the Control and Impact sites, and temporal variability operating on a large spatial scale that influences both sites similarly (Stewart-Oaten et al. 1986). By sampling at both Control and Impact on repeated surveys during the Before and After periods, the design "controls" for such natural variation. Heretofore, the standard analytical approach, using the resulting BACIPS data, was to calculate the difference between Control and Impact values (which may be transformations of the original data) on each date (henceforth termed a "delta"), and test whether the mean of these deltas changes from Before to After the perturbation (Stewart-Oaten et al. 1986, 1992, Carpenter et al. 1989).

Previous attention has focused on how the null hypothesis of no difference between the Before and After mean deltas (i.e., no effect of the perturbation)

Detecting Ecological Impacts: Concepts and Applications in Coastal Habitats, edited by R. J. Schmitt and C. W. Osenberg

should be tested when the formal statistical assumptions are violated. This has included a consideration of the effects of serial correlation in the deltas (Hurlbert 1984, Millard et al. 1985, Stewart-Oaten et al. 1986, 1992, Carpenter et al. 1989, Schroeter et al. 1993), how to test and correct for this potential problem (Stewart-Oaten 1987, 1992, Carpenter et al. 1989, Schroeter et al. 1993), and the validity of parametric and nonparametric tests to violations of distributional assumptions (Carpenter et al. 1989, Stewart-Oaten et al. 1992).

Here we suggest a change in emphasis and consideration of alternative approaches. We agree with Stewart-Oaten et al. (1992), Stewart-Oaten (Chapter 2), and Schroeter et al. (1993) that our primary goal is to obtain an estimate of how large the effect is (the effect size) along with some measure of the accuracy of this estimate (e.g., a confidence interval), and that "P-values" are of secondary importance. Although it is possible to obtain estimates of effect size using the standard approach, the estimates are based on a specific model that requires time, location, and perturbation effects to be additive. One approach to violations of the additivity assumption is to transform the nonadditive data into a form that is additive (say, by taking logarithms). As we will illustrate, however, it is possible that no simple transformation exists for which the resulting data are additive. Even if the natural time and location effects are additive after the transformation, the use of a particular transformation carries with it implicit assumptions about how the perturbation influences the Impact site. Thus, the standard approach lacks flexibility for modeling the perturbation.

In this chapter we explore an alternative "predictive" approach, where the Control value is treated explicitly as a predictor of the Impact value (Mathur et al. 1980, Stewart-Oaten et al. 1992, Stewart-Oaten, Chapter 7). This provides a natural way to include other predictors and allows us to explicitly model effects whose size can vary with environmental conditions. The predictive approach has its own assumptions and limitations, however, and it often will be hard to choose between the model used in the standard BACIPS approach and a predictive regression model. In agreement with Stewart-Oaten (Chapter 7), we think that an application of a variety of different plausible models like these can provide some indication of model uncertainty.

In three core sections below we first discuss the model underlying the standard approach, present some difficulties with this approach, and then present an application of the predictive approach. We use an example data set from studies of the effects of a nuclear generating station's discharges on giant kelp (*Macrocystis pyrifera*) in these sections. Hence, we present relevant background on the example data set in the next section before turning to the core topics.

Background on the Example Data Set

These data come from a study of the influence of the "new" Units 2 and 3 of the San Onofre Nuclear Generating Station on the marine environment. This facility is located on the southern California coast between Los Angeles and San Diego. The new units became fully operational in May of 1983. *A priori*, the cooling system of these units was predicted to adversely influence the San Onofre kelp forest by increasing particulate flux and reducing the light levels on the ocean's floor within the kelp forest (Murdoch et al. 1980, Ambrose et al., Chapter 18). Giant kelp first settle to the bottom as a microscopic stage and eventually develop and grow to lengths of tens of meters, reaching the sea's surface and forming a canopy. Successful development of the microscopic stages requires that critical levels of light reach the bottom substrate, and ambient levels of light are often near or below the critical levels (Dean and Jacobsen 1984, Deysher and Dean 1984, 1986). Flux of particulates near the bottom can have adverse effects on the early stages of giant kelp through abrasion or burial (Devinny and Volse 1978). Furthermore, successful development of giant kelp in the San Onofre area requires hard substrate, and increased settlement of particulates has the potential to bury hard substrate.

The once-through cooling system of each unit consists of a single intake in shallow water and a diffuser system for returning the seawater extending over the range of depths of the kelp forest, located immediately to the northwest of forest (see Ambrose et al., Figure 18.1). When fully operational, the combined cooling systems of the new units circulate and discharge 100 m^3 per second, and can create a turbid (dirty) plume, both by moving turbid inshore water offshore and by entraining turbid bottom water in the discharge area. The plume of the discharge tends to be moved over the San Onofre kelp forest by the predominant southeast currents. Reductions in light levels (Reitzel et al. 1995), changes in the local current pattern (Elwany et al. 1990), and increases in the particulate settlement rate (Bence et al. 1989), all apparently due to the discharge system, have been reported. Schroeter et al. (1993) report that the discharge of the generating station has led to substantial reductions in the densities of invertebrates associated with the hard bottom of the San Onofre kelp forest.

The example data collected under a BACIPS design are presented in Figure 8.1. These data were collected by side-scan SONAR in the Impact kelp forest (San Onofre) and a Control kelp forest located approximately 5 km northwest of the discharge (San Mateo). Surveys were done at (usually) 6-month intervals over a study period extending from 1978 through 1989. For each survey the side-scan records were examined and maps of different categories of "adult" giant kelp density were constructed (Murdoch et al. 1989). Here we report on areas occupied by moderate and higher density categories, corresponding to densities exceeding approximately 0.04 m^{-2}.

In the following sections these example data are analyzed and manipulated using a variety of methods. The goal of this process is to illustrate an approach for estimating the magnitude of effects due to the power plant. In our opinion, this data set is the best available for estimating such effects because of its relatively long duration of measurements related to kelp density. We stress, however, that the case for adverse effects of the generating station on the kelp forest is

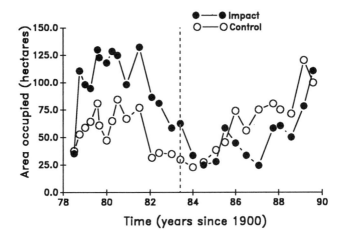

Figure 8.1. Data on areas occupied by densities of giant kelp plants exceeding approximately 0.04 m^{-2}, as determined by side-scan SONAR for the Impact (San Onofre kelp forest) and Control (San Mateo kelp forest) sites over time. Data were collected by Ecosystem Management Associates for 300 survey areas at each site (see Murdoch et al. 1989). Dashed vertical line separates Before and After periods.

based on many more data than those in the example set, including environmental measurements and mechanistic studies cited above, experimental outplants of various stages of giant kelp, other measures of kelp abundance, including counts on fixed transects and estimates from down-looking SONAR, and concomitant studies of potentially confounding factors such as sea urchin abundance and localized changes in the oceanic environment (Bence et al. 1989).

The Standard Approach –The Underlying Model and Implications

The idea that an actual population trajectory over time is a single realization of a stochastic process is central to the standard procedure, and to the alternatives suggested here. This concept is discussed extensively by Stewart-Oaten et al. (1986), and Stewart-Oaten (Chapter 7) and we will not repeat that detailed discussion here. We note that a key consequence is that the actual population value at a given place and time will usually differ from its expected value (the process mean), which itself is time (and space) dependent. Estimates (say of population abundance) can further deviate from this mean because of measurement error. Because replicate observations collected at the same time cannot provide information on the variability of the actual population about its expected value, we treat sampling dates as our level of replication. Of course, within-survey replicates might be collected to reduce sampling error, and our estimate of abundance or other parameters for that survey would then be the average or some other summary of these.

The standard approach assumes "additivity," meaning that in the absence of a perturbation effect the expected value for an observation could be expressed as the sum of time and location effects. We can write this assumption as:

$$\mu_{ij(k)} = L_i + T_{j(k)},$$

where $\mu_{ij(k)}$ is the expected value at location i (I or C for Impact or Control) and time j within the kth period (B or A for Before or After the perturbation), L is the natural location effect and T is the time effect. This additivity assumption implies that the deltas (Impact - Control values) all have the same expected value in the Before period, namely $\Delta_{j(B)} = \Delta_B = L_I - L_C$ for all j. When there is a perturbation effect (i.e., in the After period at the Impact site), we also model this as additive:

$$\mu_{Ij(A)} = L_I + T_{j(A)} + E,$$

where E is the effect of the perturbation. Therefore $\Delta_{j(A)} = \Delta_A = L_I - L_C + E$. Note that the deltas have the same expected value within periods, but these expected values differ by an amount E between periods. An alternative way of viewing the additivity assumption is to note that it implies a particular linear relationship between expected Impact and Control values, namely $\mu_{Ij(k)} = \Delta_k + \mu_{Cj(k)}$. Thus the slope stays equal to one and the perturbation changes only the intercept.

While the idea of additivity of effects is appealing, it is easy to envision cases where the untransformed data would not be additive. For example, the untransformed data might follow a multiplicative rather than additive model, so that the expected Impact value tends to be a constant multiple of the expected Control value rather than differing by a constant number. In general, failure of the additivity assumption could lead to inefficient tests or to artifactual effects (Stewart-Oaten et al. 1986, 1992). For our multiplicative example, if the Impact value tends to be half the Control value, and overall abundance increases from the Before to After period, the mean Impact - Control delta would decline, even with no effect of the perturbation. We could solve this problem and achieve additivity in this case by taking a logarithmic transformation. More generally, we could consider a class of transformations, say of the Box-Cox form $y = (x + c)^\lambda$ (Box and Cox 1964). One could then test for additivity using the Before data, say by the Tukey test for additivity (Tukey 1949), for a range of λ's and c's, and choose a transformation that passes the test.

Difficulties with the Standard Approach

Limited Flexibility of the Standard Approach

As noted above, for estimation we may need to transform the data so that (on the transformed scale) the expected Impact value increases linearly with the expected Control value, with slope one, both in the Before and After periods. This can be a restrictive requirement. Here we consider two reasons why an appropriate transformation might not exist. First, the Control and Impact sites could positively covary, and yet there may not exist any monotone transformation that makes the natural temporal and spatial effects additive. One way for this to happen is for the expected Impact value to increase to an asymptote as the expected Control value increases. In this case, the expected Impact value would exceed that of the Control for low Control values, but the opposite would be the case at high Control values. In marine systems this situation might arise when available recruits into both the Impact and Control populations respond in the same way to environmental fluctuations but with the availability of recruits always proportionally higher at the Impact site. At higher levels of recruitment, density-dependent mechanisms could operate at the Impact site and not at the Control site; in the case of a benthic marine organism this could arise because of limited substrate at the Impact site.

A second kind of difficulty is that the effect of the perturbation might not be additive on the same (transformed) scale as the natural fluctuations. For example, the abundance at the Impact site may naturally tend to be a certain percentage of the abundance at Control site, but the perturbation might cause a reduction of a certain number, rather than a constant percentage. As a result, a transformation that makes data from the Before period additive may not do so for data from the After period. The kelp data may be an example of this; the log-transformed data in the Before period appear to be additive (Table 8.1), but the same transformation fails in the After period.

In the hypothesis-testing context, using a transformation that only works in the Before period is not a problem. If the After data then are nonadditive, this implies that the effect of the perturbation is not additive on the same

Table 8.1.   Results of Additivity Tests and t-Tests for Differences between After and Before Deltas (Impact-Control Values)

Transformation	Additivity P-value	t-test P-value	Effect size
untransformed	0.02	<0.0001	−55.6 ha
$\log(x)$	0.60	<0.0001	−53.0%
$1/x$	0.45	0.0044	+0.011 ha^{-1}
$(x + 0.4)^{-2}$	0.16	0.051	+0.0015 ha^{-2}

Note: Effect size is the mean estimated effect calculated from the difference between the average After and Before deltas. Additivity P-values are attained significance level of Tukey's test for addittivity. t-Test P-values are attained significance level for the hypothesis that the Before and After deltas have the same mean, using the Welch version of the test, to allow for potentially heterogenous variances.

transformation scale that worked for the natural time and location effects. Although such a result indicates that the basic model used by the standard approach is not a realistic portrayal of the system under study, the hypothesis test for a perturbation effect remains valid. This is true because the validity of the test depends on the assumption being true only under the null hypothesis of no effect. However, estimates of effect size, and descriptions of how they vary with conditions, could be markedly off, and we think these should be of primary interest.

Choosing Transformations and Interpreting Effects

Even when a suitable transformation exists, there can still be problems in applying the standard approach. In practice it can be difficult to choose a transformation, or to interpret the effect, which is now measured in the units of the transformed variable. Table 8.1 gives results of t-tests used to test for a difference between the mean Before and After deltas for the giant kelp data for four possible treatments of the data. First, the data are untransformed. Second, the data are $\log_e$ transformed. Third, reciprocals are taken (i.e., $1/x$), and fourth we take the Box-Cox transformation $(x + 0.4)^{-2}$. In three cases the t-test for an effect is statistically significant at the 0.05 level, and in the fourth it is nearly so. The additivity assumption cannot be rejected for the three transformations, but can be rejected for the untransformed data. Estimated effect sizes are also given in the table.

For the first three treatments of the data these effect sizes have an intuitive and qualitatively consistent interpretation; a specified *decrease* in area (untransformed), a specified percentage *decrease* ($\log_e$ transform), or an *increase* in the resources required to maintain one unit area of kelp forest ($1/x$ transform). The fourth data treatment, an arbitrarily chosen Box-Cox transformation, does not have such an obvious interpretation, although the qualitative result, a reduction in kelp area, is consistent.

Each of the treatments of the data presented above imply that the effect of the perturbation acts to change the relationship between Impact and Control in a particular way. We illustrate this for hypothetical examples (Figure 8.2). In Figure 8.2a we plot relationships between the expected Impact and expected Control values under the assumption that the data follow an additive model and the perturbation causes a decline at the Impact site. We next consider the case where a log-transformation would be appropriate. In this case the expected Impact value is a constant percentage of the expected Control value and this percentage declines by a fixed amount After the perturbation. Here the perturbation causes a change in the slope of the relationship between expected Impact and Control values and the intercept is fixed at zero (Figure 8.2b). The portrayed relationships in Figure 8.2c were chosen as an example where our Box-Cox transformation $(x + 0.4)^{-2}$ would achieve additivity. In the standard approach we suspect that careful evaluation of the implicit assumption of these Impact versus Control relationships is rare.

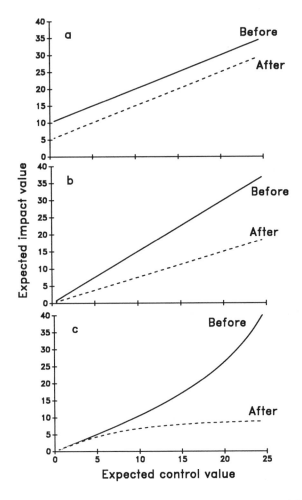

Figure 8.2. Hypothetical relationships between expected Impact and expected Control values. In each case the perturbation has caused a reduction in the expected Impact values. (a) The additive relationship assumed by the standard approach, (b) a multiplicative relationship leading to a constant percentage reduction in the Impact value for all Control values, (c) a relationship chosen so that if the data were $(x + 0.4)^{-2}$ transformed, the transformed data would be additive.

An Alternative: The Predictive Approach

The primary value of the Control is that it acts as a predictor of the Impact value, and here we consider alternative analyses where this concept is an explicit part of the method. The assessment problem can then be thought of as one of comparing two regressions, or fitted functions; the function which best predicts the Impact value from the Control value in the Before period, and the corresponding prediction function for the After period (Stewart-Oaten et al. 1992, Stewart-Oaten, Chapter 7). We could predict the Impact value as a joint function of other variables (e.g., season, current direction, sedimentation) along with the Control value, and could force certain parameters to be the same in both periods (under the assumption that the perturbation did not influence them).

The basic idea behind this approach is that we use the predictive functions to estimate the expected value at the Impact site given the Control for both the Before and After periods. The difference in the expected values, conditional on a particular Control value, is taken as an estimate of the effect of the perturbation (under one set of conditions). This approach attacks the restrictions associated with additivity in two ways. First, any function can be used to model the relationship between Impact and Control values within a period. Second, we can use different functions in the After period than in the Before period, and this allows the perturbation to influence dynamics differently than we might expect from a natural change. A critical implicit assumption is that changes in these conditional expectations (Impact given Control) occur only due to the perturbation. Often this will not be strictly true for reasons related to error in variable problems (e.g., Seber and Wild 1989) (recall that the Control value differs from the expected value of the process that generated it). This is likely to be more of a problem when the distribution of Before and After Control values differ markedly (see Stewart-Oaten, Chapter 7). Because of this difficulty we recommend that the Before and After distributions of Control values be examined to ensure they have similar ranges and variability, and that regression-based estimates be compared with estimates from other models (e.g., the standard BACIPS model).

We now illustrate this approach, again using the side-scan SONAR data on area occupied by giant kelp.

Finding Appropriate Functions

Figure 8.3 shows a plot of Impact versus Control areas (same data as Figure 8.1), with Before and After data distinguished. There is a suggestion in these data that the relationship between Impact and Control may be nonlinear in the Before period, with the Impact value reaching an asymptote. Our approach was to fit four models to the data (each separately by period): linear with assumed zero intercepts, linear with intercepts, a quadratic model, and a nonlinear logistic model. For the logistic model we assumed that the asymptote detected during the Before period remained unchanged in the After period. Thus our approach was to fit the logistic model,

$$I = \alpha/(1 + \beta e^{-kC}),$$

to the Before data, then fix the parameter determining the asymptote (α) and estimate the remaining two parameters based on the After data. We discarded the quadratic model early on because it predicts a dome-shaped relationship between Impact and Control for the Before period over the range the function needs to be

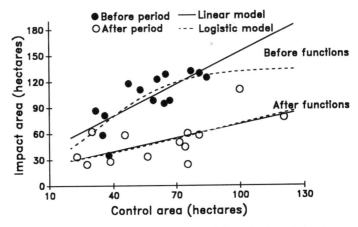

Figure 8.3. Relationship between observed Impact and Control values of giant kelp area. Solid lines indicate fitted linear model with intercepts, dashed lines indicate fitted logistic model (see text).

used in the After period. Without good evidence to the contrary, we required that as conditions continued to "improve" at the Control they also improve at Impact, so that the expected Impact value increases monotonically with the Control value.

Using the "extra sum of squares" principle (e.g., Draper and Smith 1981) we tested whether the added complexity of nonzero intercepts was warranted for the linear models. The results show that the model with nonzero intercepts fits significantly better than the simple model assuming direct proportionality between Impact and Control ($F_{2,23} = 4.91$, $P < 0.025$). The linear regression lines (with intercepts) are plotted on Figure 8.3. For any given Control value the effect size can be estimated by subtracting the Before prediction from the After prediction. The predicted relationships between Impact and Control for the logistic model are also given in Figure 8.3. For both models the residuals are plotted against the Control value for each period in Figure 8.4. Both linear and logistic models fit the data reasonably well (i.e., the residuals seem to show no patterns in relationship to the Control value), and we chose the linear model because it required the estimation of one fewer parameter.

Effect Size and Its Confidence Interval

With the predictive approach we do not assume that there is a single effect size under some appropriate measurement scale. Effect size can vary with the magnitude of the Control value, and with other predictor variables if they are included in the analysis. These effect sizes can be estimated simply by taking the difference between the predicted After ($\hat{I}_{o,A}$) and Before ($\hat{I}_{o,B}$) "Impact" values for some specified Control value C_o.

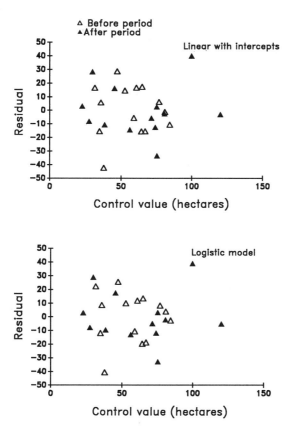

Figure 8.4. Residuals from linear model with intercepts and logistic model (see text) versus Control value. Before residuals are indicated by open triangles, After residuals by solid triangles.

We constructed approximate $(1 - \alpha)$ confidence intervals for the estimated effects at a specified Control value, $\hat{E}_0 = \hat{I}_{0,B} - \hat{I}_{0,A}$, for the linear model with intercepts as

$$\hat{E}_0 \pm t_{v_0}^{\frac{\alpha}{2}} \hat{\sigma}_0$$

where t refers to critical value of the t distribution, v_0 indicates the degrees of freedom, and σ_0 is the estimated standard deviation of the estimated effect at C_0. σ_0 was calculated as $\sqrt{(\hat{\sigma}_{0,B}^2 + \hat{\sigma}_{0,A}^2)}$ where $\hat{\sigma}_{0,B}$ and $\hat{\sigma}_{0,A}$ are the standard deviations of $\hat{I}_{0,B}$ and $\hat{I}_{0,A}$, which were calculated using standard regression approaches (e.g., Draper and Smith 1981). Taking into account that the standard deviations of the estimates of the expected Impact value can differ between the periods, v_0 can be estimated with a Satterthwaite approximation (e.g., Ames and Webster 1991):

$$v_0 = \frac{(\hat{\sigma}^2_{o,B} + \hat{\sigma}^2_{o,A})^2}{\dfrac{\hat{\sigma}^4_{o,B}}{v_B} + \dfrac{\hat{\sigma}^4_{o,A}}{v_A}}$$

where v_k is the degrees of freedom (n - number of parameters estimated) associated with period k. Note that our approach for constructing confidence intervals does not require that the prediction equations be linear; approximate confidence intervals can be calculated whenever estimates of the appropriate variances and degrees of freedom are available.

Estimated effect sizes (still for the linear model with intercepts) as a function of Control values, along with their approximate 95% confidence intervals, are plotted in Figure 8.5. Effect size increases with the Control value and the effect size as a percentage of the unaffected Impact value (i.e., the Before prediction) increases modestly with the Control value.

Although effect size can depend upon the value of the Control or other variables, the concept of an average or net effect size (and its associated confidence interval) is of practical importance. For example, some estimate of the average percent loss might be required in order to implement a mitigation plan. Then an artificial reef that is expected to provide an equal amount of kelp could be designed. In this situation some consideration needs to be given to what we mean by an "average" effect. Here we are interested in the average long-term effect, so we generate an estimate of the long-term distribution of Control values, and calculate the expected loss (both in area and as a percentage) over this distribution. Our distribution of Control values is taken as given, and is assumed to be the

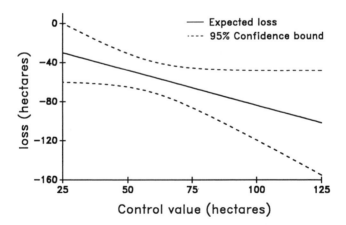

Figure 8.5. Estimated expected loss (effect size) in kelp area as a function of Control value, with 95% confidence intervals, for the linear model with intercepts.

observed set of Control values over the 11-year study including both Before and After data. Based on this set of Control values, the estimated average loss was 55.0 ha, or 52.0%. A confidence interval for the average losses was generated using a nonparametric jackknife method (e.g., Miller 1974). This method requires leaving out each of the 27 surveys one at a time, refitting the regression models with the data point deleted and calculating the average reduction leaving out the selected observation in this calculation also. The variation in these averages was used, following usual jackknife procedures, to calculate a confidence interval. The confidence intervals were (-71.1, -38.0) ha, and (-65.3, -39.2)%. This estimate of "net effect" and its confidence interval was similar to what we obtained for the standard approach using log-transformed data (effect = -53.0%, 95% confidence interval = (-65.7, -35.7)%), although we stress that this is not a guaranteed result (see Discussion).

The Independence Assumption

Although the problem of independence is not peculiar to the predictive approach, the possibility of violating this assumption needs to be considered in virtually all applications, including our example. To this point all our estimation and hypothesis testing have been done under the assumption that residual errors are uncorrelated. If there is substantial autocorrelation then (i) hypothesis tests given above are biased, (ii) the ordinary least squares estimates are not the most efficient, and (iii) we should take the true error structure into account when calculating confidence intervals. Within the context of general linear models, the usual approach is to test for first-order autocorrelation using the Durbin-Watson statistic Q. For the linear model with intercepts, the estimated first order autocorrelation was negative, small in magnitude (-0.091), and not statistically significant by the Durbin-Watson test ($Q = 1.81$, $P > 0.05$: the null distribution of the Durbin-Watson statistic was approximated by matching the first three moments of a beta distribution to $a + bQ$ for suitably chosen a and b, as suggested by Henshaw (1966) and evaluated by Durbin and Watson (1971)). Thus the available evidence indicates that first-order autocorrelation of the residuals is relatively weak for the kelp example.

Other types of violations of the independence assumption are possible. For example, residuals within a year (or some large block of time) could be correlated. Rather than performing many specific tests (none of which we have a good *a priori* basis for), we have plotted the residuals (still for the logistic model) against sampling date. There is some suggestion that the effect might be increasing during the first few surveys of the After period (i.e., the residuals are decreasing), but otherwise there are no obvious patterns (Figure 8.6).

We conclude this section by noting that there could still be weak autocorrelation that we failed to detect, or violations of the independence assumption in ways we have failed to consider. This is one more reason to treat confidence intervals only as approximate guides in interpretation.

Discussion

In the predictive approach we consider a Control to be a predictor, perhaps combined with others, of an Impact value. By comparing predictions based on data collected Before a perturbation with predictions based on data from the After period we can estimate the magnitude of an effect. We have contrasted this approach with the standard (BACIPS *t*-test) approach. The predictive approach is more flexible, in the sense that it can deal with effects of a perturbation, or natural temporal and spatial effects, that are not additive. In another sense, however, our implementation of the predictive approach is more restrictive; it assumes that changes in the expected value of Impact given the observed Control will change only due to effects of the perturbation [see Stewart-Oaten (Chapter 7) for discussion of this assumption]. Because the different models make different assumptions, and it may be hard to choose between them, we recommend that evaluations include estimates of effects based on several different models or approaches. When estimates from different methods are similar this will increase confidence in the conclusions, while large differences would indicate that model uncertainty may add greatly to the stated uncertainty obtained from any single method (Stewart-Oaten, Chapter 7).

For the kelp example, the estimated net effect is not appreciably different when using the predictive approach than for the standard approach on log-transformed data. This suggests that the formal estimates of effects and their confidence intervals are not highly dependent on arbitrary properties of a particular model. In large part, this favorable result occurred because the predictive approach yielded Impact-Control relationships that were similar to those implied

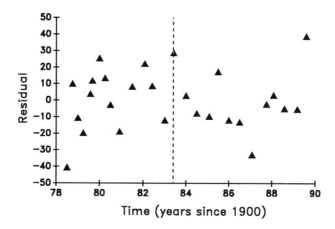

Figure 8.6.   Residuals from linear regression (with intercepts) between Impact and Control giant kelp areas plotted against time. Dashed vertical line separates Before and After periods.

by the standard BACI t-test model applied to log-transformed data (compare Figures 8.2b and 8.3). This similarity, of course, is not guaranteed.

Use of the predictive approach may help us rule out some models. When using the predictive approach we are explicit about the fact that the Control is a predictor. This naturally leads us to think about whether our data fit the regression (or other model) we use, and whether our estimates of effects in the After period require us to predict outside the bounds of the Before data. This is possible to some degree when using a t-test, but there is nothing inherent in the test to promote plotting Impact versus Control, or to look at whether the relationship changes at extreme values. In addition, the course of action when lack of fit is observed may not be obvious for the t-test. This is still tricky in regression analyses but is a standard part of the approach (e.g., Draper and Smith 1981, Carroll and Ruppert 1988, Seber and Wild 1989).

We know of two examples in the literature where a form of the predictive approach has been used. Mathur et al. (1980) tested for differences in zooplankton abundance between Before and After operations of a power plant's cooling system started by analysis of covariance, using temperature, stream water flow and abundance at a Control site as covariates. Reitzel et al. (1995) analyzed irradiance data from the San Onofre study; they stratified the data on the basis of current direction and followed the standard approach (t test on deltas) within the strata. This stratification by current direction allowed them to detect effects whose sign depended upon current direction. Both of these studies incorporated elements of the predictive approach, which we think provided insight beyond what could have been obtained based only on a test comparing the means of the Before and After deltas. We encourage an even more flexible and exploratory approach.

We conclude this largely statistical chapter by stressing that statistics can be only part of the equation. We recognize that any analysis will rest on assumptions, not all of which can be adequately tested (Schroeter et al. 1993, Stewart-Oaten, Chapter 7). In the end, decisions about the reality and importance of an apparent effect should depend upon the weight of all the available evidence–including the results of mechanistic studies, consideration of potential alternative explanations, and consistency among different sets of data–not just the P-value from a single test, or even estimates of the effect and associated confidence intervals.

Acknowledgments

R. Ambrose and W. Murdoch provided comments as our work progressed and R. Schmitt and C. Osenberg provided input on earlier drafts of this chapter. This work was supported in part by the Minerals Management Service, U.S. Department of the Interior, under MMS Agreement No. 14-35-0001-30471 (The Southern California Educational Initiative) and by the Marine Review Committee. The side-scan SONAR data were collected by Ecosystems Management Associates under a cooperative agreement between the Marine Review Committee and Southern California Edison. The views and interpretation contained in this document are those of the authors and should not be interpreted as necessarily representing the official policies, either express or implied, of the U.S. government or of the Marine Review Committee.

References

Ames, M. H., and J. T. Webster. 1991. On estimating approximate degrees of freedom. The American Statistician **45**:45–50.

Bence, J. R., S. C. Schroeter, J. D. Dixon, and T. A. Dean. 1989. Technical report to the California Coastal Commission. K. Giant kelp. Marine Review Committee, Inc.

Box, G. E. P., and D. R. Cox. 1964. An analysis of transformations. Journal of the Royal Statistical Society, B **26**:211–252.

Carpenter, S. R., T. M. Frost, D. Heinsey, and T. K. Kratz. 1989. Randomized intervention analysis and the interpretation of whole-ecosystem experiments. Ecology **70**:1142–1152.

Carroll, R. J., and D. Ruppert. 1988. Transformation and weighting in regression. Chapman and Hall, London, England.

Dean, T. A., and F. R. Jacobsen. 1984. Growth of *Macrocystis pyrifera* (Laminariales) in relation to environmental factors. Marine Biology **83**:301–311.

Devinny, J. S., and L. A. Volse. 1978. Effects of sediments on the development of *Macrocystis pyrifera* gametophytes. Marine Biology **48**:343–348.

Deysher, L. E., and T. A. Dean. 1984. Critical irradiance levels and the interactive effects of quantum irradiance and dose on gametogenesis in giant kelp, *Macrocystis pyrifera*. Journal of Phycology **20**:520–524.

Deysher, L. E., and T. A. Dean. 1986. In situ recruitment of sporophytes of giant kelp, *Macrocystis pyrifera* (L.) C. A. Agardh: effect of physical factors. Journal of Experimental Marine Biology and Ecology **103**:41–63.

Draper, N. R., and H. Smith. 1981. Applied regression analysis. Wiley, New York, New York.

Durbin, J., and G. S. Watson. 1971. Testing for serial correlation in least squares regression. III. Biometrika **58**:1–19.

Elwany, M. H. S., J. Reitzel, and M. R. Erdman. 1990. Modification of coastal currents by power-plant's intake and thermal discharge systems. Coastal Engineering **14**:359–383.

Henshaw, R. C. 1966. Testing single equation least squares regression models for autocorrelated disturbances. Econometrica **34**:646–660.

Hurlbert, S. J. 1984. Pseudoreplication and the design of ecological field experiments. Ecological Monographs **54**:187–211.

Mathur, D., T. W. Robbins, and E. J. Purdy. 1980. Assessment of thermal discharges on zooplankton in Conowingo Pond, Pennsylvania. Canadian Journal of Fisheries and Aquatic Sciences **37**:937–944.

Millard, S. P., J. R. Yearsley, and D. P. Lettenmaier. 1985. Space-time correlation and its effect on methods for detecting aquatic change. Canadian Journal of Fisheries and Aquatic Sciences **42**:1391–1400.

Miller, R. G. 1974. The jackknife - a review. Biometrika **61**:1–15.

Murdoch, W. W., R. C. Fay, and B. J. Mechalas. 1989. Final report of the Marine Review Committee to the California Coastal Commission. Marine Review Committee, Inc.

Murdoch, W. W., B. J. Mechalas, and R. C. Fay. 1980. Report of the Marine Review Committee to the California Coastal Commission: Predictions of the effects of San Onofre Nuclear Generating Station, and recommendations. Part I: Recommendations, predictions, and rationale. Marine Review Committee, Inc.

Reitzel, J., M. H. S. Elwany, and J. D. Callahan. 1994. Statistical analyses of the effects of a coastal power plant cooling system on underwater irradiance. Applied Ocean Research **16**:373–379.

Schroeter, S. C., J. D. Dixon, J. Kastendiek, R. O. Smith, and J. R. Bence. 1993. Detecting the ecological effects of environmental impacts: a case study of kelp forest invertebrates. Ecological Applications **3**:331–350.

Seber, G. A. F., and C. J. Wild. 1989. Nonlinear regression. John Wiley and Sons, New York, New York.

Stewart-Oaten, A. 1987. Assessing effects on fluctuating populations: tests and diagnostics. *in* ASA/EPA conferences on interpretation of environmental data: III - Sampling and site selection in environmental studies (May 14–15, 1987). Publication EPA-230-08-88-035. U.S. EPA, Office of Policy, Planning, and Evaluation, Washington, D.C.

Stewart-Oaten, A., W. W. Murdoch, and K. R. Parker. 1986. Environmental impact assessment: "psuedoreplication" in time? Ecology **67**:929–940.

Stewart-Oaten, A., J. R. Bence, and C. W. Osenberg. 1992. Assessing effects of unreplicated perturbations: no simple solutions. Ecology **73**:1396–1404.

Tukey, J. W. 1949. One degree of freedom for non-additivity. Biometrics **5**:232–242.

CHAPTER 9

ON BEYOND BACI

Sampling Designs That Might Reliably Detect
Environmental Disturbances[1]

A. J. Underwood

There are increasing needs for reliable detection of environmental disturbances due to human activities. There are also needs for ecological research to
become more concerned with problems of anthropogenic influence on natural
systems at spatial and temporal scales of relevance to the organisms and habitats
affected (e.g., Peters 1991). Unfortunately, there are often problems in the use of
appropriately valid experimental and sampling designs and replication for detection of unnatural disturbances to biological variables and for identifying a causal
relation between an observed effect and the putative anthropogenic cause.

Here, I briefly review some of the problems of some of the most widely used
procedures. Then, I discuss sampling designs that are appropriate to detect a variety of environmental perturbations. Finally, I consider some of the ecological
research programs (sensu Lakatos 1974) which are needed for the future development of practices of environmental sampling and assessment.

Throughout, I have chosen to consider populations as the appropriate units for
investigation. There is a lack of coherence as to the definition of some more com-
pound units, such as communities (Underwood 1986) and a lack of predictive
power of many suborganismal variables (Underwood and Peterson 1988). In
principle, however, any univariate measurements are analyzable in the ways suggested here. There may be multivariate analogs of these procedures, but they
have not yet appeared in the relevant literature. There are procedures for simpler
situations (Warwick 1986, Clarke and Green 1988, Warwick and Clarke 1991),
but they cannot handle the temporal patterns of change described here.

There are many practical problems of detection of human influences on
abundances of populations, but two are paramount in designing sampling

---

[1]This chapter was previously published in a modified form in Ecological Applications 4:3–15.
© 1994 by the Ecological Society of America.

programs. First is the large temporal variance of many populations, so that their abundances are very "noisy." Second, many populations show a marked lack of concordance in their temporal trajectories from one place to another (e.g., Osenberg et al., Chapter 6). This results in considerable statistical interaction between changes in mean abundance from time to time and differences from place to place.

Sampling must therefore be sufficient to identify unusual patterns of change in a very interactive and very variable measurement. This is not a property of some of the most widely used procedures (see examples in Underwood 1991).

Problems with Current Sampling Designs

There are several problems with current, widely-used environmental sampling designs (see also reviews in Underwood 1991, 1992, Osenberg and Schmitt, Chapter 1). The principle on which they are based is Green's (1979) BACI (Before-After-Control-Impact) design. In this, a sample is taken Before and another sample taken After a possible impact, in each of the putatively disturbed (hereafter called the Impact) and undisturbed Control locations. If there is an environmental disturbance that affects a population, it would appear as a statistical interaction between the difference in mean abundance of the sampled population between Control and potentially Impacted locations Before the disturbance and that difference After the disturbance.

As pointed out by Hurlbert (1984), this design is confounded. Any difference from Before to After the potential disturbance may occur between two times of sampling, but may not be related to (caused by) the human activity. Thus, the design was extended by Bernstein and Zalinski (1983) and by Stewart-Oaten et al. (1986) to have several replicated times of sampling (Figure 9.1a).

Stewart-Oaten et al. (1986) discussed the need for appropriate temporal replication, preferably in a nonregular frequency of sampling to avoid coincidences with natural cycles. They also considered, in detail, how to analyze the data, using t-tests on the differences between two locations Before and After the potential impact. All details of the rationale and procedure are available in their paper.

While this solved some of the problems of lack of temporal replication, it did not solve problems caused by the lack of spatial replication. The comparison between a single Impact and a single Control location is still confounded by any other cause of different time-courses of abundances in the two places that is not due to the identified human activity. It is not unusual for populations to have different temporal trajectories in different locations and temporal interaction among places is not at all uncommon.

Stewart-Oaten et al. (1986) discussed some aspects of such interaction. They concluded that if there were a trend or temporal change in the difference between the two locations before the putative impact, the variable being sampled would not be usable for assessment of the impact. Obviously, if the difference in mean

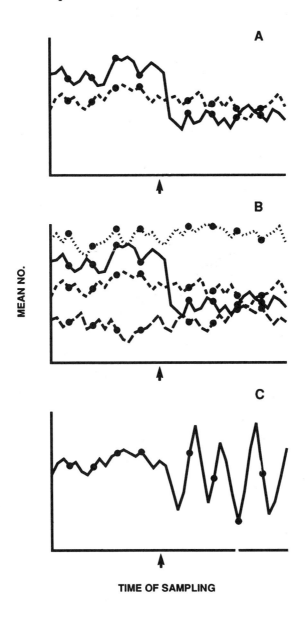

Figure 9.1. Sampling to detect environmental impacts. (A) BACIPS design; replicated samples are taken at several times in a single control (dashed line) and the potentially impacted location (solid line) Before and After a planned disturbance (at the time indicated by the arrow). (B) Sampling three Control locations to provide spatial replication. (C) An impact that has no effect on long-run mean abundance, but causes greater temporal variation.

abundance between two locations already varied from time to time before an impact, with some consistent temporal pattern of change, it would be expected to differ after an impact. Therefore there is no defined null hypothesis for any test so that an impact could be detected. Because many populations can be expected to interact in their patterns of abundance from time to time and location to location, this is a serious restriction on the usefulness of the proposed sampling design.

The problem of confounding (or "pseudoreplication," Hurlbert 1984) caused by comparing abundances in two locations, one potentially disturbed and the other a Control, should be overcome by having several replicated disturbed and several Control locations. This is, after all, the solution to the same problem in any routine field experiment (Hurlbert 1984, Underwood 1986). Of course, there will rarely be replicated planned developments causing the possible disturbance to a population. There is, however, no reason not to have replicated Control locations.

Thus, a randomly chosen set of locations is needed with the appropriate features of physical characteristics, mix of species, abundance of the target species, etc., as previously dictated the choice of a single Control location. Note that there is no need to attempt to choose places with identical characteristics or abundances of the population. Not only is this impractical, it is unnecessary. The set of locations chosen to serve as Controls must simply represent the range of habitats of the one that might be disturbed (the Impact location). This is not a conceptually more difficult chore than the choice of two similarly representative locations, one Impact and one Control.

Obviously, the Control locations (as with a single Control in BACI designs) must be a representative sample of places of the same general habitat as that in which the impact is expected. So, if an outfall is to be placed on a rocky headland with steep cliffs and fast local currents, Controls must be placed in similar locations. It can (and has) been argued that selecting such controls may be very difficult or impossible. Indeed, it may be. It is, however, equally difficult to select a single Control location. Where there cannot be any Control (because, for example, the potentially impacted area is very large or is unique), there would be no comparative study comparable to a BACI or an asymmetrical, multi-control design. This is not a new problem for the designs considered here.

There may, nevertheless, be cases where multiple controls are not available and BACI procedures are all that can be available. Often, replication of Control locations is not a more difficult task than choosing one Control. Clear-headedness about the appropriately relevant locations that might be chosen is an identically valuable commodity for both types of design.

Now, there is a sample of locations, representing a population of locations in which the monitored species can be found. The variance among mean abundances of a species from location to location represents the variance among any set of such locations (i.e., the set being sampled is a sample of locations from a population of many such locations). Locations represent a random factor in the sampling design (Snedecor and Cochran 1967, Winer 1971, Underwood 1981). The only constraint on the set of locations chosen for sampling is that it must (obviously!) include the location that might be disturbed by a planned human activity (the Impact location).

As explained in Underwood (1992), an environmental impact affecting the abundance of a sampled population in the Impact location must cause the temporal pattern of abundance in that location to differ from the range of patterns in the set of Control locations. There will probably be differences from location to location in the patterns of abundance from time to time. Nevertheless, to have an effect, the disturbance must cause more change in the Impact location than occurs at the Control locations.

An impact must also be detectable as a different pattern of statistical interaction from Before to After it starts, between the Impact and Control locations than occurs among the Control locations. This is illustrated in Figure 9.1b and described in full in Underwood (1992). Asymmetrical analyses of variance of sets of data from such sampling will be described below. Use of several Control locations and asymmetrical analyses also allows a solution to Stewart-Oaten et al.'s (1986) problem of not being able to detect impacts in populations which have spatial and temporal interactions in their abundances. This will also be summarized below (and see Underwood 1992).

There are other problems with the BACI designs recommended by Bernstein and Zalinski (1983) and Stewart-Oaten et al. (1986). For their methods of analysis of the data, it is necessary for both locations (Control and Impact) to be sampled at the same time. This is not always possible because of logistic constraints, weather, time taken to sample, etc. Use of asymmetrical analyses of variance of the type recommended here can partially overcome this problem, but this is not considered further here. Full details are available in Underwood (1991, 1992).

Also, as illustrated in Underwood (1992), the temporally replicated BACI design assumes that the spatial scale of a putative environmental impact is known before it occurs. For example, a proposed outfall putting warm water into an estuary may have only a local effect on the surrounding few hundred square meters of mudflat. To detect this, appropriate Control locations would be areas of a few hundred square meters elsewhere in the estuary. If, however, the estimated scale of impact is wrong and the outfall causes a change in abundance of some population over the whole estuary, such sampling will not detect it, because all the Controls would also be affected.

To cover this possibility, Control locations must also be examined in other estuaries along the coast, independent of any possible effects of the warm water from this outfall. Now, the impact would be detectable as a difference (from Before to After the outfall's construction) in the abundance in the disturbed as opposed to the Control locations. Such an effect will not be matched by any change in the Controls. If the smaller scale were correctly predicted, the impact would be detected among the locations in the single estuary where the outfall is constructed. If this is wrong, there will be a different change in the Controls in that estuary from that occurring in Control locations elsewhere.

Because the spatial scale of such a disturbance might not be well defined in advance, sampling at two scales (estuaries and sites within estuaries) would be necessary. The Bernstein and Zalinski (1983) and Stewart-Oaten et al. (1986) design cannot be extended to analyze such data. In contrast, the asymmetrical analyses of variance reviewed here can be modified for this. Complete examples are provided in Underwood (1992), where I also discussed other advantages for the demonstration of causality of sampling several spatial scales.

As a final problem, there is a whole class of environmental disturbances that cannot be detected at all by a BACI design. These are disturbances that do not affect long-run mean abundances of a population, but instead, alter the temporal pattern of variance in abundance (Figure 9.1c). Analysis of appropriate sampling for these is not considered in detail in this review, having been described in full in a previous paper (Underwood 1991).

The final part of this discussion is a consideration of the ecological research programs necessary to replace specific regimes of sampling to detect particular environmental impacts. These are required to solve the usual problem of lack of sufficient time to sample before a possible impact and the lack of statistical power in many sampling designs used to detect impacts.

Asymmetrical Sampling Design to Detect Environmental Impacts

First, consider the BACI design advocated by Bernstein and Zalinski (1983) and Stewart-Oaten et al. (1986). The analysis of variance of this design is summarized in Table 9.1a. As described in Underwood (1991), the F-ratio indicated in Table 9.1a is the t-test recommended by these authors. Also in Table 9.1b is the same analysis, extended to compare abundances in more than two locations.

In Table 9.2, the asymmetrical analysis is described, using a modeled set of data illustrated in Figure 9.2a. Now, the useful contrasts of the Impacted versus Control locations and its interactions with time can be extracted from the variation among all locations and its interaction with time.

An environmental impact should now be evident, in the simplest case, as an interaction between the difference between the mean abundance in the Impacted location and that in the Control locations Before compared to After the disturbance began (i.e., B × I in Table 9.2; see Figure 9.3 and Table 9.3 for examples). Alternatively, if the impact is not sustained, nor sufficient to alter the mean abundance in the Impacted location over all times of sampling After the disturbance, it should be detected in the pattern of statistical interaction between the times of sampling and the contrast of the Impacted and Control locations (i.e., T(Aft) × I in Table 9.2). This is explained in full in Underwood (1992) and illustrated below.

Thus, a difference is sought between the time-course in the putatively Impacted location and that in the Controls. Such a difference would indicate an unusual event affecting mean abundance of the population in the single disturbed location, at the time the disturbance began, compared with what occurred

Table 9.1.   Analyses of Variance in Sampling Designs to Detect Environmental Impacts

Source of variation[a]	Degrees of freedom	F-ratio versus	Degrees of freedom for F-ratio
Before versus After = B	1	T(B)	
Control versus Impact = C	1	T(B)	
B × C	1	T(B) × C [b]	$1, 2(t-1)$
Times (Before or After) = T(B)	$2(t-1)$	Residual	
T(B)× C	$2(t-1)$	Residual [b]	$2(t-1), 4t(n-1)$
Residual	$4t(n-1)$		

Source of variation[c]	Degrees of freedom	F-ratio versus	Degrees of fredom for F-ratio
Before versus After = B	1	No test	
Among Locations = L	$l-1$	T(B) × L	
B × L	$l-1$	T(B) × L	$(l-1), 2(t-1)(l-1)$
Times (Before or After) = T(B)	$2(t-1)$	T(B) × L	
T(B) × L	$2(t-1)(l-1)$	Residual	$2(t-1)(l-1),$
Residual	$2lt(n-1)$		$2lt(n-1)$

[a]Relevant F-ratios are calculated from expected values of Mean Squares as in Underwood (1981). B and C are fixed factors Times are a random factor, nested in either Before or After. BACI design data are collected in two locations (Impact and Control) at t randomly-chosen times Before and After a planned disturbance. n replicate samples are taken at each time in each location.

[b]This test is identical to the t test recommended by Stewart-Oaten et al. (1986).

[c]Similar design, but there is a total of l locations sampled; locations are a random factor, otherwise details are as above. There is no formal test for comparing Before versus After. This is irrelevant because an impact must cause an interaction (B × L or T(B) × L); see text for details.

elsewhere in undisturbed Controls. The impact will either be detected as a different pattern of interaction among the times of sampling or at the larger time scale of Before to After the disturbance.

The patterns of significance in such analyses under different types of responses to disturbance are illustrated below.

Patterns in Analyses to Detect Environmental Impacts

Pulse Effects

It is informative to consider disturbances of two types—pulse and press (Bender et al. 1984). The former are not sustained; the disturbance is removed after a period, although effects may be longer term. The latter are sustained disturbances. In environmental disturbances, the construction of, say, a marina

Table 9.2. Asymmetrical Analysis of Variance of Model Data from a Single Impact (I) and Three Control Locations, Sampled at 6 Times Before and 6 Times After a Disturbance That Causes No Impact (See Figure 9.2a)

Source of variation	Degrees of freedom		Mean square	F-ratio	F-ratio versus
Before versus After = B	1		331.5		
Among Locations = L	3		25114.4		
[a]Impact versus Controls = I	1		3762.8		
[a]Among Controls = C	2		35790.2		
Times (B) = T(B)	10		542.0		
B × L	3		375.0		
[a]B × I	1		454.0	1.51	Residual
[a]B × C	2		335.5	1.12	Residual
T (B) × L.	30		465.3		
[a]Times (Before) × L	15		462.2		
[a]T(Bef) × I		5	515.6		
[a]T(Bef) × C		10	435.9		
[a]Times (After) × L	15		468.2		
[a]T(After) × I		5	497.3		
[a]T(After) × C		10	453.6	1.51	Residual
Residual	192		300.0		

[a]Represent repartitioned sources of variation to allow analysis of environmental impacts as specific interactions with time periods (B × I or T(Aft) × I; see text for more details).

will cause pulse disturbances while it is being built. For example, there will be chemical and sedimentary changes to the habitat while building is done. This is a pulse disturbance—it ends when the marina is built. There is, however, also a press disturbance after the marina is finished, involving the altered water flow around its walls, any hydrocarbon pollutants or antifouling leachates from the boats, etc. Either or both types of disturbance may have consequences for nearby populations of organisms.

There is some blurring of distinction between the two categories because they are not independent of the time scales of life histories of the organisms being affected (Underwood 1991), but they require different approaches, mechanisms and interpretations (Bender et al. 1984).

So, what happens in a location subject to a pulse disturbance that reduces the mean abundance of a population, which recovers very rapidly once the disturbance is removed? This situation is illustrated in Figure 9.2B and 9.2C, for two different magnitudes of response. In each case, only the population in the Impact location was reduced.

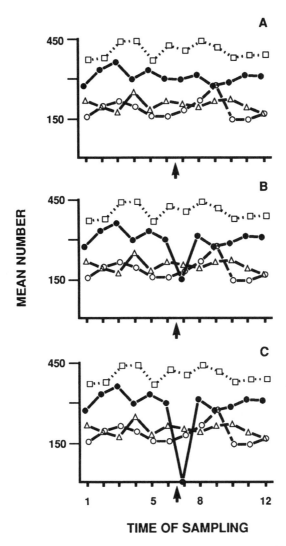

Figure 9.2. Simulated environmental disturbances in one location (solid line), with three Controls, all sampled six times Before and After the disturbance (at the time indicated by the arrow). These data are analyzed in Table 9.3. (A) No effect of the disturbance; (B) a pulse reduction to 0.5 of the original mean; (C) a pulse reduction to 0.

The analysis of such data is modeled in Table 9.3, assuming that residual variances (i.e., among samples at each time and location) were not altered by the disturbance. To simulate the impact, the mean number of animals at the seventh time of sampling (the first after the impact) was reduced by the requisite proportion.

The effect on the analysis of the data is obvious in Table 9.3. There is now a significant interaction between the difference between Impact and Control

Table 9.3. Asymmetrical Analyses to Detect Environmental Impact

Source of variation	df	(a) MS	(a) F	(b) MS	(b) F	(c) MS	(c) F	(d) MS	(d) F	(e) MS	(e) F	(f) MS	(f) F	(g) MS	(g) F
Before versus After = B	1	526.2		901.7		341.7		4212.3		9687.5		4277.2		4228.5	
Among Locations = L	3														
[a]Impact versus Controls = I	1	2108.2		980.6		3593.9		1881.2		10149.1		1953.2		1899.1	
[a]Among Controls = C	2	35790.2		35790.2		35790.2		35790.2		35790.2		35790.2		35790.2	
Y(B)	10	635.0		908.1		581.6		552.5		560.8		580.9		531.2	
B x L	3														
[a]B x I	1	1125.6		2340.2		492.2		12625.6	26.55**	29370.3	63.65**	12825.0	11.40**	12675.3	21.14**
[a]B x C	2	335.5		335.5		335.5		335.5	0.75	335.5	0.75	335.5	0.75	335.5	0.75
T(B) x L	30														
σT(Bef) x L	15														
σT(Bef) x I	5	515.6		515.6		515.6		515.6		515.6		515.6		515.6	
σT(Bef) x C	10	435.9		435.9		435.9		435.9		435.9		435.9		435.9	
σT(Aft) x L	15							435.3							
[a]σT(Aft) x I	5	1041.4	3.47**	2666.0	8.89**	1785.0	5.95**	453.6	1.45	407.2	1.36	1734.2	5.78**	683.3	2.28*
[a]σT(Aft) x C	10	453.6	1.51	453.6	1.51	453.6	1.51	300.0	1.51	453.6	1.51	453.6	1.51	453.6	1.51
Residual	192	300.0		300.0		300.0		300.0		300.0		300.0		300.0	

Note: Four locations are sampled, each with five random, independent replicates, at each of 6 times Before and 6 times After a putative impact starts in one location; there are 3 Control locations; locations represent a random factor and every location is sampled at the same time; times represent a random factor nested in each of Before or After. Simulated disturbances are: (a), (b) pulses causing a brief reduction in mean abundance to 0.5 and 0, respectively; (c) temporal variance increased (standard deviation $\times$ 5) without altering mean abundance; (d), (e) press disturbance decreasing mean abundance to 0.5 or 0.2, respectively; (f) combination of altered temporal variance (standard deviation $\times$ 5) and press reducing mean to 0.5; (g) as (f), standard deviation $\times$ 0.5, press to 0.5.

[a]Repartitioned sources of variation; impacts can be detected by tests of B $\times$ I and/or T(Aft) $\times$ I (see text for details).

*P <0.05
**P <0.01

locations and the times of sampling After the disturbance (T(Aft) × I in Table 9.3a and 9.3b). This is easily interpretable; the pulse caused a much more radical change from one time to another in the Impacted location than in any of the Controls. Thus, compared to the situation Before the impact, there is now greater (and significant) interaction than Before (T(Bef) × I in Table 9.3a and 9.3b was not significant) and there is no corresponding change among Controls (T(Aft) × C is not significant). As explained in Underwood (1992), this test assumes no Type II error in pooling (Winer 1971, Underwood 1981) the non-significant variance component T(Aft) × C for testing for impact as T(Aft) × I.

Impacts Affecting Temporal Variance

The second sort of impact is one that does not alter the mean abundance, but causes greater oscillations in numbers from time to time (Underwood 1991 and Figure 9.1c). In the first case, temporal variance was increased. This was modeled by calculating the deviation of the mean abundance in the Impact location at each time of sampling from the mean abundance in that location over all times of sampling After the disturbance. The deviations were then multiplied by the desired amount to alter the temporal standard deviation in that location. Results are illustrated in Figure 9.3a and 9.3b and analyzed in Table 9.3c. Again, this has the effect of causing a significant interaction in the difference between the mean abundance in the Impact and Control locations and the time of sampling (T(Aft) × I in Table 9.3c). This is caused, again, by the greater fluctuations in that location than occurred Before the disturbance or than occurred in the Control locations After the disturbance.

Impacts on temporal variance can also cause decreased variation from time to time in the disturbed location, as shown in Figure 9.3b. Here, there will again be a difference in the pattern of temporal interaction in the analyses, but the variance associated with such interactions will be smaller than in the Control location and than occurred Before the disturbance. This can still be identified in appropriate tests, which are described in Underwood (1992) and not dealt with in this summary.

Press Disturbances

Press disturbances cause a different pattern in the data. If the press simply causes a sustained increase or decrease in mean abundance (as in Figure 9.3c and 9.3d) it also causes an increase in the variation due to the interaction between the mean abundances in Impact and Control locations from Before to After the disturbance (Table 9.3d and 9.3e). This is, of course, the change detectable by the standard BACI design (Green 1979; Bernstein and Zalinski 1983, Stewart-Oaten et al. 1986, Underwood 1992). In the situations simulated here, the press disturbances were detectable as a significant pattern in the data (B × I was significant in Table 9.3d and 9.3e). Note that B × C was not significant and has been ignored in the analysis (see Underwood 1992 for details).

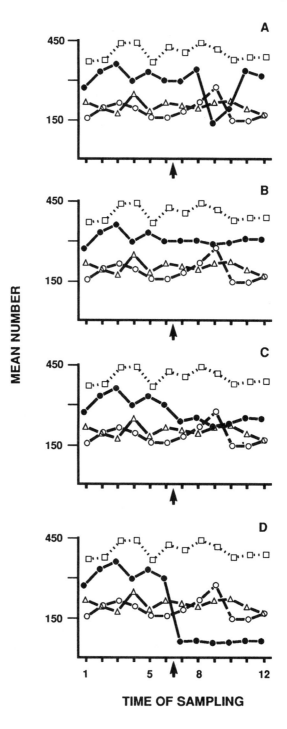

Combinations of Effects

It is possible for a sustained series of disturbances to cause both a continuing impact, affecting numbers in the population, and a simultaneous change in temporal variance. For example, an outfall pipe may cause overall reductions in survival of a population of fish, reducing mean abundance. The discharges from the outfall may, however, be intermittent and only coincide occasionally with episodes of recruitment. If recruits arrive when the outfall is operating, abundance will be pushed to very small numbers. If, however, recruits arrive when the outfall is inactive, the numbers entering the population will cause its abundance to rise dramatically. Thus, around the reduced mean abundance there will be greater temporal variance than in Control locations. Several examples are illustrated in Figure 9.4.

As can be seen in Table 9.3, both effects were detected as significant interactions. The altered mean abundance After the disturbance caused an increased interaction between the difference from Impacted to Control locations Before, compared with After, the disturbance (B × I is significant in Table 9.3f and 9.3g). In addition, the altered temporal variation caused interactions between the Impacted and Control locations in their temporal changes After the disturbance (T(Aft) × I is significant in Table 9.3f and 9.3g).

Thus, the combination of different types of impact is detected as a combination of the patterns of interactions they cause. The analyses in Table 9.3 (f and g) demonstrate combinations of the patterns of significance of press effects (Table 9.3d and 9.3e) and altered temporal variation (Table 9.3c).

This not only indicates the usefulness of this design of sampling, but also shows that considerable information can be gained from the analysis about the nature of the biological responses to the disturbance. This would be of great benefit for predictions about effects of future proposed developments, prevention of their deleterious biological effects, or remedial action to reconstruct damaged ecosystems.

Temporally Interactive Populations

As discussed fully by Stewart-Oaten et al. (1986), in standard BACI procedures, data cannot be used for populations in which, Before a disturbance, there is already an interaction among locations in the differences in abundance from time to time. The present designs can sometimes solve this problem, provided that impacts are large (because the tests are generally not likely to be powerful).

Figure 9.3.   Simulated environmental disturbances in one location (solid line), with three controls, all sampled six times Before and After the disturbance (at the time indicated by the arrow). These data are analyzed in Table 9.3. (A-B) The impact is an alteration of temporal variance after the disturbance; temporal standard deviation × 5 in (A) and × 0.5 in (B). (C-D) A press reduction of abundance to 0.8 (C) and 0.2 (D) of the original mean.

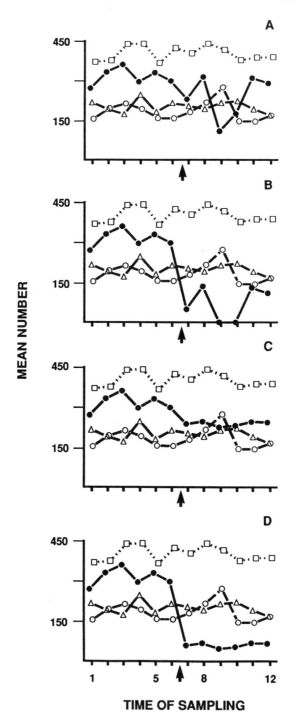

MEAN NUMBER

TIME OF SAMPLING

The logic of this procedure is described in full in Underwood (1992). If there exists, Before the disturbance, a pattern of interaction among the locations, it will be detected as one or other, or both, of the F-ratio tests on these interactions. Thus, in Table 9.3, T(Bef) × I, T(Bef) × C, or both, will be significant, depending on whether there is a different temporal pattern in the Impact location from that in the Controls, or whether there is, instead, or additionally, such a difference among the Controls.

An environmental impact will result in a changed magnitude of interaction between the Impact and Control locations that is not matched in the Controls. The interaction among times between the Control and Impact locations must alter relative to what occurred Before the disturbance (T(Aft) × I must differ from T(Bef) × I in Table 9.3). Furthermore, there must be a larger difference in such interaction than now occurs among the Controls (T(Aft) × I must be significantly larger than T(Aft) × C in Table 9.3). Finally, in order to be sure that such patterns are not part of a more general environmental change, coincident with the disturbance, there must be similar amounts of interaction among Controls After compared with Before the disturbance (i.e., T(Aft) × C and T(Bef) × C should be similar). Otherwise, the pattern identified as significant for the Impact location is also present in the Controls and cannot be identified as being due to the disturbance.

These tests are illustrated in Table 9.4 for the analyses in Table 9.3. Some of the statistical tests are two-tailed because a change in temporal interaction can result in a smaller or a larger Mean Square in an analysis of variance. Interactions among times of sampling After the disturbance can therefore be larger (if temporal variance increases) or smaller (if it decreases) relative to what occurred Before the disturbance (see also Underwood 1991).

The data in the simulations would have been identified as interactive if, for example, their residual variances had been much smaller. Many of the interactions Before the disturbance would then have been significant. Under these circumstances, impacts would have been detected only for the most extreme conditions because the tests have few degrees of freedom and are not particularly powerful. Here, only a pulse to zero (a temporary extinction) would be unambiguously identifiable as an impact (only (b) in Table 9.4 has the appropriate pattern of significant results). This is, however, still a considerable advance on the lack of any test for such populations in the BACI procedure.

There is one other possible interactive structure in abundances of populations. That is an interaction over the longer time scale, from Before to After the

---

Figure 9.4. Simulated environmental disturbances in one location (solid line), with 3 controls, all sampled 6 times before and after the disturbance (at the time indicated by the arrow). These data are analyzed in Table 9.3. In each case, disturbance causes a press response and, simultaneously, alters temporal variation as in Figure 9.3. (A) temporal S.D. × 5, press to 0.8; (B), S.D. × 5, press to 0.2; (C), S.D. × 0.5, press to 0.8; (D), S.D. × 0.5, press to 0.2.

Table 9.4.   *F*-ratios for Detecting Environmental Impacts When There Are Temporal
Interactions as Described in Full in Underwood (1992)

Test	Df	Simulated disturbance						
		(a)	(b)	(c)	(d)	(e)	(f)	(g)
1-tailed *F*-ratios in analysis of variance in Table 9.3								
B x I versus B x C	1, 2	3.35	6.97	1.47	37.62*	87.54*	38.23*	37.78*
2-tailed *F*-ratios of Mean Squares from analyses in Table 9.3								
T(Aft) x I versus T(Aft) x C	5,10ᵃ	2.30	5.88**	3.93**	1.04	1.11	3.82*	1.51
T(Aft) x I versus T(Bef) x I	5, 5	2.02	5.17*	3.46	1.18	1.27	3.36	1.33
T(Aft) x C versus T(Bef) x C	10, 10	1.04	1.04	1.04	1.04	1.04	1.04	1.04

Note: Some tests are 2-tailed because either Mean Squaare can be greater in response to different
  disturbances. Mean Squaares and abbreviations are in analyses (a) to (g) in Table 9.3.
ᵃor vice versa, depending on which Mean Squaare is larger in the 2-tailed test.
*$P < 0.05$
**$P < 0.01$

disturbance, rather than among times Before or among times After. If some general environmental change is happening while the disturbance occurs, the Control locations may have different temporal patterns at a later time (i.e., coincidentally after the disturbance). There will then be an interaction among the locations in their differences After compared to Before the disturbance (B × C will be significant in Table 9.3).

If this occurs, the only way the disturbance could have an effect is to cause a different pattern in the abundance in the Impacted location. Thus, the interaction from Before to After in the difference between the Impact and Control locations must differ from that among Controls (and B × I will differ from B × C in analyses in Table 9.3). The *F*-ratio test to detect this is shown in Table 9.4. Even though there are minimal degrees of freedom, in the simulated data there was virtually no longer-term interaction among Control locations (B × C in Table 9.3). The tests for press effects of disturbance which caused a major Before to After interaction in the Impacted location were still significant (Table 9.4d–9.4g). Thus, even where there are few locations, if short-term interactions (among times of sampling) are absent, even nonparallel changes in abundance among locations do not prevent the detection of an environmental impact.

Power of Tests

It is not realistic in this overview to consider the general properties of power of the tests (principles reviewed by Winer 1971 and Cohen 1977; see Mapstone, Chapter 5; Osenberg et al., Chapter 6). Nevertheless, it is useful to examine the particular sets of disturbances simulated here.

For each simulated disturbance described, I calculated the power of statistical tests to detect the effects. In each case, I examined the altered variance (Mean Square) associated with sources of variation affected by the disturbance (as in Table 9.3). I then calculated the power of each F-ratio test used (one-tailed tests in Table 9.3 and two-tailed tests in Table 9.4) to detect (as significant) that amount of variation. For example, a pulse to half the previous abundance caused the Mean Square for the interaction between differences among times of sampling After the disturbance and the average mean difference between the Impact and Control locations (T(Aft) × I in Tables 9.2 and 9.3a) to change from 497.3 to 1041.4. The power of the F-ratio to detect the latter (increased) Mean Square is 0.95 (Table 9.5).

I calculated the power of all relevant tests on all simulated sets of data to gain some insight into types of environmental disturbance that were more likely to be detectable. Obviously, this is *not* an estimate of power of the tests in real sets of data.

Power was calculated using the method for random sources of variation (Variance components or Type II models) as described by Winer (1971). All tests were done with probability of Type I error (i.e., α) at 0.05, so power was calculated with this probability. I also calculated power with $\alpha = 0.10$, i.e., relaxing the stringency of tests and making detection of environmental impacts far less conservative. This was based on the premise that choice of $\alpha = 0.05$ is conventional, but not particularly well founded on any principle of the relative costs of Type I as opposed to Type II errors (Mapstone, Chapter 5).

Considering first the tests for impacts in populations that do not have temporal interactions when undisturbed, the one-tailed F-ratios described earlier (as in Table 9.3) were powerful for any large change to temporal variance or a very large pulse. Power was greater than 0.90 in tests for such conditions (T(Aft) × I versus Residual in Table 9.4). For smaller pulses (to 0.5), power remained at 0.80, for $\alpha = 0.05$.

Press effects were less likely to be detected. The power of the tests was in the range 0.6 to 0.7, even for an 80% (to 0.2) reduction of mean abundance (B × I/T(B) × I in Table 9.5).

Press effects in combination with changes of temporal variance were generally more likely to be detected; power of the tests (T(Aft) × I/Residual in Table 9.5) was mostly somewhat larger for these disturbances ((iv) in Table 9.5).

If the residual variance had been smaller, making some of these tests more powerful, the interactions among times of sampling Before the disturbance would probably also be found to be significant. The populations would then be considered temporally interactive and the two-tailed F-ratios in Table 9.4 would be relevant identifiers of impacts. The power of these was generally not great; they all had power in the range 0.5 to 0.8. Note that power of 0.5 implies that an impact, even of the magnitude of a sustained reduction to a fraction of the normal population's abundance, would only be detectable in one out of every two assessments.

Table 9.5. Power of Tests Used to Detect Environmental Impacts

Simulated disturbance	Analysis in Table 9.3	Data in Fig.	df / α	B × I/T(B) × I 1, 10 .05	.10	T(Aft) × I/Residual 5, 192 .05	.10	B × I/B × C 1, 2 .05	.10	T(Aft) × I/T(Aft) × C 5, 10 .05	.10	T(Aft) × I/T(Bef) × I 5, 5 .05	.10
(i) Pulse to:													
0.5	(a)	9.2B		—	—	0.8	0.9	—	—	—	—	—	—
0	(b)	9.2C		—	—	0.95	0.95	—	—	0.7	0.8	0.5	0.6
(ii) Temporal S.D.×													
5	(c)	9.3A		—	—	0.95	0.95	—	—	0.5	0.6	—	0.5
2				—	—	—	—	—	—	—	—	—	—
(iii) Press to:													
0.5	(d)	9.3C		0.6	0.7	—	—	0.5	0.6	—	—	—	—
0.2	(e)	9.3D		0.7	0.8	—	—	0.6	0.7	—	—	—	—
(iv) Temporal S.D.×/press to:													
5/0.5	(f)			0.6	0.7	0.95	0.95	0.5	0.6	0.5	0.6	—	—
5/0.2	(g)	9.4B		0.8	0.9	0.9	0.9	0.6	0.7	—	—	—	—
2/0.5				0.6	0.7	0.7	0.8	0.5	0.6	—	—	—	—
2/0.2				0.8	0.9	0.7	0.8	0.6	0.7	—	—	—	—
0.5/0.5				0.6	0.7	—	—	0.5	0.6	—	—	—	—
0.5/0.2		9.4D		0.8	0.9	—	—	0.6	0.7	—	—	—	—

Note: Calculations are from simulations as discussed in the text. Disturbances are: (i) pulses causing a brief reduction (as specified) in mean abundance for the first time of sampling after the disturbance; (ii) fluctuating, causing changes to temporal variance without altering the mean (multiplying temporal standard deviation by the amount specified); (iii) press disturbances causing sustained reduction (as specified) in mean abundance; (iv) combinations of (ii) and (iii), a press disturbance also causing altered standard deviation of temporal differences. Power is given if ≥0.50, for α (Probability of Type I error) = 0.05 and 0.10.

As discussed below, the lack of power of programs of environmental sampling is a major problem (see also Osenberg et al., Chapter 6). In contrast to the gloom, however, highly interactive populations are, by definition, characterized by large natural fluctuations from time to time and natural differences from place to place. Thus, to persist at all, they must probably be resilient and able to recover from nonanthropogenic disturbances, natural fluctuations in recruitment and mortality, etc.

In such populations, for human impact to be biologically meaningful, it must be very large, or it will not move the abundance out of its natural range of fluctuations. Therefore, in interactive populations, the only effects of human disturbances that are going to matter are very large ones. These sampling programs may well turn out to be powerful enough to detect such effects in real situations, but more work is needed to optimize the designs to maximize their power and efficiency.

Discussion

Scales of Sampling

The sampling designs considered above are necessary for detecting environmental impacts, because they include temporal and spatial replication to take into account natural variation. In many cases, the spatial or temporal scale of the possible effects of a disturbance are unknowable before the disturbance happens. The choice of scales for sampling to detect the disturbance is then very difficult, if not impossible (e.g., see Chapters in Section III). Under these circumstances, hierarchical sampling of different spatial scales (Green and Hobson 1970, Underwood 1992) is appropriate and the asymmetrical designs can be extended to cover more than one spatial scale. I provided examples of such sampling in an earlier paper (Underwood 1992).

When time scales are uncertain, these designs are also extendible. Underwood (1991) illustrated their use for two different time scales to detect pulse and press disturbances that affect temporal variance, but not mean abundance. The use of hierarchical temporal sampling should probably be much more widespread in ecology (including environmental sampling) than is currently the case.

It is often assumed that differences seen from time to time in a monitoring program can be interpreted, for example, as indicating seasonal trends in mean abundance. Without knowledge of any fluctuations at shorter intervals, this is not a logically valid conclusion. The samples are confounded or pseudoreplicated (Hurlbert 1984) because there is only one set of replicates at one time in each season. This is illustrated in Figure 9.5, where a single set of replicate samples, taken once in each of four seasons, reveals an apparent seasonal cycle of abundance. In fact, there is no cycle, but a slight general increase of abundance through time (Figure 9.5c). If sampling were at two time scales, in this case three

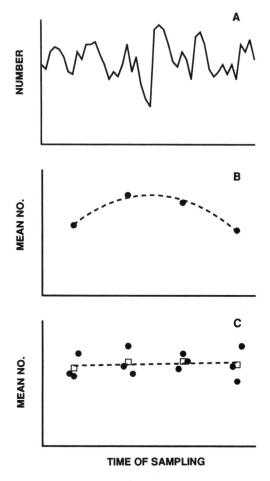

Figure 9.5. Consequences of sampling without hierarchical temporal replication. (A) The real abundance, through time, of the sampled population; (B) samples, however well-replicated, once in each season show an apparent seasonal trend; (C) sampling in each season is replicated at three times and there is (correctly) no cyclic trend in seasonal means.

short intervals during each season, the lack of seasonal patterns of mean abundance would be revealed. The two time scales are then seasonal (as originally required) and shorter periods in each season. The alternative, of course, would be to demonstrate consistent seasonal patterns over several years (cycles). This would not only take longer, but would also be a different study. The time scale would be several years, which is presumably appropriate for some tests of hypotheses. It is, however, irrelevant for hypotheses about a particular (and appropriately defined) period, such as one year.

Whatever the case, sampling at more than one temporal or spatial scale will be more expensive and time consuming, but may be mandatory. Cost-benefit procedures for optimizing the allocation of resources to each scale of sampling can be used where there are hierarchically structured spatial and/or temporal scales of sampling. Such procedures are well known (e.g., Cox 1958, Cochran 1963, Snedecor and Cochran 1967, Underwood 1981) and examples of their use are available in ecological field studies (e.g., Kennelly and Underwood 1985).

Improving Designs to Detect Impacts

The power of environmental sampling to detect impacts is often poor and there are serious deficiencies of capability of the designs discussed here to detect even quite large environmental disturbances (Table 9.5). Two conclusions are evident from this.

First, there must be more research on sampling procedures to optimize the designs and attempt to increase their power to detect unusual changes. There has been recent discussion on some of the related issues in relevant contexts (Andrew and Mapstone 1987, Clarke and Green 1988, Eberhardt and Thomas 1991). Some work is needed to determine the relative efficiency of increasing the numbers of times of sampling, number of Control locations and the numbers of replicate samples in each location at each time. There will obviously be different gains in precision for different types of effects. Note that only an increase in number of Control locations will have any dramatic effect on detection of press disturbances in temporally interactive systems (Tables 9.4 and 9.5). A further consideration is rapid procedures for determining when temporally independent samples can be taken. If a population is sampled too frequently, the data from time to time are not independent, leading to major problems in any statistical analyses (e.g., Cochran 1947, Eisenhart 1947, Green 1979, Underwood 1981, Stewart-Oaten et al. 1986). Much more needs to be known about detection of non-independence and the frequencies with which temporally independent samples could be taken of various types of organisms with different life histories (see Frank 1981, Connell and Sousa 1983, Keough and Butler 1983).

This is not as pessimistic as it seems. These designs are quite powerful for some types of disturbance, even with few Control locations, times of sampling and numbers of replicates (three, six and five here). In the arbitrary case modeled, a number of disturbances would have been examined with power in excess of 0.90 (Table 9.5). This was also the case for real data in an extensive monitoring exercise as part of the CSIRO Jervis Bay Marine Environmental Projects (CSIRO 1991), where power to detect realistic sizes of disturbance was 0.90 or better for a number of populations monitored (Underwood personal observation).

Nevertheless, long time-courses, particularly before a planned disturbance, are not the norm in environmental investigations. Logistical and financial constraints will continue to conspire to prevent adequate sampling unless the necessity for appropriately powerful sampling becomes a legal requirement of impact assessment. Alternative approaches are urgently needed, as discussed below.

Estimating Spatial and Temporal Variances: Removing the Need for Before Data

To overcome the problem of lack of time and resources for powerful sampling before many impacts, research is needed to determine the rates of temporal change and the magnitudes of spatial differences for various populations. This could be done for key habitats in which repeated disturbances are planned. For example, there have been numerous disturbances of similar sorts in many habitats (such as those due to forestry, salmon fishing, agriculture, foreshore developments in mangrove forests, mussel and oyster farming, building marinas, boat-ramps, outfalls from different types of industries, cooling-water outlets, etc.).

Because these will recur repeatedly, it makes sense to choose some suitable species and to monitor now in a range of undisturbed, randomly chosen replicated habitats. If these are randomly chosen and the times of sampling are randomly assigned, the variance among times, locations and their interactions are objective measures of the variances from a population of such locations and times. This is the essence of random factors in multifactorial analyses of variance (e.g., Scheffe 1959, Winer 1971, Underwood 1981, 1989). Thus, variances estimated in such a sampling program would serve as Before data for any other set of locations at some subsequent time. They could be used to contrast against data collected After for a set of locations chosen as Controls and including one location in which a disturbance is occurring. This would obviate the need for specifically acquired Before data and could be a very powerful data set, based on numerous times of sampling and many locations (see also Underwood 1989).

Such a research program (sensu Lakatos 1974) needs to be implemented for those habitats and populations where there is reasonable expectation that a regime of sampling in some period of time will be representative of a subsequent period and for which there is an expectation of repeated, planned environmental assaults. Unless the latter is realistic, there is no point in collecting the information.

All of this begs the questions of which species (e.g., Paine 1974, Underwood and Peterson 1988, Jones and Kaly, Chapter 3) and how to pay for the research. The former requires its own research. The latter requires a shift in emphasis of payment from developers. If this all works, the cost of individual EIS's would be smaller, because of the reduced need for location- and time-specific Before data. Also, such sampling would better protect developers from (and the prohibitive legal costs of defending) an accusation of causing environmental damage when none has occurred. While sampling continues to be in only two locations, many instances must arise where differences between the Control and the putatively Impacted location are not due, as claimed, to the identified human disturbance. Both should be political selling points for a change of strategy.

Experimental Disturbances

The final new research program needed will be experimental impacts. As argued elsewhere (Underwood 1989), these should be of two kinds. First is the use of existing human disturbances. For example, Hilborn and Walters (1981) have made a convincing case for using existing impacts as experiments. Far more can be done to evaluate what actually happened in places where human influences have occurred. This can be particularly powerful when repeated instances of similar disturbances have occurred (and can be used as replicates to contrast against a set of undisturbed Controls). Examination of such events will provide two types of information.

First, we would learn much more about the magnitudes (and directions) and temporal scales of changes caused by particular sorts of human disturbances. This would not only allow greater predictive accuracy for future planned development than is currently available from EIS procedures, but would also provide much of the information needed to optimize future environmental sampling in relation to future developments. Knowledge of actual sizes of effects of disturbances would go a long way towards solving problems in the calculation of power in sampling designs (e.g., Cohen 1977, Peterman 1990, Osenberg et al., Chapter 6).

Second, such studies will also provide evaluations of the worth of the predictions made before the development occurred. Thus, treating the development as an experiment will provide information to determine whether the prior EIS's were realistic in their predictions. Or, we could discover whether they erred because of failure of understanding the biology of the system, inadequacy of the data gathered before the disturbance, incompetence on the part of the ecologists concerned, or a complete failure of the EIS process itself. Such environmental auditing can be useful (Buckley 1991), but needs to be more widespread (Ambrose et al., Chapter 18) and to make use of existing and previous disturbances as experiments (Hilborn and Walters 1981).

The second kind of experiment is a deliberate attempt to simulate the magnitudes of disturbances as they affect populations. Small-scale examples of these have proven very useful for estimating the effects of anthropogenic disturbances (e.g., McGuinness 1990) and are the bread and butter of much ecological research (e.g., Connell 1983, Schoener 1983, Hurlbert 1984, Hairston 1989, Peters 1991). What is needed is a shift of emphasis and perhaps of scale so that the experiments are recast to test explicit hypotheses about environmental disturbances. Some discussion of protocols, problems and interpretations is in a previous paper (Underwood 1989) so I will not consider it further here.

Conclusion

Improved procedures for detection and interpretation of environmental impacts are needed. These must be coupled more tightly with logic and the sort of professionalism that field ecologists claim to use in more academic field experiments (although there are still too many problems for complacency about this; Underwood 1981, 1986, Hurlbert 1984). For too long, a perceived lack of rigor in environmental monitoring as opposed to "academic" ecological experimentation has been correlated with an argument that there is a difference between applied and pure science. It is important to get the debate back onto the distinction between good and bad science, regardless of its "purity" or its "application". Otherwise, decisions about environmental disturbances will continue to be dominated by (often) more expensive, more stochastic processes of law (Lester, Chapter 17) instead of being responsive to the successes of modern ecology.

Acknowledgments

The preparation of this chapter was funded by the Australian Research Council, the Institute of Marine Ecology and the Research Grant of the University of Sydney. I am grateful for extensive discussions with and help from M.G. Chapman, K.R. Clarke, P.G. Fairweather, M.P. Lincoln Smith, K.A. McGuinness, and C.H. Peterson and for the comments of two anonymous referees. M. G. Chapman and A. Underwood helped considerably with the preparation of the manuscript (twice!).

References

Andrew, N. L., and B. D. Mapstone. 1987. Sampling and the description of spatial pattern in marine ecology. Oceanography and Marine Biology Annual Review 25:39–90.

Bender, E. A., T. J. Case, and M. E. Gilpin. 1984. Pertubation experiments in community ecology: theory and practice. Ecology 65:1–13.

Bernstein, B. B., and J. Zalinski. 1983. An optimum sampling design and power tests for environmental biologists. Journal of Environmental Management 16:35–43.

Buckley, R. 1991. Auditing the precision and accuracy of environmental impact predictions in Australia. Environmental Monitoring and Assessment 18:1–24.

Clarke, K. R., and R. H. Green. 1988. Statistical design and analysis for a 'biological effects' study. Marine Ecology Progress Series 46:213–226.

Cochran, W. G. 1947. Some consequences when the assumptions for the analysis of variance are not satisfied. Biometrics 3:22–38.

Cochran, W.G. 1963. Sampling techniques. Wiley, New York, New York.

Cohen, J. 1977. Statistical power analysis for the behavioral sciences. Academic Press, New York, New York.

Connell, J. H. 1983. On the prevalence and relative importance of interspecific competition: evidence from field experiments. American Naturalist 122:661–696.

Connell, J. H., and W. P. Sousa. 1983. On the evidence needed to judge ecological stability or persistence. American Naturalist 121:789–824.

Cox, G. 1958. The planning of experiments. Wiley, New York, New York.

Eberhardt, L. L., and J. M. Thomas. 1991. Designing environmental field studies. Ecological Monographs 61:53–73.

Eisenhart, C. 1947. The assumptions underlying the analysis of variance. Biometrics 3:1–21.

Frank, P. W. 1981. A condition for a sessile strategy. American Naturalist 118:288–290.

Green, R. H. 1979. Sampling design and statistical methods for environmental biologists. Wiley and Sons, New York, New York.

Green, R. H., and K. D. Hobson. 1970. Spatial and temporal structure in a temperate intertidal community, with special emphasis on *Gemma gemma* (Pelecypoda: Mollusca). Ecology **51**:999–1011.

Hairston, N. G. 1989. Ecological experiments: purpose, design, and execution. Cambridge University Press, Cambridge, UK.

Hilborn, R., and C. J. Walters. 1981. Pitfalls of environmental baseline and process studies. Environmental Impact Assessment Review **2**:265–278.

Hurlbert, S. J. 1984. Pseudoreplication and the design of ecological field experiments. Ecological Monographs **54**:187–211.

Kennelly, S. J., and A. J. Underwood. 1985. Sampling small invertebrates on natural hard substrata in a sublittoral kelp forest. Journal of Experimental Marine Biology and Ecology **89**:55–68.

Keough, M. J., and A. J. Butler. 1983. Temporal changes in species number in an assemblage of sessile marine invertebrates. Journal of Biogeography **10**:317–330.

Lakatos, I. 1974. Falsification and the methodology of scientific research programmes. Pages 91–196 *in* I. Lakatos and A. E. Musgrave, editors. Criticism and the growth of knowledge. Cambridge University Press, Cambridge, United Kingdom.

McGuinness, K. A. 1990. Effects of oil spills on macro-invertebrates of saltmarshes and mangrove forests in Botany Bay, New South Wales. Journal of Experimental Marine Biology and Ecology **142**:121–135.

Paine, R. T. 1974. Intertidal community structure: experimental studies on the relationship between a dominant competitor and its principal predator. Oecologia **15**:93–120.

Peterman, R. M. 1990. Statistical power analysis can improve fisheries research and management. Canadian Journal of Fisheries and Aquatic Sciences **47**:2–15.

Peters, R. H. 1991. A critique for ecology. Cambridge University Press, Cambridge, UK.

Scheffe, H. 1959. The analysis of variance. Wiley, New York, New York.

Schoener, T. W. 1983. Field experiments on interspecific competition. American Naturalist **122**:240–285.

Snedecor, G. W., and W. G. Cochran. 1967. Statistical methods, 6th Edition. University of Iowa Press, Ames, Iowa.

Stewart-Oaten, A., W. W. Murdoch, and K. R. Parker. 1986. Environmental impact assessment: "psuedoreplication" in time? Ecology **67**:929–940.

Underwood, A. J. 1981. Techniques of analysis of variance in experimental marine biology and ecology. Annual Review of Oceanography and Marine Biology **19**:513–605.

Underwood, A.J. 1986. The analysis of competition by field experiments. Pages 240-268 *in* J. Kikkawa and D. J. Anderson, editors. Community ecology: pattern and process. Blackwells, Melbourne, Australia.

Underwood, A.J. 1989. The analysis of stress in natural populations. Biological Journal of the Linnaean Society **37**:51–78.

Underwood, A.J. 1991. Beyond BACI: experimental designs for detecting human environmental impacts on temporal variations in natural populations. Australian Journal of Marine and Freshwater Research **42**:569–587.

Underwood, A.J. 1992. Beyond BACI: the detection of environmental impacts on populations in the real, but variable, world. Journal of Experimental Marine Biology and Ecology **161**:145–178.

Underwood, A. J., and C. H. Peterson. 1988. Towards an ecological framework for investigating pollution. Marine Ecology Progress Series **46**:227_234.

Warwick, R. M. 1986. A new method for detecting pollution effects on marine macrobenthic communities. Marine Biology **92**:557–562.

Warwick, R. M., and K. R. Clarke. 1991. A comparison of some methods for analysing changes in benthic community structure. Journal of the Marine Biology Association of the United Kingdom **71**:225–244.

Winer, B. J. 1971. Statistical principles in experimental design, 2nd Edition. McGraw-Hill, New York, New York.

EXTENSION OF LOCAL IMPACTS TO LARGER SCALE CONSEQUENCES

DETERMINING THE SPATIAL EXTENT OF ECOLOGICAL IMPACTS CAUSED BY LOCAL ANTHROPOGENIC DISTURBANCES IN COASTAL MARINE HABITATS

Peter T. Raimondi and Daniel C. Reed

Determining the spatial scale of ecological change that results from a local (i.e., point source) anthropogenic disturbance is the focus of many ecological impact assessment studies. Frequently, measures of physical parameters are used to evaluate the potential maximum spatial extent of ecological change resulting from an impact (Spellerberg 1991). Such an approach assumes that changes in the physical environment (e.g., changes in seawater temperature, salinity, or chemistry) are comparable to, or provide indirect measurements of, impacts on ecological parameters (e.g., population density and community structure). This is unlikely to be the case unless the spatial extent of the physical and ecological impacts are the same.

There are compelling reasons to expect that the spatial extents of physical and ecological impacts are not always the same in coastal marine habitats. Many species typically display a nonlinear or threshold response to changes in the physical environment. This is because various physiological mechanisms allow many organisms to tolerate some degree of changing environmental conditions associated with a disturbance (e.g., increased levels of toxicants) without undergoing any appreciable change in their demographic rates. Moreover, because marine organisms vastly differ in physiology, resource requirements, and life-history characteristics, their tolerances to a given perturbation (and the corresponding spatial extent of the ecological impact that results from it) are likely to vary among species as well as among different ecological parameters within a species. This type of phenomenon can cause the spatial extent of the ecological impact to be partially or even completely unrelated to alterations in the physical environment caused by a perturbation. Unfortunately there are few published data sets that provide information on the spatial relationships among different

Detecting Ecological Impacts: Concepts and Applications in Coastal Habitats, edited by R. J. Schmitt and C. W. Osenberg

physical and ecological variables affected by a local impact. The scarcity of such information further complicates the task of determining the spatial scale of ecological change that results from a local disturbance. In this chapter we examine spatial relationships among various physical/chemical and biological variables used to measure the spatial extent of impacts caused by an industrial outfall in a coastal marine habitat.

Of equal concern in ecological impact assessment studies is the potential for ecological impacts to spread to distant locations far from the area that is physically disturbed. Such spatial separation between physical and ecological impacts is possible in nearshore habitats because many marine species reproduce via planktonic propagules (e.g., gametes, spores, and larvae) that are capable of dispersing substantial distances (tens to hundreds of kilometers). This causes renewal rates in local populations of many species to be largely dependent on the transport and arrival of propagules produced elsewhere. Thus, a local perturbation that alters demographic rates (e.g., birth, death, immigration, emigration) of nearby populations can potentially affect the structure and dynamics of distant populations.

If the physical and ecological effects caused by human activities covary in space and time, then assessment of ecological change, while often laborious, should not be overly complex. However, if physical and ecological effects of a disturbance are separated, ecological assessment can become very difficult. This is because when ecological effects of a local impact are dispersed to distant sites they usually become diluted making it more difficult to detect ecological change at any particular site. Furthermore, natural variation in population parameters coupled with the high costs of increased sampling effort serves to reduce the statistical power of detecting small changes caused by a distant perturbation. For these reasons, efforts to monitor ecological impacts that are spatially separated from physical impacts are almost surely destined to fail if typical protocols (including the BACIPS design, Chapters 6–8) are used. Nevertheless, local impacts can in theory substantially influence the dynamics and structure of more distantly located populations (Nisbet et al., Chapter 13).

Designing effective and logistically-tractable ecological assessment studies hinges critically on understanding the possible relationships between the spatial scale of physical and ecological impacts. The most fundamental cause of spatial uncoupling between physical and ecological impacts is the ability of individuals to move from one location to another. If movement of individuals is severely limited, then ecological impacts are likely to be restricted to the area of physical impact, or some subset of that area. In contrast, if organisms are highly mobile the effects of a local perturbation can spread to distant areas, far removed from the site of local activities causing physical and ecological impacts to be either partially or fully separated in space.

While organismal mobility seems to be a necessary condition for the extension of ecological effects beyond the area of physical impact, other attributes may make such separation between physical and ecological effects more or less likely to occur. Thus, in addition to examining spatial relationships between physical and biological variables we also discuss how various life-history attributes of marine organisms may cause ecological impacts to become spatially separated from physical impacts.

Spatial Relationships among Physical and Ecological Variables Following a Local Disturbance

Detecting the spatial extent of ecological impacts resulting from local physical disturbances is often difficult. A major complication arises when physical measures of impact, which are often used to estimate biological impacts, do not correlate well with ecological ones. This may happen when: (i) the physical variables measured are not those that cause ecological effects (or are not correlated with the causal agents), or (ii) physical variables are not adequately measured (e.g., instrumentation and/or sampling design are inadequate to detect changes caused by disturbance). That individual physical measures of impact may not correlate with other physical measures and individual ecological measures of impact may not correlate with other ecological measures further complicates the task of assessing the spatial extent of local impacts. If physical and ecological measures of impact are poorly correlated, then the spatial extent of an impact can be meaningful only relative to the measured parameter. This would reduce our ability to predict and determine the extent and type of ecological effects that result from localized physical disturbances.

To provide insight into the degree of correlation among different physical and ecological measures of impact we examined results from a group of companion studies designed to assess the impact of a produced water (a byproduct of oil and gas development) discharge on the nearshore marine environment. The outfall is located off the coast of Carpinteria, California, USA (34°23'N, 119°30'W) on a flat sandy bottom at a depth of 11 m. This area has been characterized as an open coast high energy environment (Osenberg et al. 1992). The effluent is discharged ~ 300 m offshore through a series of small diameter (4–5 cm) diffuser ports. The average discharge of effluent during the course of the investigations (1989–1992) was 2.64 million liters/day (Osenberg et al. 1992). A discussion of the chemical nature of the effluent can be found in Higashi et al. (1992).

Surveys and experiments were done to estimate the spatial extent of the physical and biological/ecological impacts resulting from the discharge of produced water. Data were collected at a series of locations, along an 11 m isobath, arranged in a gradient away from the diffusers. Physical variables most likely to be affected by the outfall were chosen and used to estimate physical extent of the impact (Table 10.1). Water clarity (estimated from secchi disk measurements) and sediment grain size were measured because of the belief that the rapid discharge of effluent to a sand-bottom habitat would stir up sediments and

Table 10.1. Estimates of Spatial Extent of Impact at Carpinteria for Various Physical/Chemical and Biological Variables

	Distance (meters) from outfall where measured variables = background or control levels	Source[*]
Physical/chemical variables		
Barium in the shell of *Mytilus edulis*	500–1000	a
Barium in the shell of *Mytilus californianus*	500–1000	a
Barium in sediments	10–20	b
Salinity	5–10	c
Water clarity	0	d
Sediment grain size	0	a
% Organic matter	0	a
Biological variables		
Nematode density	50–100	e
Polychaete density	50–100	e
Mytilus edulis		
Tissue production	500–1000	e
Mytilus californianus		
Tissue production	500–1000	e
Strongylocentrotus		
Male gonadal mass	50–100	c
Female gonadal mass	50–100	c
Fertilization	50–100	c
Sperm attack rate	50–100	c
Haliotis rufescens - precompetent larvae		
Settlement	100–1000	f
Survivorship	100–1000	f
Haliotis rufescens - competent larvae		
Settlement	100–500	f
Survivorship	0	f
Metamorphosis	50–100	f
Swimming	100–500	f
Macrocystis pyrifera		
Fecundity (spore production)	0–5	g
Sporophyte recruitment	5–50	g

[*]Sources: a = Osenberg and Fan (personal communication); b = Higashi and Fan (personal communication); c= Krause (1993); d = Reed (unpublished data); e = Osenberg et al. (1992); f = Raimondi and Schmitt (1992); g = Reed et al. (1994).

increase turbidity near the diffusers. Produced water discharged at Carpinteria is hyposaline (about 17 ppt) and contains numerous hydrocarbons and heavy metals, the most conspicuous being barium which has an elevated concentration of 6–19 ppm (compared to ambient control levels of <0.002 ppm). Therefore, water-column salinity, and barium and organic matter in the sediments were measured to map the extent of the outfall plume. In addition, barium was measured in the shells of the mussels *Mytilus californianus* and *M. edulis*. Measuring constituents of the effluent in sediments and mussels can be useful in determining the spatial extent of impacts because sediments and mussels potentially integrate the temporally variable supply of effluent over time. To measure the biological/ecological response to the produced water discharge, 16 variables from 8 species were examined (Table 10.1).

Less than 50% (3/7) of the physical/chemical variables measured at the Carpinteria site were significantly correlated with proximity to the diffusers, an event we consider to be evidence for an impact (Table 10.2). By contrast, a greater percentage, 75% (12/16), of the biological variables examined were correlated with distance from diffusers. A concern about the validity of comparing correlations involving physical/chemical variables to those from biological variables is that the results might be affected by selection of variables. We agree, but note that all physical/chemical variables which as a group were less correlated with distance from diffusers, were studied because of specific predictions about the physical results of the impact (e.g., likely increase in salinity and decrease in barium concentration as a function of distance from the diffusers; as above).

The spatial extent to which each variable was affected by the outfall was estimated by determining the distance from the diffusers where levels of each variable were no different from background or control levels (Table 10.1). In general, the distances over which the biological variables were affected were ~ an order of magnitude greater than those over which the physical/chemical variables were affected with the exception of barium in the shells of the mussels, *M. californianus* and *M. edulis*.

We investigated the relative value of different biological and physical/chemical variables in predicting the spatial extent of ecological impacts by examining the degree to which: (i) biological measures of impact were correlated with each other and, (ii) physical/chemical measures of impact were correlated with biological measures. Pairwise correlations were done on all combinations of variables by arranging variable values by distance from the diffusers. The significance levels for all correlations are given in Table 10.2 and the ratios of statistically significant to nonsignificant correlations involving each variable are summarized in Table 10.3 along with whether the variables themselves were significantly correlated with distance from the diffusers.

Results indicate that certain variables were better than others at predicting the spatial pattern of the ecological impact (here defined as the area over which any biological variable was affected: Table 10.3). For example, polychaete number was not significantly correlated with any of the other 15 biological variables. By

Table 10.2. Significant Levels of Correlation between Variables

Physical / Chemical Variables — Biological Variables

	Barium in mussel shells (M. edulis) BARME	Barium in mussel shells (M. californ.) BARMC	Barium in sediments BARIUM	Salinity SAL	Water Clarity VIS	Sediment Grain Size SED	% Organic Matter POM	Infaunal density — Nematodes NEMAT	Infaunal density — Polychaetes POLY	Mussel tissue production (M. edulis) EDU	Mussel tissue production (M. californ.) CAL
BARME	0.004										
BARMC	0.354	0.371									
BARIUM	0.262	0.301	(0.000)								
SAL	0.484	0.544	0.448	0.575							
VIS	0.659	0.476	0.474	0.468	0.142						
SED	0.969	0.650	0.509	0.509	0.533	0.416					
POM	0.135	0.195	0.006	0.461	0.603	0.361	0.200				
NEMAT	0.308	0.531	0.220	(0.001)	0.975	0.919	0.487	0.126			
POLY	(0.018)	(0.024)	0.420	0.202	0.292	0.843	0.837	0.152	0.698		
EDU	(0.010)	(0.025)	0.384	0.308	0.632	0.978	0.737	0.173	0.508	0.055	
CAL	0.158	0.062	0.367	0.293	0.102	0.295	0.666	0.300	0.743	0.043	0.336
MGON	0.107	0.014	0.539	0.368	0.269	0.298	0.206	0.480	0.775	0.152	0.118
FGON	(0.008)	(0.001)	0.204	0.539	0.079	0.332	0.538	0.173	0.882	(0.028)	0.055
SPERM	(0.012)	(0.002)	0.131	0.217	0.073	0.354	0.561	0.161	0.932	0.027	0.061
FERT	(0.038)	(0.027)	0.471	0.108	0.093	0.566	0.674	0.104	0.874	0.006	0.066
PSET	0.077	(0.030)	0.065	0.131	0.413	0.453	0.194	0.436	0.971	0.219	0.032
PSUR	0.051	0.094	0.863	0.471	0.816	0.839	0.973	(0.043)	0.703	0.034	0.017
CSET	0.881	0.658	0.173	0.065	0.979	0.131	0.817	0.065	0.710	0.667	0.646
CSUR	(0.007)	(0.004)	0.317	0.863	0.582	0.349	0.778	0.931	0.912	0.018	0.050
CSWIM	(0.002)	0.227	0.074	0.173	0.693	0.341	0.773	0.090	0.970	0.045	0.030
MFEC	0.372	0.186	0.163	0.317	0.102	0.337	0.859	0.285	0.752	0.108	0.609
MSPOR	0.051	0.074		0.074	0.279	0.677	0.677	0.094	0.515	0.283	0.105
LD	(0.016)	(0.000)		0.163	0.651	0.592	0.846	(0.020)	0.377	0.041	0.007

Biological Variables

	Strongylocentrotus purpuratus (Sea Urchin) — Gonad Size Males MGON	Strongylocentrotus purpuratus — Gonad Size Females FGON	Strongylocentrotus purpuratus — Fertilization FERT	Strongylocentrotus purpuratus — Sperm Activity SPERM	Strongylocentrotus purpuratus — Pre-competent larvae Settlement PSET	Strongylocentrotus purpuratus — Pre-competent larvae Survival PSUR	Haliotis rufescens (Red Abalone) — Competent larvae Settlement CSET	Haliotis rufescens — Competent larvae Survival CSUR	Haliotis rufescens — Metamorph CMET	Haliotis rufescens — Swimming CSWIM	Macrocystis pyrifera (Kelp) — Fecundity (spore production) MFEC	Macrocystis pyrifera — Sporophyte Recruitment MSPOR
FGON	0.002											
FERT	(0.006)	(0.000)										
SPERM	(0.006)	(0.017)	0.000									
PSET	(0.043)	0.057	0.011	0.006								
PSUR	0.239	(0.016)	0.041	0.048	0.090							
CSET	0.201	0.146	0.053	0.047	0.017	0.102						
CSUR	0.281	0.695	0.713	0.693	0.762	0.919	0.892					
MET	0.059	(0.049)	(0.001)	0.000	0.005	0.064	0.005	0.527				
CSWIM	0.103	(0.048)	0.001	0.002	0.016	(0.038)	0.009	0.589	0.000			
MFEC	(0.013)	0.290	0.231	0.204	0.177	0.589	0.366	0.368	0.221	0.312		
MSPOR	0.589	0.409	0.170	0.199	0.270	0.239	0.157	0.657	0.167	0.107	0.642	
LD	(0.019)	(0.006)	0.000	0.000	0.024	0.007	0.006	0.792	0.005	0.003	0.433	0.237

Note: Significant ($P \leq .05$) positive correlations are indicated by bold type (with light shading), and significant negative correlations are indicated by parentheses and bold type type (with light shading). LD = Log of distance from diffusers.

Table 10.3.   Relationships among Variables

	P/B	B/B	Correlated with distance from diffusers?
Physical/chemical variables			
Barium in the shell of *Mytilus edulis*	9/16		Yes
Barium in the shell of *Mytilus californianus*	9/16		Yes
Barium in sediments	1/16		No
Salinity	1/16		Yes
Water clarity	0/16		No
Sediment grain size	0/16		No
% organic matter	0/16		No
Biological variables			
Nematode density		1/15	Yes
Polychaete density		0/15	No
Mytilus edulis			
Tissue production		7/15	Yes
Mytilus californianus			
Tissue production		4/15	Yes
Strongylocentrotus purpuratus			
Male gonadal mass		7/15	Yes
Female gonadal mass		7/15	Yes
Fertilization		8/15	Yes
Sperm attack rate		9/15	Yes
Haliotis rufescens - precompetent larvae			
Settlement		7/15	Yes
Survivorship		5/15	Yes
Haliotis rufescens - competent larvae			
Settlement		7/15	Yes
Survivorship		0/15	No
Metamorphosis		8/15	Yes
Swimming		9/15	Yes
Macrocystis pyrifera			
Fecundity (spore production)		1/15	No
Sporophyte recruitment		0/15	No

Note: Ratios are the number of significant correlations ($P \leq .05$) over the total number correlations (significant and nonsignificant). P/B = Physical/chemical vs biological variables; B/B = Biological vs other biological variables. See Table 10.2 for description of data.

contrast, sea urchin fertilization was significantly correlated with 9 of 15 biological variables. Interestingly, the same variable measured in different ways yielded different results. Concentration of barium in the shells of mussels (a physical/chemical variable) was as good as any biological variable in predicting the spatial extent of ecological impacts, while barium in the sediments was correlated with only one biological variable. Like barium in the sediments, other physical/chemical variables were generally poor predictors of ecological impacts. There were large differences in ability to predict the response of other variables even among those biological and physical/chemical variables affected by the impact (i.e., those significantly correlated with distance from the diffusers). The range in number of significant correlations for such variables was from 1 to 9 out of 15. The magnitude of this range should act as a precautionary warning about the importance of selecting indicator variables (species) that are truly representative of other ecologically important variables (see also Chapter 3).

Finally, one piece of information that is often missing from studies of local impacts in the marine environment is an assessment of the spatial distribution of the causative agent. This information is potentially important for at least two reasons. First, laboratory assays cannot be used to predict performance in the field without knowledge of the concentration gradient of the causative agent in the environment. Second, the relationship between causative agent and biological performance can be obtained by comparing the spatial distributions of performance and causative agent(s). Obviously, the best method for assessing the spatial distribution of a causative agent is to directly measure it in the vicinity of the impact. However, this is often complicated by uncertainty about the identity of the causative agent, or concern that multiple agents could be producing the effect. For example, a large amount of information (Table 10.1) suggests that the discharge of produced water is toxic to a wide variety of marine organisms, yet there is no consensus as to the identity of the specific toxin (Higashi et al. 1992). Furthermore, even if the causative agent is known, environmental processes can cause it to be extremely variable in its occurrence, making it difficult to detect unless variables that integrate over time are used.

An alternative to measuring the spatial distribution of causative agents in the field is to predict it from performance under known concentrations in the laboratory and at known distances from the effluent source in the field. We did this for the Carpinteria study using data on abalone settlement. Abalone settlement was chosen because it was as sensitive (as measured by spatial extent of effect) as any other biological variable examined at Carpinteria (Table 10.1) and because it was a variable that could easily be examined in laboratory and field. We estimated the concentration of produced water as a function of distance from the diffusers from the observed settlement of abalone larvae at known concentrations of produced water in the laboratory (Raimondi, *personal observation*) and from data on settlement in the field.

The estimated concentration of produced water near the diffusers was greater than 100% (at the 5 m Site it is >1000%) and gradually declined to ~0.0001% at

1000 m from the diffusers (Figure 10.1). Estimates of field concentrations greater than 100% were produced because at distances close to the diffusers performance was much worse than ever found using lab experiments; here field concentrations were extrapolated. These results indicate that the discharge of produced water was much more toxic than could have been predicted from laboratory bioassays.

This approach can be used to determine which physical/chemical variables best describe the distribution of the causative agent in the field. In Figure 10.2 we compare the estimated concentration of produced water with the physical/chemical variables that showed a significant correlation with distance from the diffusers (salinity and barium in mussel shells) and with a similar variable that did not correlate with distance (barium in sediments). Only spatial pattern of barium in the shells of the two mussel species was similar to that of produced water estimated from abalone settlement. Interestingly, the spatial pattern of barium in the sediments differed substantially from that of barium in mussel shells. The dissimilarity between the spatial distribution of barium in the sediments and produced water probably reflects the inability of sediments to accumulate toxicants in an open coast environment.

Life-History Attributes and the Dispersal of Ecological Impacts

Spatial patterns of ecological impacts may be completely uncorrelated with those of physical impacts if individuals have the ability to move from one location to another. Most marine organisms have such mobility at some stage of development. Populations of many pelagic species are sustained by the extensive

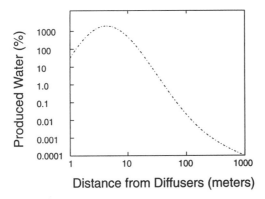

Figure 10.1. Estimated concentration of produced water as a function of distance from produced water diffusers. Estimate is based upon comparison of laboratory and field studies assessing the settlement of abalone larvae (see text). The curve was generated using a nonlinear curve fitting routine (Systat Inc. 1992).

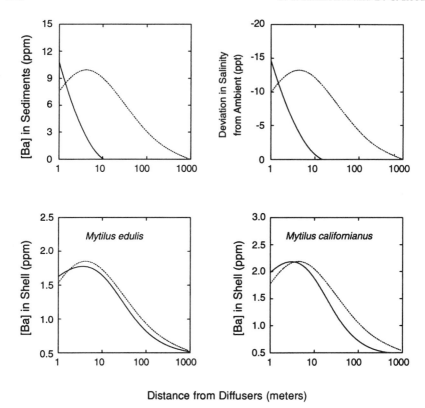

Figure 10.2. Relationships between physical/chemical variables (solid lines) and distance from produced water diffusers. Sources of data are given in Table 10.1. Curves were generated using a nonlinear curve fitting routine (Systat Inc. 1992). For comparison, the curve describing the estimated concentration of produced water (dotted line; see Figure 10.1) is included on each figure.

movement of juveniles and adults that travel in search of food or breeding sites. In contrast, the majority of organisms living in coastal marine habitats have sessile or weakly mobile juveniles and adults. Widespread movement in these species is confined to planktonic propagules, which in many cases disperse considerable distances causing local populations to be strongly reliant on replenishment by propagules produced elsewhere (Scheltema 1971, Keough 1988, Roughgarden et al. 1988). Such movement by individuals is critical in maintaining species distributions and genetic diversity (Scheltema 1971, Slatkin 1987) and can profoundly influence population dynamics and community structure (Connell 1961, Sale 1977, Underwood and Denley 1984, Roughgarden et al. 1985, Raimondi 1990). Thus, localized activities that alter the rates by which individuals (adults or propagules) emigrate from locally disturbed areas or immigrate to more distant nondisturbed sites have the potential to influence population dynamics and community structure over much larger spatial scales (e.g., Keough and Black, Chapter 11, Nisbet et al., Chapter 13). Below we discuss the biological attributes of organisms that are most likely to influence rates of movement and how human activities that affect these attributes might cause local ecological impacts to be dispersed to more distant locations.

Changes in Emigration Rates from Locally Disturbed Sites

Highly mobile species may choose to leave locally disturbed areas, thus causing an increase (usually temporary) in their abundance at some distant location (Hose et al. 1983, Ebeling et al. 1985, Stouder 1987, Breitburg 1992). Most coastal marine species, however, are much less mobile and the extension of local impacts to distant locations occurs via processes that influence propagules, which are the primary source of emigrants. Perhaps the most direct way that localized activities can reduce propagule production at a site is by causing a decrease in adult density (provided that a reduction in the abundance of propagule-producing adults is not fully compensated for by an increase in the fecundity of survivors). Local perturbations that cause a decrease in population size may also indirectly affect rates of propagule production (and hence emigration) by limiting the incidence of syngamy. Species most likely to be affected in this manner include sessile organisms that have internal fertilization (e.g., barnacles that do not self fertilize). Although there can be substantial variation in the length of intromittent organs among individuals, there are presumably limits to the distances separating individuals in which copulation is still possible. Disturbances that increase adult spacing beyond these limits will prevent fertilization from occurring regardless of how many gametes an individual produces. A reduction in population density can also adversely affect fertilization rates in species that broadcast their gametes into the water column. Recent studies indicate that under a wide range of hydrographic conditions, the vast majority of fertilization in broadcast spawners that are sessile or have limited mobility occurs between closely spaced individuals (Pennington 1985, Denny and Shibata 1989, Reed 1990).

Sublethal effects of anthropogenic activities also have the potential to affect emigration rates from locally disturbed sites. Although most marine organisms have some capacity for tolerating increased levels of toxicants, emigration rates can be affected if toxicants alter per capita gamete production and/or viability. Pollutants are known to cause a reduction in both the quality (Hogan and Brauhn 1975, Westernhagen et al. 1981, Fan et al. 1992) and quantity (Linden 1976, Osenberg et al. 1992) of gametes produced by adults living in disturbed areas. Typically, nonpolar organic compounds cause the most harm to gametes because of their tendency to bioaccumulate (i.e., be taken up and concentrated by living systems) in gonads and other body tissues that are high in lipid (highly polar compounds are less likely to enter the body system, and when they do, tend to be quickly excreted: Manahan 1992). Not surprisingly, elevated levels of polar compounds (e.g., chlorinated hydrocarbons) in gonads have been listed as a cause of serious abnormalities in gametes and embryos of a variety of species. Notable examples that have resulted in reproductive failure include eggshell thinning in birds (Ratcliffe 1967, Anderson et al. 1975) and premature births in marine mammals (De Long et al. 1973, Addison 1989). Some organisms may actually cleanse tissues sequestering toxins by shunting the toxins to the fatty tissues of eggs. Not surprisingly, several studies have found major differences between males and females in the bioaccumulation of environmental toxicants; males continuously accumulate contaminants with age while females periodically reduce their contaminant levels when reproducing (Gaskin 1983, Britt and Howard 1983).

The effects of toxicants on gametes are not necessarily restricted to processes that occur *in vivo*. Numerous studies have shown that spawned gametes may be particularly sensitive to waterborne pollutants (Allen 1971, Kobayashi 1980, Krause et al. 1992). For obvious reasons, fertilization rates of species that broadcast spawn are more likely to be affected by toxicants in the water column than species that transfer gametes via intromittent organs.

Even if the toxicity of released substances is low, the potential for fertilization rates to be significantly reduced still exists. This is because fertilization in virtually all organisms is dependent on chemical recognition (sperm is guided to the egg via a specific pheromone) and changes in water column chemistry that arise from a local impact have the potential to disrupt the chemical communication system between male and female gametes. In many brown algae, for example, a pheromone is produced by the female plant that not only guides the sperm to the egg but also triggers the release of sperm from male plants (Müller et al. 1985). Reed et al. (1994) found that an effluent of 1% produced water hampered the ability of male gametophytes of the giant kelp *Macrocystis pyrifera* to detect female pheromone. This in turn caused a threefold reduction in the quantity of sperm released by these males relative to that of male plants placed in seawater lacking the effluent. Contaminants can also change the sexual behavior of adults. Linden (1976) found that concentrations of crude oil ≥.002% completely suppressed copulatory behavior in the amphipod *Gammarus oceanicus*. Such chemical disruption that alters sexual behavior of gametes or adults can greatly influence emigration rates in species whose propagules disperse after fertilization.

The number and caliber of gametes that an individual produces are generally highly dependent on its nutritional status. Consequently, local anthropogenic activities can *indirectly* affect propagule emigration rates (via changes in per capita fecundity) if they alter the quality and/or quantity of an organism's food resources. When fecundity is altered via changes in food, the effect and extent of the impact should vary among species with different nutritional requirements. For example, ocean discharges that cause an accumulation of organic matter on the bottom might be expected to positively influence suspension feeders and deposit feeders that are able to use the material for food. Changes in the abundance of these secondary consumers may in turn cause corresponding increases in the abundance and fitness of species at higher trophic levels. In contrast, species that do not rely on heterotrophic nutrition may be negatively influenced by the increased flux of organic matter. Plants use light energy to fix gaseous carbon into organic carbon and increased turbidity resulting from the discharge of organic matter may reduce primary production (both in the water column and on the bottom) by decreasing irradiance. Substantial reductions in plant biomass may in turn negatively influence the fitness of primary consumers that actively graze micro- and macroalgae. Obviously the nature of the impact will determine which species are affected. Discharges of municipal wastes that are rich in nutrients may increase primary production to a level where macrophytes exclude sessile invertebrates (Banner 1974, Gameson 1975, Rastetter and Cooke 1979), whereas the discharge or spillage of petroleum products may favor chemotrophic organisms (Atlas et al. 1978, Cowell et al. 1978, Spies and Davis 1979, Bakke et al. 1982).

Such patterns were observed near a coastal nuclear generating station in San Onofre, California. Widespread increases in the abundance of benthic invertebrates living in or on the sediments were observed near the power plant and were attributed to an increase in organic matter on the bottom that fell out of the discharge plume (Murdoch et al. 1989). Increases related to the discharge generally extended 3–6 times farther for species that fed on and just above the surface of the sediments compared to species that fed in the sediments, presumably because the latter are affected mainly by heavier particles that fall out of the discharge plume earlier, while surface feeders are more influenced by lighter particles that fall out of the plume over a larger distance downcurrent. Similar increases in the abundance of more mobile benthic feeding fish were also observed. In contrast, adult and subadult densities of giant kelp, the lone autotroph examined in this study, were reduced ~60% below the densities expected in the absence of the power plant (Bence et al., Chapter 8). The local reduction in kelp was attributed to an increase in turbidity and a decrease in bottom irradiance caused by the discharge plume, which suppressed the development and survival of microscopic stages of kelp. Significant declines in sea urchins and many species of fish that use the kelp bed for food and shelter were also observed (Ambrose et al., Chapter 18).

Changes in Immigration Rates to Distant Undisturbed Sites

Local anthropogenic activities that influence mortality of mobile stages during transport or cause sublethal effects that render an individual incapable of settling and/or fully developing upon its arrival to a new site will potentially alter immigration rates at distant locations, thus extending local impacts over much greater spatial scales. Most marine organisms have multiple free-living life-history stages. Perturbations affecting any stage can cause ecological impacts to be extended in space. Because different stages typically use different resources, live in different habitats, are of different size and shape and may even prey on one another, their susceptibility to a given perturbation is likely to vary. There is considerable evidence that propagules (gametes, spores and larvae) are among the most sensitive life stages to hazardous wastes (see above) and other changing environmental conditions that may arise from anthropogenic activities (Conner 1972, Rosenberg 1972, Wilson 1977, Weis and Weis 1989).

Propagules of many marine organisms undergo several different developmental stages during dispersal (e.g., barnacle larvae typically have six distinct naupliar larval stages that feed in the plankton, followed by a nonfeeding cyprid stage, which ultimately attaches to a hard surface and makes the metamorphic transition to a sessile, feeding adult). Evidence from field experiments indicates that different larval stages of the same species may be affected differently by the same perturbation. Raimondi and Schmitt (1992) found that the produced water outfall off Carpinteria differentially influenced the survivorship of precompetent and competent larvae of the red abalone, *Haliotis rufescens*; survivorship of precompetent larvae was negatively affected while that of competent larvae was unaffected. Although encountering the discharge plume did not prove fatal to competent larvae while in the water column, it did significantly impair their ability to settle and metamorphose into the sessile adult form relative to larvae that did not encounter the plume. Such effects could potentially influence immigration rates to areas down current outside the zone of physical impact. Metamorphosis in red abalone larvae typically occurs 1 to 3 weeks after competency is achieved and during this period individuals remain as swimming larvae. This is a sufficient amount of time for affected larvae to be transported considerable distances along an open coastline.

Not all anthropogenic causes of propagule mortality necessarily result from the discharge of toxic compounds. The dispersal potential of spores and larvae of many species depends on the availability of exogenous energy sources and there is some evidence that food limitation during the planktonic dispersal phase may influence recruitment success in a wide variety of marine organisms, including fishes (Richards and Lindman 1987), invertebrates (Connell 1961, Olson and Olson 1989, Morse and Morse 1991) and algae (Reed et al. 1990). Localized activities that influence water-column productivity, therefore, may affect the dispersive potential of propagules by adversely affecting their food supply. The effect of a local reduction in propagule food supply on the dynamics and structure of distantly located populations has not been examined. However, such effects are likely related to the ratio of the length of time a propagule spends in the area of reduced food to the total duration of its planktonic feeding phase.

Presently, there are very few types of anthropogenic impacts that cause a detectable reduction in the planktonic food supply over large spatial areas. Thus, unless hydrographic features increase retention time in the impact area (see Keough and Black, Chapter 11), propagules are not likely to be subjected to limited food supplies for lengthy periods of time (indeed, significant ecological impacts would not be expected at distant sites in situations where propagules are retained). It is unlikely that a temporary reduction in propagule food supply will lead to significant ecological changes in distant populations for species with propagules that spend relatively long periods feeding in the plankton. For such species, the ecological consequences of a temporary reduction in food may be ameliorated by renewed and lengthy feeding outside the zone of food reduction. Moreover, a lengthy planktonic phase greatly reduces the probability that propagules from a local source will disperse to and settle in the same area, thereby decreasing the odds they would contribute significantly to any one population outside the zone of food reduction. This may not be the case for species with propagules that feed for shorter periods in the plankton. For such species, there is less chance that a temporary reduction in food will be compensated for and a greater probability that propagules of similar origin will disperse to and settle in the same area. Moreover, the odds that a greater proportion of propagules will be adversely affected by a temporary reduction in food increases as the ratio of the duration of the food limitation to the duration of planktonic feeding period increases. For these reasons, it may be easier to detect ecological changes in distant (i.e., outside the zone of physical impact) populations that arise from a local reduction in propagule food supply for species that have propagules with relatively short feeding durations.

Sublethal effects of local activities may affect the behavior of mobile stages, and thereby influence their ability to disperse and settle at distant locations. Stimulation of a neuro- or chemoreceptor initiates most behaviors (Pawlik 1992) and local perturbations may interfere with both the receptor (direct effect) and the stimulus or cue (indirect effect). Of the two, direct effects have been investigated the most thoroughly in the marine environment. There have been a number of studies that have demonstrated detrimental effects of various anthropogenic activities on mating and other behaviors of vertebrates (Anderson 1978, Garshelias and Garshelias 1984, Sokolov et al. 1990). Much less is known concerning how such activities affect invertebrate behavior. Whether or not the interference of behavior at one location causes effects to be manifested at more distant locations probably depends in part on whether exposure to a perturbation permanently alters behavior or simply changes it during the period of exposure. If the behavior is permanently altered there may be effects resulting from exposure at other times or places where "normal" manifestation should have occurred. If the behavior is altered only during exposure then the potential for local impacts to be dispersed is less likely, unless the affected behavior decreases the probability that the individual will leave the locally affected area.

There is some evidence from the field that local activities can influence larval behavior in ways that could alter immigration rates to physically undisturbed sites. Raimondi and Schmitt (1992) found that some behaviors of the larvae of red abalone were permanently altered (e.g., settlement behavior) and others temporarily so (e.g., swimming behavior) following exposure to a produced water-discharge plume; larvae observed outside the plume showed no such changes. Both alterations in behavior could cause ecological effects at distant locations. In the case of settlement behavior, large numbers of individuals that were temporarily exposed to produced water were unable to settle, even when they were outside the physically affected area. Such behavior could lead to reduced settlement and subsequent adult numbers in locations down current of the plume. The change observed in swimming behavior was a complete suppression in swimming by individuals during exposure to the discharge. Although this effect was sometimes reversible (i.e., ~50% of affected larvae resumed swimming after being placed in nonplume water), it too could cause an ecological impact at distant locations if a loss of swimming ability reduces the odds that an individual will leave the area affected by the discharge. Such may be the case for species like abalone that have negatively buoyant larvae. These larvae quickly sink to the bottom once they stop swimming, thus greatly reducing the chance of their being transported substantial distances by prevailing currents.

Any changes in population dynamics and community structure that arise from reductions in rates of propagule immigration are more likely to be detected in systems that are limited by the recruitment of new individuals rather than by interactions among established residents. The likelihood that a population will be limited by recruitment rates may vary with particular life-history characteristics. For example, because clonal organisms can also reproduce asexually, the dynamics of their populations might be expected to be much less dependent on rates of propagule immigration than those of aclonal organisms (i.e., species that can only reproduce sexually). Consequently, local impacts that reduce rates of propagule immigration to more distant sites may have a greater effect on populations of aclonal species than clonal species. Similar arguments can be made concerning ecological dominance. The percent of the substrate that a species occupies should be more tightly linked to rates of propagule immigration in aclonal species than clonal species simply because of the latter's potential for unlimited clonal growth.

Conclusions

A goal in the field of impact assessment is to be able to predict general ecological impacts from investigation of a few variables (see also Jones and Kaly, Chapter 3). This would result in a tremendous savings in the cost and effort required to accurately assess impacts resulting from anthropogenic disturbances. Our analysis of results from studies of a local impact at Carpinteria, California indicates that it may be difficult to attain this goal. The effectiveness of using biological variables to predict general ecological impacts is likely to be limited by differences in responses among species, ontogenetic stages and life-history attributes. Some physical/chemical variables may be well correlated with the maximum extent of ecological impacts, such as we found for barium in mussel shells. However, unless it can be determined *a priori* that a particular physical/chemical variable correlates well with biological variables, its usefulness for predicting the extent of ecological impacts is limited.

Detecting the extent to which ecological impacts are dispersed to locations removed from the area of physical impact is perhaps the most difficult challenge facing scientists responsible for ecological impact assessment in coastal marine habitats. Tracking individuals (especially small propagules) in a habitat as vast and complex as the nearshore marine environment can be extremely difficult. Assessing the contribution of a particular population affected by an impact to the ecology of populations and communities at distant sites is typically confounded by complex circulation patterns and other numerous, but unaffected, source populations.

Due to the difficulty of detection there is little empirical evidence to date that implicates local impacts as the causative agent producing significant ecological change at distant locations. Examples in which local impacts have been charged with producing larger-scale effects are restricted primarily to modeling studies, which are forced to make inherent assumptions that are difficult to validate (e.g., the assumption that compensatory mechanisms are not important in the plank-tonic propagule stage: Nisbet et al., Chapter 13). However, there should be little disagreement that local impacts have the potential to significantly alter the com-position and dynamics of nearshore marine communities over larger spatial scales.

Detecting effects dispersed in complex habitats such as the coastal marine environment is critically dependent on the ability to make accurate predictions. Detection, hopefully, will become easier if there is a clear sense of what charac-teristics of species contribute to ecological impacts being dispersed over larger spatial scales. In this chapter, we present some ideas as to which attributes of species are the most likely to lead to broad-scale population changes as a result of a localized disturbance. Many of the ideas we discuss still need to be rigor-ously tested. Even if verified, however, knowledge of which species characteris-tics lead to dispersed effects will not, by itself, allow dispersed impacts to be detected in most situations. Equally important is knowing *where* impacts are likely to be dispersed to. Details of oceanographic transport and larval behavior under different transport regimes are critical to the accurate prediction of where dispersed impacts will most likely occur (Keough and Black, Chapter 11). Complementary information on "sources and sinks" (sensu Pulliam 1988) of population replenishment will aid in predicting where and how much change can be expected from a given local impact. Such information may require the use of modern techniques in molecular biology and in the analyses of stable isotopes to develop genetic and environmental markers for distinguishing source and sink populations. Moreover, population effects may not be limited to numerical changes; changes in source populations can alter the genetic makeup of sink populations. Such genetic alterations can result in delayed effects by reducing the stability and resiliency of a population during times of environmental change (Soulé and Simberloff 1986, Gilpin 1991). All such information that serves to improve the accuracy of predicting the scale over which localized effects are dispersed will undoubtedly aid in the detection of such effects.

Acknowledgments

We thank J. Bence and C. Osenberg for comments on the manuscript and T. Fan, R. Higashi, P. Krause, and C. Osenberg for allowing us to use their unpublished data. Support during the preparation of this manuscript was funded by the Mineral Management Service, U.S. Department of Interior under MMS Agreement No. 14-35-0001-3071, the UC Toxics Program and the National Science Foundation (OCE-9201682). The views and conclusions in this chapter are those of the authors and should not be interpreted as necessarily representing the official policies, either express or implied, of the U.S. Government.

References

Addison, R. F. 1989. Organochlorines and marine mammal reproduction. Canadian Journal of Fisheries and Aquatic Sciences **46**:360–368.

Allen, H. 1971. Effects of petroleum fractions on the early development of a sea urchin. Marine Pollution Bulletin **2**:138–140.

Anderson, D. W., J. R. Jehl, R. W. Risebrough, L. A. Woods, L. R. Deweese, and W. G. Edgecomb. 1975. Brown pelicans: improved reproduction off the southern California coast. Science **190**:806–808.

Anderson, S. S. 1978. Scaring seals by sound. Mammal Review **8**:19–24.

Atlas, R. M., A. Horowitz, and M. Busdosh. 1978. Prudhoe crude oil in Arctic marine ice, water and sediment ecosystems: degradation and interactions with microbial and benthic communities. Journal of Fisheries Research Board of Canada **35**:585–590.

Bakke, T., T. Dale, and T. F. Thingstad. 1982. Structural and metabolic responses of a subtidal sediment community to water extracts of oil. Netherlands Journal of Sea Research **16**:524–537.

Banner, A. H. 1974. Kaneohe Bay, Hawaii: urban pollution and a coral reef ecosystem. Pages 685–702 in A. M. Cameron, B. M. Campbell, A. B. Cribb, R. Endean, S. J. Jell, O. A. Jones, P. Mather and F. H. Talbot, editors. Proceedings of the 2nd international symposium on coral reefs., Brisbane, Australia.

Breitburg, D. L. 1992. Episodic hypoxia in Chesapeake Bay: interacting effects of recruitment, behavior, and physical disturbance. Ecological Monographs **62**:525–546.

Britt, J. O., and E. B. Howard. 1983. Tissue residues of selected environmental contaminants in marine mammals. in E. B. Howard, editor. Pathobiology of marine mammal diseases. CRC Press, Boca Raton, Florida.

Connell, J. H. 1961. The effects of competition, predation by *Thais lapillus*, and other factors on natural populations of the barnacle, *Balanus balanoides*. Ecological Monographs **31**:61–104.

Conner, P. M. 1972. Acute toxicity of heavy metals to some marine larvae. Marine Pollution Bulletin **3**:190–192.

Cowell, R. R., A. L. Mills, J. D. Walker, P. Garcia-Tello, and V. Campos-P. 1978. Microbial ecology studies of the Metula spill in the straits of Magellan. Journal of Fisheries Research Board of Canada **35**:573–580.

De Long, R. L., W. G. Gilmartin, and J. G. Simpson. 1973. Premature births in California sea lions: association with organochlorine pollutant residue levels. Science **181**:1168–1170.

Denny, M. W., and M. F. Shibata. 1989. Consequences of surf-zone turbulence for settlement and external fertilization. American Naturalist **134**:859–889.

Ebeling, A. W., D. R. Laur, and R. J. Rowley. 1985. Severe storm disturbances and the reversal of community structure in a southern California kelp forest. Marine Biology **84**:287–294.

Fan, T. W.-M., R. M. Higashi, G. N. Cherr, and M. C. Pillai. 1992. Use of noninvasive NMR spectroscopy and imaging for assessing produced water effects on mussel reproduction. Pages 403–414 in J. P. Ray and F. R. Engelhardt, editors. Produced water: technological/environmental issues and solutions. Plenum Press, New York, New York.

Gameson, A. C. H. 1975. Discharge of sewage from sea outfalls. Pergamon Press, Oxford, England.

Garshelias, D. L., and J. A. Garshelias. 1984. Movements and management of sea otters in Alaska. Journal of Wildlife Management 48:665–678.

Gaskin, D. E. 1982. The ecology of whales and dolphins. Heineman, Exeter, New Hampshire.

Gilpin, M. 1991. The genetic effective size of a metapopulation. Biological Journal of the Linnaean Society 42:165–175.

Higashi, R. M., G. N. Cherr, C. A. Bergens, and T. W.-M. Fan. 1992. An approach to toxicant isolation from a produced water source in the Santa Barbara Channel. Pages 223–233 in J. P. Ray and F. R. Engelhardt, editors. Produced water: technological/environmental issues and solutions. Plenum Press, New York, New York.

Hogan, J. W., and J. L. Brauhn. 1975. Abnormal rainbow trout fry from eggs containing high residues of PCB (Aroclor 1242). Prog Fish Culture 37:229–230.

Hose, J. E., T. D. King, K. E. Zerba, R. J. Stoffel, J. S. J. Stephens, and J. A. Dickinson. 1983. Does avoidance of chlorinated seawater protect fish against toxicity? Laboratory and field observations. Pages 967-982 in R. J. Jolley, editor. Water chlorination: environmental impact and health effects Vol 4(1), chemistry and water treatment. Ann Arbor Science Publications, Ann Arbor, Michigan.

Keough, M. J. 1988. Benthic populations: is recruitment limiting or just fashionable? Proceedings of the 6th International Coral Reefs Symposium 1:141–148.

Kobayashi, N. 1980. Comparative sensitivity of various developmental stages of sea urchins to some chemicals. Marine Biology 58:163–171.

Krause, P. R. 1993. Effects of produced water on reproduction and early life stages of the purple sea urchin (Strongylocentrotus purpuratus): field and laboratory tests. Ph.D. Dissertation. University of California Santa Barbara.

Krause, P. R., C. W. Osenberg, and R. J. Schmitt. 1992. Effects of produced water on early life stages of a sea urchin: state-specific responses and related expression. Pages 431–444 in J. P. Ray and F. R. Englehardt, editors. Produced water: technological/environmental issues and solutions. Plenum Press, New York, New York.

Linden, O. 1976. Effects of oil on the reproduction of the amphipods Gammarus oceanicus. Ambio 5:36–37.

Manahan, S. E. 1992. Toxicological chemistry. Lewis Publishers, Ann Arbor, Michigan.

Morse, D. E., and A. N. C. Morse. 1991. Enzymatic characterization of the morphogen recognized by Agaricia humulis (scleractinian coral) larvae. Biological Bulletin 181:104–122.

Müller, D. G., I. Maier, and G. Gassman. 1985. Survey on sexual pheromone specificity in Laminariales (Phaeophyceae). Phycologia 24:475–484.

Murdoch, W. W., R. C. Fay, and B. J. Mechalas. 1989. Final report of the Marine Review Committee to the California Coastal Commission. Marine Review Committee, Inc.

Olson, R. R., and M. H. Olson. 1989. Food limitation of planktotrophic marine invertebarate larvae: does it control recruitment success? Annual Review of Ecology and Systematics 20:225–247.

Osenberg, C. W., R. J. Schmitt, S. J. Holbrook, and D. Canestro. 1992. Spatial scale of ecological effects associated with an open coast discharge of produced water. Pages 387–402 in J. P. Ray and F. R. Englehardt, editors. Produced water: technological/environmental issues and solutions. Plenum Press, New York, New York.

Pawlik, J. R. 1992. Chemical ecology of the settlement of benthic marine invertebrates. Oceanography and Marine Biology Annual Review 30:273–335.

Pennington, J. T. 1985. The ecology of fertilization of echinoid eggs: the consequences of sperm dilution, adult aggregation, and synchronous spawning. Biological Bulletin 169:417–430.

Pulliam, H. R. 1988. Sources, sinks, and population regulation. American Naturalist 132:652–661.

Raimondi, P. T. 1990. Patterns, mechanisms, consequences of variability in settlement and recruitment of an intertidal barnacle. Ecological Monographs 60:283–309.

Raimondi, P. T., and R. J. Schmitt. 1992. Effects of produced water on settlement of larvae: field tests using red abalone. Pages 415–430 in J. P. Ray and F. R. Englehardt, editors. Produced water: technological/environmental issues and solutions. Plenum Press, New York, New York.

Rastetter, E. B., and W. J. Cooke. 1979. Responses of marine fouling communities to sewage abatement in Kaneohe Bay, Oahu, Hawaii. Marine Biology 53:271–280.

Ratcliffe, D. A. 1967. Decrease in egg shell weight in certain birds of prey. Nature 215:208–210.

Reed, D. C. 1990. The effects of variable settlement and early competition on patterns of kelp recruitment. Ecology 71:2286–2296.

Reed, D. C., R. J. Lewis, and M. Anghera. 1994. Effects of an open coast oil production outfall on patterns of giant kelp (Macrocystis pyrifera) recruitment. Marine Biology 120:25–31.

Reed, D. C., C. D. Amsler, and A. W. Ebeling. 1990. Dispersal in kelps: factors affecting spore swimming and competency. Ecology 73:1577–1585.

Richards, W. J., and K. C. Lindman. 1987. Recruitment dynamics of reef fishes: planktonic processes, settlement and demersal ecology and fisheries analysis. Bulletin of Marine Science 41:392–410.

Rosenberg, R. 1972. Effects of clorinated aliphatic hydrocarbons on larval and juvenile Balanus balanoides. Environmental Pollution 3:313–318.

Roughgarden, J., Y. Iwasa, and C. Baxter. 1985. Demographic theory for an open marine population with space-limited recruitment. Ecology 66:54–67.

Roughgarden, J., S. D. Gaines, and H. Possingham. 1988. Recruitment dynamics in complex life cycles. Science 241:1460–1466.

Sale, P. F. 1977. Maintenance of high diversity in coral reef fish communities. American Naturalist 111:337–359.

Scheltema, R. S. 1971. Larval dispersal as a means of genetic exchange between geographically separated populations of shallow-water benthic marine gastropods. Biological Bulletin 140:284–322.

Slatkin, M. 1987. Gene flow and the geographic structure of natural populations. Science 236:787–792.

Sokolov, V. E., A. Krivolutsky, A. I. Taskaev, and V. A. Shevchenko. 1990. Bioindications of the ecological after-effects of the Chernobyl nuclear power station accident in 1986-1987 as exemplified by terrestrial ecosystems. in D. A. Krivolutsky, editor. Bioindications of chemical and radioactive pollution. Mir Publishers, Moscow, Russia.

Soulé, M. E., and D. Simberloff. 1986. What do genetics and ecology tell us about the design of nature reserves? Biological Conservation 35:19–40.

Spellerberg, I. F. 1991. Monitoring ecological change. Cambridge University Press, Cambridge, England.

Spies, R. B., and P. H. Davis. 1979. The infaunal benthos of a natural oil seep in the Santa Barbara channel. Marine Biology 50:227–237.

Stouder, D. J. 1987. Effects of a severe-weather disturbance on foraging patterns within a California surfperch guild. Journal of Experimental Marine Biology and Ecology 114:73–84.

Systat, Inc. 1992. Evanston, Illinois.

Underwood, A. J., and E. J. Denley. 1984. Paradigms, explanations and generalizations in models for the structure of intertidal communities on rocky shores. Pages 151–180 in D. R. Strong, D. S. Simberloff, L. G. Abele and A. Thistle, editors. Ecological communities, conceptual issues and the evidence. Princeton University Press, Princeton, New Jersey.

Weis, J. S., and P. Weis. 1989. Effects of environmental pollutants on early fish development. Reviews in Aquatic Sciences 1:46–73.

Westernhagen, H. von, V. Dethlefsen, W. Ernest, U. Harms, and P. D. Hansen. 1981. Bioaccumulating substances and the reproductive success in Baltic flounder Platichthys flesus. Aquatic Toxicology 1:85–99.

Wilson, K. W. 1977. Acute toxicity of oil dispersants to marine fish larvae. Marine Biology 40:65–74.

CHAPTER 11

PREDICTING THE SCALE OF MARINE IMPACTS
Understanding Planktonic Links between Populations

Michael J. Keough and Kerry P. Black

The vast majority of marine organisms disperse through plankton at some stage of their lives, and this dispersal (or lack of it) divides species into a series of local or subpopulations, and provides links between those subpopulations. The spatial extent of those local populations and the rates at which they receive immigration from other local areas will influence long-term population dynamics. The identification of local stocks has been a long-standing focus of fisheries research (Sinclair 1988, Fogarty et al. 1991). Larval dispersal and its effects on population dynamics has also been a source of considerable debate in the mainstream ecological literature in the past decade (Underwood and Denley 1984, Keough 1988). Planktonic dispersal has been incorporated relatively recently into conceptual models used for environmental impact assessment (Fairweather 1991), and there is no reason to expect that it will be any less important for populations subject to anthropogenic interference than to "natural" ones.

Many anthropogenic effects on marine environments are localized in space (and sometimes time). For example, effluent discharges into receiving waters typically are through single points or relatively short diffusers, and construction activities tend to induce localized sedimentary changes. Impacts may have acute or chronic (sublethal) effects, both of which may alter the output of larvae from a local population. Acute effects may reduce the pool of reproducing adults, and alter the number of larvae produced. Chronic effects may influence the fecundity of organisms directly, or indirectly by reducing their growth rates. The important questions are whether any such changes in local larval output have local or regional consequences to population dynamics (Nisbet et al., Chapter 13).

This question is most topical in the large number of places in which toxicant concentrations, particularly in mussels and other bivalves, are used as indicators of ecological impact (National Research Council 1990b; Viarengo and Canesi

1991). In most cases, accumulation of toxicants is presumed to reflect declining health of individual organisms, which may in turn influence fecundity. It is then generally assumed that declines in health reflect actual or probable population level changes. For mussels, the link between individual health and population dynamics has rarely been demonstrated (Viarengo and Canesi 1991), yet mussels, especially *Mytilus edulis*, remain the most commonly used bioindicator world-wide (see also Jones and Kaly, Chapter 3).

Other "applied" problems involve links between local populations and regional events. Recent attempts at enhancement of marine fish stocks use the concept of Harvest Refugia (Russ and Alcala 1989, Alcala and Russ 1990), in which small reserves provide protection for large, fecund individuals of exploited species. Populations within refugia then export recruits, which increase fishery yields in exploited areas (e.g., Alcala and Russ 1990).

These examples all require some estimate of the exchange of propagules between populations. Traditional methods of measuring exchange use population genetic techniques, particularly electrophoresis, but these methods are directed at identifying levels of genetic change that control population differentiation, rather then ecologically meaningful rates. More direct methods, such as direct observation of dispersing larvae (Olson 1985, Young 1986, Davis and Butler 1989; see also review by Levin 1990), are not practical for most species. Recent marine studies have attempted to model the dispersal of larvae between populations, incorporating information about patterns of water movement and the biology of larval stages (Shanks 1983, Gaines and Roughgarden 1987, McShane et al. 1988, Black et al. 1991b). A separate area of vigorous discussion has concerned the relative roles of passive transport and active larval behavior in determining larval dispersal (Butman 1987, Mullineaux and Butman 1991). These studies, combined with the long history of investigation into the biology of propagules of marine organisms (reviewed by Young 1990), are making it possible to develop sophisticated models of larval transport.

We believe that these ideas have not been implemented yet in the assessment of impacts of human activity (henceforth EIA), and we describe our current understanding of dispersal processes, and explore information needs for particular EIA cases.

Predicting dispersal requires five sets of information:
  (i) knowledge of the relationship between the localized impact and the number of larvae exported from the local population;
 (ii) the numbers of propagules received from outside the region;
(iii) an understanding of the hydrodynamic processes that transport propagules;
(iv) knowledge of the relative contribution of larval biology (including behavior) to dispersal, and quantification of relevant aspects of that biology;

(v) an understanding of the relationship between fertilization, larval density, settlement, and recruitment/population dynamics.

The variability in the physical environment challenges our capacity to create a generalized framework to characterize larval dispersal. Tidal currents are often fast in a bay, especially near the entrance. Wind and wave-driven turbulence may reduce the potential importance of active larval behavior along an open shoreline, or in shallow depths. The vertical structure of the flow and the vertical and horizontal mixing will vary between regions and different techniques are often required for their treatment. In addition, a number of scales have to be treated, from the microscale (<10 m) to the horizontal regional scale (km). The bottom and surface boundary layers both present special cases, particularly if larvae are released within these layers.

The effects of localized impacts, particularly input of toxicants, have been described for many species and many kinds of impacts. The methodologies are well developed, and we will not discuss them in detail here. The relationship between larval density and population dynamics can be dealt with experimentally, and there is a substantial body of recent theoretical (e.g., Roughgarden et al. 1985, Menge and Sutherland 1987, Possingham and Roughgarden 1990) and empirical (Connell 1985, Roughgarden et al. 1985, Keough 1986, 1988, Davis 1987) studies attempting to solve this problem. We will therefore focus on issues (ii), (iii), and (iv)—larval transport processes and larval biology.

Larval Transport Processes

Larval dispersal occurs in a wide range of environments, ranging from exposed ocean coasts to sheltered bays. It is important to know whether larval transport processes are specific to individual environments, or if the same general processes apply everywhere. There have been substantial recent advances in understanding how large-scale oceanographic processes influence larval transport (Kingsford 1990, Farrell et al. 1991, Kingsford et al. 1991). These processes often operate at spatial scales of kilometers. In most EIA activity, the spatial scale of the impact is smaller; even for relatively large impacts, Control and Impact sites are often separated by less than 10 km (e.g., Murdoch et al. 1989, National Research Council 1990a, 1990b). Models are less well developed at this and smaller spatial scales, but we believe that recent developments have considerable potential.

We will discuss models of transport processes from four different areas, which represent a broad cross-section of the world's marine environments. The individual cases also vary in the degree of complexity that is necessary to produce accurate descriptions of patterns of water movement and in the nature of the major influences on that movement. Our aim here is to provide broad descriptions of the state of development, complexity, and some results from the models developed for each of these examples. Specifically, we will consider processes in four main regions where larval transport occurs (Figure 11.1).

These are:
- coral reefs—on the offshore edge of the continental shelf in the western Pacific Ocean
- an ocean strait—linking the Tasman Sea to the Southern Ocean
- a continental shelf—on an exposed, high-energy coastline in south-eastern Australia
- a more sheltered bay—linked to the ocean strait by a narrow entrance

The models developed for circulation in the examples detailed below involve a series of partial differential transport equations, which incorporate information about boundary conditions, bathymetry, coastal currents, benthic topography, prevailing winds and the currents that they drive, and Coriolis effects. The general structure of these models has been described in detail elsewhere by Black (e.g., Black et al. 1991b). In all cases, the same model code was applied. The bathymetry and boundary conditions only were altered.

The numerical models follow the paths taken by particles through a spatial grid, so they are particularly useful for examining larval dispersal. Supported by a prior analysis of field measurements of circulation to understand the regional hydrodynamics, the models are then tested using these current and sea level measurements for confirmation of model behavior. The calibration of the hydro-dynamics is usually a much more tenable problem than' calibration of the dispersal.

Propagules, toxicants and other materials are transported by two processes: advection and diffusion. The advection is normally handled deterministically using either current meter measurements or the predictions of the models, while a probability function describes the diffusion. Shear dispersion arises when ver-tical or horizontal variations in currents cause the material in a patch to diverge or converge. Being an advection process, the shear dispersion is not related to random turbulence. A characteristic of the environments to be described below is the tendency for shear dispersion to mostly govern the dispersal of larvae.

Oscillating wave or tidal currents can induce a net circulation due to topo-graphic effects or Stokes' drift. However, their most important role relates to the creation of extra turbulence. Notably, current mixing depends most strongly on depth, currents, and magnitude of bed roughness, while the wave-induced mixing depends strongly on the water depth and wavelength. Short period waves in deep water may be inconsequential except in the surface boundary layer, when the waves are breaking.

Figure 11.1. Study regions on the east and southeast coast of Australia. Key areas are the Great Barrier Reef, Bass Strait, southeast Gippsland and Port Phillip Bay.

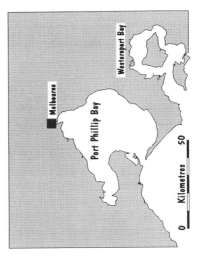

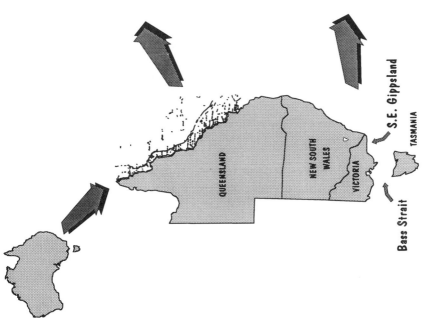

John Brewer Reef

Reef edge into deep water

Reef

Lagoon

Melbourne

Port Phillip Bay

Westernport Bay

Kilometres

0   50

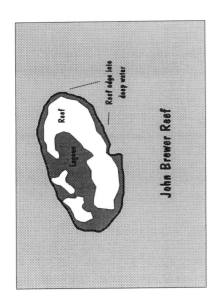

QUEENSLAND

NEW SOUTH WALES

VICTORIA

TASMANIA

S.E. Gippsland

Bass Strait

In the following sections, we use examples from our work on the numerical simulation of marine physical systems undertaken for scientific and applied applications. We draw on a series of references by Black and co-authors to illustrate the important processes for larval dispersal and the types of investigations done for an EIA. We do not intend to review the oceanography, but use these case studies to illustrate the oceanographic processes that repeatedly play a role.

Great Barrier Reef

Some of the best understood water movements are those around various reefs on the Great Barrier Reef, which have been described at two spatial scales: one concerned with mesoscale effects, and the other at the level of single reefs or groups of reefs. This latter scale is most relevant to the spatial scale on which most EIA occurs. These models have been used to understand the recruitment or larval dispersal of primarily corals and echinoderms. The Great Barrier Reef is adjacent to the western boundary of the Pacific Ocean. Currents that travel across much of the Pacific and throughout the Coral Sea, local wind-driven flows and tidal circulation, occur simultaneously (Burrage et al. 1991). Success with the models cannot be achieved without the prerequisite of a detailed data acquisition and analysis program to be used to define the boundary conditions. In general, a model provides a convenient means to unify and expand on these existing measurements. The characteristics of the systems are:

- Propagules considered thus far have weak locomotory ability - coral eggs, planulae, and Crown-of-Thorns larvae (but see Kingsford et al. 1991)
- Strong tidal currents, up to 50 cm/s, but averaging around 20 cm/s
- Longshore currents (East Australian Current and wind forcing combined), commonly around 20 cm/s but averaging about 10 cm/s
- Reefs are surrounded by relatively deep water, presenting a bluff body on the scale of kilometers
- There is relatively little vertical stratification of the water column, although subsurface intrusions of Coral Sea water have been recorded (Andrews and Furnas 1986), and a thermocline occurs in some reef lagoons in summer
- Tides, local wind, and the East Australian Current account for greater than 90% of current strength

The models have led to some important conclusions:

(1) Larval retention around reefs is quite possible, even with high tidal flows and larvae with a long planktonic period (Black et al. 1990). Simulations suggest that larvae could easily be retained for 10 days around individual reefs. Retention is produced by the recirculating tidal currents and eddies around reefs. The examples in Figure 11.2 are taken from John Brewer Reef on the central Great Barrier Reef, and they show tidal flows through half of a tidal cycle. At slack water, most velocities are <10 cm/s, but toward midtide, velocities away from the reef reach 30 cm/s, and large eddies develop along the side of the reef. Velocities within

these eddies are <10 cm/s, and there is no net movement of water. The eddies disappear toward slack water, and currents away from the reef reverse as the tide turns. At peak flows again, similar eddies redevelop, this time on the opposite side of the reef. At all times, velocities within the main lagoon remain low.

The eddies can generate and maintain larval patchiness (Black 1993). On a within-reef scale, some areas will retain more larvae or toxicants than others. In the example from John Brewer Reef, locations on the east and west sides of the reef are typically low retention, while the areas in which the eddies form are high retention zones. The extent of retention seems to be surprising, and may lead to an increased focus on processes that produce larval retention. The attractive feature of these models is that they can make quantitative predictions about the amount of larval retention, and if neighboring sources of larvae are included (e.g., Figure 11.3), can predict the relative inputs from local and foreign populations (Black 1993). An understanding of the oceanography can also be useful for suggesting appropriate locations for discharges, notably areas of rapid diffusion and low retention.

(2) The patterns predicted here do not depend greatly on the shape of the reefs—they are more a consequence of reefs that slope steeply upward from 40 to 50 m deep to the intertidal. This model could therefore be adapted readily for modeling transport around any new Impact reef.

(3) Within the lagoon, even more retention of particles is possible. Particles can move around the lagoon several times before eventually moving out of the area. Breaking waves on the reef flat generate currents that contribute to within-lagoon circulation, and can significantly increase retention times for larvae.

(4) Two- and three-dimensional versions of the model have been developed, with the aim of comparing their predictions (Black et al. 1991a). In general terms, the models do not differ in their predictions for vertically well-mixed larvae. However, as we note below, this finding is not relevant to exposed coasts near the shoreline or bays and estuaries.

(5) The predictions made by this model have been tested by contrasting predictions to actual propagule transport or recruitment. At Bowden Reef, the model predicted various localitions of high and low larval concentrations. For four such areas, there are measurements of abundance of coral larvae (Willis and Oliver 1988). There was good agreement, with the rank order of sites as predicted by the model.

This model was also used to predict areas of larval retention for Crown-of-Thorns larvae (Black and Moran 1991). Subsequent field surveys were used to identify areas of high recruitment and primary outbreaks of Crown-of-Thorns. All locations of primary outbreaks were in areas of predicted high larval retention, and no area of predicted low advection had an outbreak. Sammarco and Andrews (1988) also found a correspondence between coral recruitment and circulation patterns around a single reef, using the same model and modeling technique.

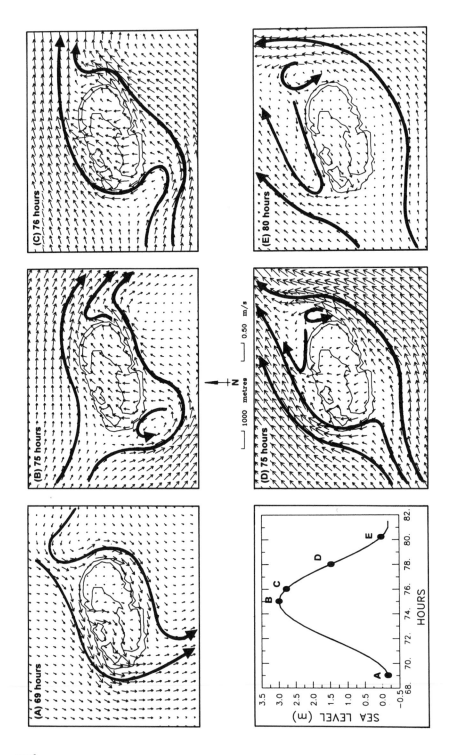

The models also have limitations:

(1) Even though the Great Barrier Reef has been the subject of some of the most detailed studies, current reef-scale models incorporate mainly unstratified waters. If there are situations in which stratification is likely to be important, additional hydrodynamic information is required. One situation where stratification is likely is at point discharges that involve water at temperatures or salinities different from those of receiving waters.

(2) The models discussed above use grid sizes ranging from 200 to 2000 m (Black 1993). There are very few models to work on smaller spatial scales, although this situation is being remedied with some current measurements and modeling at spatial scales of tens of meters (Eckman 1990, Benzie, Moran, and Black, personal communication). The greater the importance of local retention, the greater the need for small-scale models.

(3) In most cases, time-averaged values are used for wind-driven currents, principally to simplify the system so that a better understanding of the processes can be achieved. When local weather conditions vary at times of larval release, the generalized model is not appropriate. Variation in wind strengths will cause local currents to vary, and consequently the relative contributions of local versus foreign larvae will vary. More detailed modeling is used in these cases (Black et al. 1991b, Gay et al. 1991).

(4) There is a trade-off between the size of the grids into which the area is divided and the details of transport. Large grids mean that less calibration is required, but these models produce less detailed structure, and cannot accurately predict some of the major circulatory phenomena, especially the tidal eddies. In the case of reefs on the Great Barrier Reef, the model has been run with the size of each grid cell being 200, 500, 2000, and 4000 m, for a reef approximately 4000×4000 m. The eddies are predicted only when there are at least 20 cells over the reef, and, more importantly for dispersal, the coarse grid model (4000 m) predicted that 0.01% of larvae would be retained after 10 days, while 6% remained when the grid size was 200 m.

Extrapolation to Other Geographic Areas

One of the things making this modeling so successful at the mesoscale is the result that shapes of reefs on the Great Barrier Reef have little effect on those flow patterns that determine reef-scale larval retention. The reef waters experience a high level of mixing due to shear dispersion. Thus, the simulation of larval dynamics near the reef is simplified, as random turbulence and reef shape play a

---

Figure 11.2. Circulation patterns around John Brewer Reef on Australia's Great Barrier Reef. The currents are shown at selected times over the tidal cycle. The bold lines show the flow stream-lines. The lower left panel shows a tidal cycle, with the points on the curve indicating the corresponding circulation pattern.

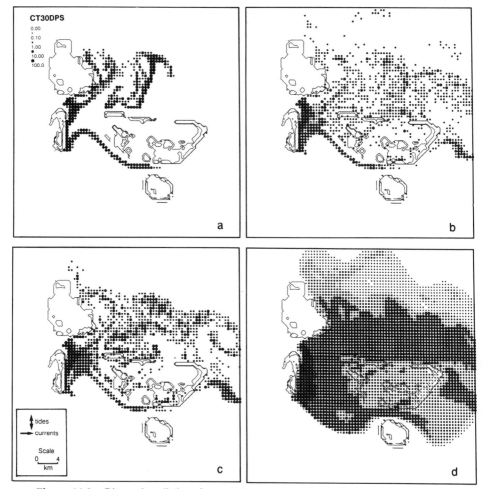

Figure 11.3. Dispersal predictions from a point source near Davies Reef on the Great Barrier Reef. The particles spread throughout the reef group within 100 h. The last panel shows the "integrated abundances" after 400 h, i.e., the positions visited by particles, with the number of visits being proportional to the shading intensity.

secondary role (Black et al. 1990). Most temperate studies generally deal with nearshore coastal environments, which are more gently sloping, and have irregular reefs, headlands, and occasional offshore islands. These areas often have coverings of macrophytes, which may effectively divide the area into two quite different hydrodynamic regimes. Similar variations also occur in coral reef benthic structure at the microscale. The boundary layer thus formed is potentially very important in all environments (e.g., Eckman and Duggins 1991). It may also vary between habitats, depending on the morphology of the particular macrophytes. Smaller kelps, such as *Ecklonia* in southern Australia, or *Eisenia* in North America, will have dramatically different effects from very large plants, such as *Macrocystis* or *Durvillea*.

Bass Strait

Bass Strait is a large water body between 60 and 80 m deep. Deep ocean to the east and west, and land to the north and south form natural boundaries (Figure 11.1). Its hydrodynamic behavior can be likened to a semi-enclosed water body on a large scale (Black 1992). Its behavior is also influenced by the dynamics of the adjacent Southern Ocean and Tasman Sea, so a much larger physical region than just Bass Strait must be considered. The area is also characterized by frequent strong winds. In this case, the physical size and spatial and seasonal variability are major concerns for an EIA.

The area has fast tidal currents at the east and west of the Strait, which fall by an order of magnitude in the central regions. Additional variations in wave action, wind curl, and remote forcing (e.g., Middleton and Viera 1991) have meant that useful models have taken several years to establish. This time scale may prove difficult for many EIA procedures, which often have little lead time.

The first hydrodynamic models were established in the early 1980s after dedicated data acquisition programs were established to measure currents and sea levels, particularly across the western entrance (Fandry et al. 1985). A number of isolated small studies and three large programs have been conducted more recently: a fisheries application (Craig et al. personal communication), an effluent dispersal assessment (Black et al. 1993), and a research study of coastal-trapped waves (Middleton, Black and Symonds, personal communication). As part of the investigation of coastal-trapped waves, integrated unsteady models of the Strait in two and three dimensions were established (Middleton and Black 1994) to supplement prior tidal and wind-driven simulations (Black 1992). The models currently in use represent the cumulative effect of a number of independent studies. They have produced a number of important results:

1. Two-dimensional models do not produce an accurate description of the dynamics. Three-dimensional models of the Strait show wind-driven circulation varying significantly with distance below the surface (Figure 11.4). Because wind acts at the surface only, wind forcing always creates a complex three-dimensional structure. In a region where prevailing winds strongly favor an on/offshore direction, the coastline can be characterized by almost constant upwelling or downwelling, and such phenomena can have strong influences on larval transport (Farrell et al. 1991).

Velocity shear between layers will spread material residing throughout the depth over a wide area; incorporation of larval behavior into the model is essential. Variations in buoyancy or active swimming behavior over long time periods will significantly alter study outcomes.

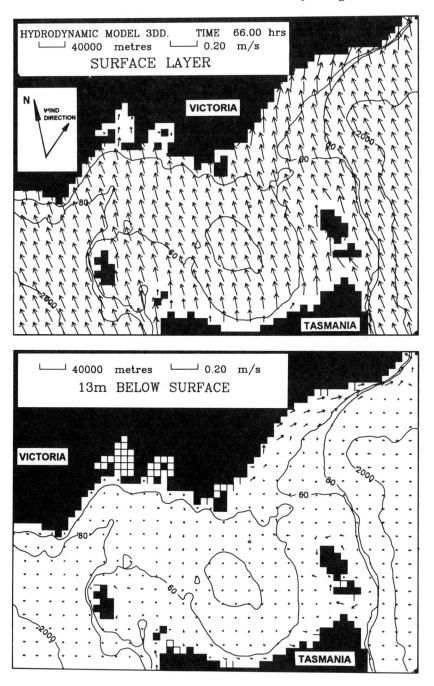

Figure 11.4. Wind-driven circulation in Bass Strait for an 8 m/s southwest wind. The upper panel shows the surface currents, while the lower panel shows currents at 13 m depth.

2. Spatial variation in vertical shear may mean that the relative importance of hydrodynamics and larval behavior is not consistent through the whole region. Vertical shear and turbulence levels in strong tidal currents near the entrances to Bass Strait will be much greater than in slower central regions. Jonsson et al.'s (1991) experiments indicated that swimming ability of larvae is impaired above a threshold shear stress, due to a torque across the organism in a shearing current. In such a case, EIA studies near the center of Bass Strait would require different biological data from those done at either end.

Horizontal shear is also common. The low frequency circulation under coastal-trapped waves and wind forcing can be spatially complex (Figure 11.5). Eddies form due to the combined effects of topography and unsteadiness of the forcing. Adjacent clouds of larval material may travel in quite different directions in these conditions during the 10–14 d cycles associated with these low-frequency phenomena.

3. In locations close to the shore, exposure to wave attack must also be considered. Deep ocean swell can still be "felt" on the seafloor at around 40 m depth,

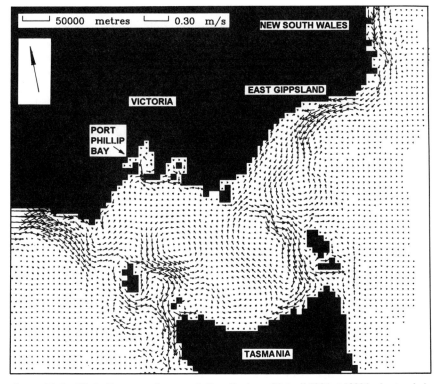

Figure 11.5. Vertically-averaged currents in Bass Strait, on 23 April 1984 at 1200 h, due to winds and coastal-trapped waves.

although the actual depth depends on the size and wavelength of the swell. This depth is reflected in the mud depositional limit, which occurs on the exposed west coast of Tasmania at about 45 m (Jones and Davies 1983) and 42 m in eastern Bass Strait (Black et al. 1993). This compares with about 50 m off the Washington and Oregon shelves (Smith and Hopkins 1972). The boundary is reflected by changes to the character of the benthos, the nature of the sea bed sediments and their mobility, with resulting consequences for the suitability of the substrate for recruitment.

4. The degree of vertical mixing of the water column depends strongly on local weather conditions. In the water column, very high vertical eddy diffusivities of up to $E_z=0.3$ m²/s for storm wind speeds of 20 m/s are possible (Davies 1985). Vertical mixing would occur within minutes in shallow water and within hours in Bass Strait depths in these conditions. In calm conditions, eddy diffusivities may be 3 orders of magnitude less and mixing will occur very slowly. Temperature stratification may become established, creating a two-layer system with very little exchange between the layers (Munk and Anderson, 1948). Again, well-mixed and temperature-stratified waters require very different models.

There is always a temptation to "simplify" the problem or the system for an EIA. The results, however, may be misleading. Early studies of Bass Strait, for example, treated tides and local winds only. More recent investigations of coastal-trapped waves have confirmed that much of the circulation in the strait is related to these remotely forced phenomena. Coastal-trapped waves entering the strait from the west excite currents throughout Bass Strait and up the east coast of Australia (Figure 11.5). With 10–14 d periods, these low frequency events may influence pelagic dispersal of many species more than the local wind or tidal conditions in some locations. To model all of Bass Strait and incorporate the vertical mixing requires an unusually large data set incorporating factors such as density structure, wave action, sea levels, and wind strength, all at a number of locations. Much of the necessary data has been collected, although further research is needed to attempt to more effectively link the individual physical components into a fully unified model.

Continental Shelf

Continental shelf models of the hydrodynamics and dispersal were initially developed for an EIA of the likely dispersal of effluent from a proposed pulp mill. However, the open coast physical environment was challenging and the program was extended to incorporate a research study of wave/current interaction on natural shorelines (Black, Symonds, Simons, Pattiaratchi and Nielsen, personal communication). By combining these two large investigations, fourteen current meters deployed through the water column at three sites and nine tide gauges deployed at five sites recorded over two 6-week periods at an exposed coast in southeastern Australia (Figure 11.1). A detailed bed sediment survey and *in situ* measurements of suspended sediment on the continental shelf were made (Black et al. 1993).

Both two- and three-dimensional numerical models of this open coast environment were established to compare the predictions. They were calibrated against the current meter and sea level measurements. The tidal currents were slow in this region (<10 cm/s); the currents primarily responsible for the dispersal of water-borne material were driven by local winds. The three-dimensional model could treat coastal upwelling and downwelling and unsteady wind and tide conditions.

Three factors distinguished the three-dimensional model from its less sophisticated counterpart. In the three-dimensional model, surface water brought towards the coast by onshore winds downwelled and flowed offshore at the bed (Figure 11.4). Second, the up/downwelling associated with the Ekman spiral had a small significant impact on the outcomes near the coast. Third, vertical shear was properly represented in the three-dimensional model. All of these factors are advection phenomena and combined to significantly increase the level of dispersal without altering the eddy diffusivity due to random turbulence. Consequently, the dispersal patterns were very different in the two models. Material spread over a far wider region in the three-dimensional model (Figure 11.6). Notably, "average" conditions would not accurately represent the dispersal resulting from variation in the weather and current patterns in this case, and short-term variation in these phenomena may produce rapid alterations in dispersal profiles.

The spatial scale of the likely dispersal at this open coast site was very large. Longshore currents of up to 0.8 m/s were recorded and 0.3 m/s was common. The latter current represents an excursion of 25 km/day or 180 km/week. In winter, the net currents were directed to the east for more than 96% of the 6-week deployment. The excursion associated with this flow was about 544 km. Thus, an EIA would need to consider a very broad regional scale to treat the biological implications. This scale may be too large for an accurate assessment. In an EIA, the two-dimensional model could lead to positioning of Control sites well within the "Impact" zone, in the mistaken belief that larvae and effluents would not be transported to these Controls. The hydrodynamic studies indicate that the physical scale of the potential larval inputs is large in this case.

Port Phillip Bay

Port Phillip Bay in southern Australia (Figure 11.1) is an interesting environment and an important case to consider. This bay, like others around the world, responds to a wide variety of physical processes. Some events enter through the entrance which is linked to Bass Strait and which, in turn, is subject to ocean-scale circulation. Other responses are local, such as the effect of local winds or internal temperature and salinity gradients.

Australia's second largest city of Melbourne lies on Port Phillip Bay's northern shores. A number of environmental issues were being raised, each of

a

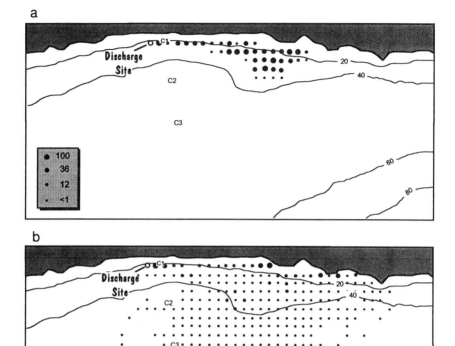

b

Figure 11.6.   Dispersal patterns from a point source at C1 in east Gippsland, predicted by the (a) two-dimensional model (upper panel) and the (b) three-dimensional model (averaged through depth; lower panel). Sizes of symbols are proportional to the concentration of particles.

which could be answered with an accurate model, so a three-year program of measurements was established to record currents and sea levels throughout the bay. Some 44 deployments of instruments, each for 6 weeks, were made (Rosenberg et al. 1992). This program depicted the circulation in the bay and provided the necessary data for establishment of two- and three-dimensional models.

The calibration in two dimensions centered on the determination of bed friction and the need to identify the appropriate local winds to represent the wind over water. In three dimensions, the value of the vertical eddy viscosity was the primary additional input and field measurements of currents through the vertical dimension were needed to estimate this parameter. Further work is being undertaken to refine these estimates. The winds at this latitude are variable, rotating anticlockwise. Winds from the northwest quadrant are common in winter, while southerlies predominate in summer. The former are offshore along the Bass Strait shoreline near the entrance to the Bay (see Figure 11.1).

Tidal currents are rapid near the entrance, but these reduce quickly inside the bay where the dispersal is mostly governed by wind forcing (Black et al. 1992). The strong longshore coastal currents present in the nearby Strait do not occur in the bay because of its enclosed nature. Instead, the bay is characterized by a highly three-dimensional wind-driven circulation pattern dominated by vertical shear. Net circulation is therefore a function of the wind, which disperses and mixes material throughout the bay.

The three-dimensional model of Port Phillip Bay showed that the heading and magnitude of currents varied with depth at a particular location (Figure 11.7). Thus, larvae with different vertical positions were expected to travel in different directions. Applying this model to a toxicant input in the southwest part of the Bay showed material heading both upwind and downwind during a severe storm. After one day, material initially released on a linear transect was found throughout much of the northern bay (Black et al. 1992). Notably, the material was initially and subsequently vertically well mixed. The observed spread in the model was due to the advection differences through the vertical layers and was not associated with density stratification, although the latter would enhance the tendency for material in the upper and lower layers to separate.

The movement of the center of mass of the cloud followed the track depicted by the two-dimensional model, i.e., downwind along the bay margins and upwind through the central deeper regions (Figure 11.6). However, the three-dimensional model, which accounts for shear between layers of material, showed a much greater spread. A simple parameterization of this difference would be very difficult to achieve. The effective horizontal eddy diffusivity associated with the spreading material was found to be about 600 m^2/s, but this decreased again after the land restrictions became effective.

The outcomes of these sorts of simulations will be different if larvae mix vertically quickly or remain at a particular level. Where shear through the vertical is prevalent, the time scale of vertical mixing is a fundamental parameter which establishes the rate and spatial extent of the dispersal. The shear dispersion in storm conditions monopolized the random diffusion. In calm conditions, random diffusion or vertical stratification can dictate the outcomes. Larval behavior would also need to be considered.

Near the entrance, low frequency oscillations (10–14 d period) coming from Bass Strait shift the balance of the currents, such that the net current is directed into the bay over one half of the 10–14 d cycle and out of the bay during the other half. This may result in pulses of recruitment coming from Bass Strait (Jenkins and May 1994). Coastal up/downwelling at the entrance, which will alternate in sequence with the rotating wind directions at these latitudes, may also alter the quantities of available larvae at the entrance in a cyclic way. Similarly, buoyant

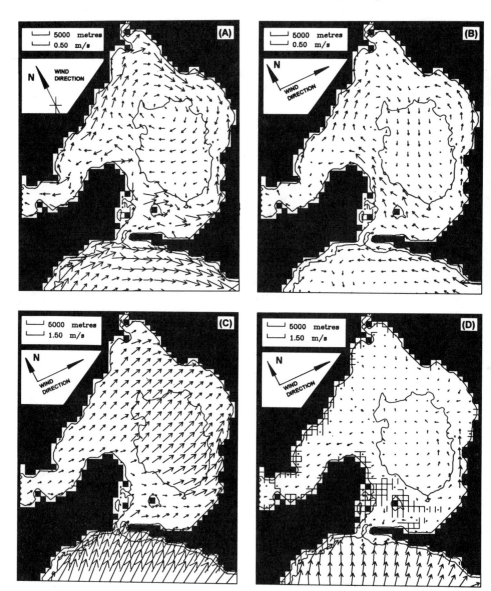

Figure 11.7. Wind-driven currents in Port Phillip Bay. The Bay is saucer-shaped in the central regions, reaching 24 m depth. The integrated velocity through all the layers matches the currents predicted by the two-dimensional model. (A) Vertically-averaged current for a 16 m/s south wind. (B) Vertically-averaged current for a 22 m/s west wind. (C) Near-surface currents for a 22 m/s west wind. (D) Current in the layer 4.5 to 7.5 m deep for a 22 m/s west wind.

larvae will be swept shoreward during onshore winds which are more common in summer. The current direction along the coast near the entrance will also change with the wind, or due to coastal-trapped waves. Cyclic pulses of larval input could also be produced by these processes if the larvae come preferentially from one side of the entrance, or if there is any periodicity in spawning.

We can identify other, indirect consequences of these circulation patterns that may be relevant to larval transport. The nature of the sea bed in the bay reflects these hydrodynamic processes: sandy beds predominate around the margins where wave action dominates, while muds are common in deeper or more sheltered waters. Medium sands and "clean" shell predominate in the highly mobile beds under the fast tidal currents near the entrance. In an estuary, Black et al. (1989) identified a strong correspondence between the net currents and sea bed type. Shell lag predominated where net currents were accelerating, causing the bed to be depleted of readily moved material. The shells had a much higher entrainment threshold with the result that only an armor of shell detritus remained in the high flows. In general, the type of substrate and therefore the suitability for settlement are a function of the current speeds and the availability of sediments (Black 1992).

Many of the laboratory experiments on larval settlement have been conducted in relatively slow flows of 0.05 to 0.1 m/s (Jonsson et al. 1991, Mullineaux and Butman 1991). The currents in the cases cited above were commonly 0.2 m/s or more in the water column, although the currents very near the bed will be slower and the sheltering offered for settlement will increase with bed roughness (Eckman 1990). Medium noncohesive sands will be transported along the bed once the current at 1 m above the bottom exceeds about 0.35 m/s. Thus, the sea bed sediments may be regularly moving at many sites. Velocities above the bottom may be high enough to modify the behavior of larvae and unstable sediments may have important effects on settlement.

Any modification of circulation patterns by construction activities may transport larvae to unsuitable settlement habitats or alter an existing habitat, rendering it unsuitable for settlement. Similarly, if a discharge alters the biology of propagules, their modified dispersal may transport them to habitats with substrates that have different suitabilities for settlement and early survival.

Larval Attributes Contributing to Dispersal

Most invertebrate (and vertebrate) larvae are not capable of swimming very fast, with swimming speeds typically at least one order of magnitude below even weak tidal currents or wind-driven circulation (Chia et al. 1984, Keough 1988). Even relatively fast-swimming larvae, such as late larval stages of some fish, may only be able to swim in weak currents. Active swimming by larvae may be most important for adjustment of their vertical position, allowing them to move between bodies of water traveling at different velocities, or to investigate settlement sites once they approach suitable habitat. Recent literature, reviewed by Butman (1987), has focused on whether larvae act as passive particles or whether their swimming contributes to dispersal. Butman considered that larval swimming was potentially important. In a later paper, Mullineaux and Butman (1991) have also suggested that transport of larvae to settlement sites may be a passive process, but that behavioral responses are important for choosing settlement sites. Black et al. (1991b) have also questioned whether weakly-swimming larvae, such as those of *Acanthaster planci*, are capable of maintaining or changing their vertical positions in the water column in moderate currents. This area of research remains an area of some confusion, with a shortage of good field data.

In the following section, we use examples taken from our own work on larval bryozoans to illustrate components of larval biology that are potentially important for dispersal, and the kinds of data that may be obtained. We do not intend to review comprehensively the larval biology of marine invertebrates; that topic has been dealt with elsewhere in great detail (e.g., Thorson 1950, Scheltema 1974, Crisp 1976).

Morphology (Passive Attributes)

Larvae vary in their buoyancy, ranging, for example, from positively buoyant coral eggs and planulae to negatively buoyant cyprids (e.g., Chia et al. 1984, Willis and Oliver 1988). Some propagules undergo changes in buoyancy through their lifetime. For example, coral eggs may be initially positively buoyant, but change to negative buoyancy as larvae. Passive changes in buoyancy are likely to be important. Vertically stratified flow, seasonal temperature changes, freshwater run off, and coastal upwelling can all contribute to variations in the density of the medium. However, the densities of the larvae are expected to change more significantly, at least over the spatial and temporal scales of the planktonic lives of the larvae.

Within a species, there may also be considerable variation in sinking rate. Figure 11.8 shows the distribution for the bryozoan *Bugula neritina*. For these approximately spherical, heavily ciliated larvae, there was a threefold range in sinking rate. Larvae of the same species swim at rates of approximately 1 mm/s in the absence of a strong cue, and reach maximum speeds of 3 mm/s when swimming towards a strong cue, light (Boxshall personal communication). This variation in sinking and swimming rates is potentially as important as ontogenetic variation.

Duration of the Planktonic Phase

The duration of the planktonic period has an obvious relationship with hydro-dynamic processes: if the planktonic period is very short, there can be little dispersal. Marine animals and plants vary widely in the duration of their (usually) planktonic dispersal stage, ranging from as much as a year (Scheltema 1971, Scheltema and Williams 1983) to as little as a few minutes (Olson 1985, Davis and Butler 1989). Some species bypass the planktonic period completely.

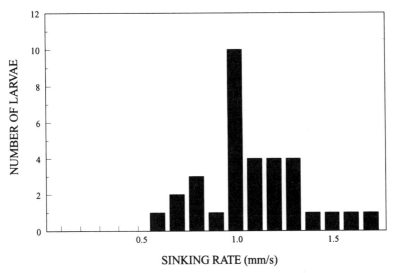

Figure 11.8. Sinking rates of freshly-killed larvae of the bryozoan *Bugula neritina*. Rates were measured in a 25 mm-diameter cylinder. All larvae were obtained from a single batch of colonies, which were kept in the dark for > 12 h, then exposed to strong light.

The duration of the planktonic period, coupled with numerical modeling of the currents, can provide an estimate of the maximum or potential dispersal of propagules (see also Raimondi and Reed, Chapter 10).

In species with propagules that undergo planktonic development, there is generally a time during which they are not competent to settle and metamorphose. For example, in red abalone, *Haliotis rufescens*, the precompetent period lasts 5–7 days, and the competent period for 2–5 weeks (Morse 1990). The length of the precompetent period provides a minimum estimate of the duration of the planktonic phase. After reaching competency, larvae settle in response to a variety of cues, but most larvae are capable of deferring settlement in the absence of such cues. In some cases, larvae may become less selective as the length of this delay increases (Knight-Jones 1953; see also review by Pechenik 1990). Larvae may or may not continue to feed after the onset of competency.

Larvae of species lacking planktonic development are capable of settling very soon after they are released. They, too, settle in response to various cues, and are capable of delaying settlement. Lacking feeding apparati, their capacity to delay metamorphosis may be limited by physiological reserves. Some larvae, however, are capable of taking up dissolved organic material directly from seawater (Manahan 1986), providing them with a means to extend their planktonic period.

The simplest estimate of the duration of the planktonic phase comes from laboratory experiments depriving larvae of settlement cues. Larvae are observed until they settle or die. Even under these controlled laboratory conditions, however, larval swimming times are highly variable. We have examined the settlement behavior of various species of bryozoans, which produce nonfeeding planktonic larvae. In the laboratory, larvae are typically released when colonies are exposed to strong light after a period of darkness. Settlement occurs when the larvae attach themselves to a substrate and begin to metamorphose. In most species that we have studied, there is little settlement soon after release, then a peak, followed by a long tail (Figure 11.9; see also Keough 1989, Keough and Klemke 1991). The example shows also that in seven populations with very different mean duration of the planktonic phase, the level of variability was similar. In other nonfeeding larvae, the peak in settlement may occur within 2 days of release, with the tail extending at least to 17 d (Keough and Klemke 1991). In all cases, larvae settling at the extreme times were still capable of metamorphosing into apparently healthy juveniles.

There is no reason to suspect that larvae of species producing feeding larvae will show any less variation. Various studies have shown that it is possible for these species to delay metamorphosis by over a year, although laboratory-based delays of 15 –50 d are more common (reviewed by Pechenik 1990). Variation over this time scale may have profound effects on passive transport.

The other important feature is that there may be considerable intraspecific variation in planktonic periods (Raimondi and Keough 1990). In *Bugula neritina*, a species that has been well studied, the peak in settlement may vary dramatically between populations (Figure 11.9 and see Raimondi and Keough 1990). In conditions of rapid water flow, even a difference between 1 and 6 h may be important. This variation has important practical consequences, requiring that estimates of planktonic duration be obtained for local populations, rather than simply using values from other literature.

While attention has been focused recently on the potential for passage through toxicant plumes to induce deformities (Kingsford and Gray, Chapter 12) or interfere with settlement responses of propagules (Raimondi and Schmitt 1992), it is also possible that toxicants could modify the duration of the planktonic period.

Larval Behavior Contributing to Transport

Most larvae respond during their swimming period to some stimuli that might influence their dispersal. The most obvious responses are those to light or gravity, both of which could result in larvae moving between strata of different velocities in the water column. Some organisms show such behavior for much of their larval period (reviewed for crustaceans by Cronin and Forward 1986). Many species also show ontogenetic changes in behavior through their larval period. For example, larvae of most bryozoans are positively phototactic when they are released, and many subsequently become negatively phototactic (Ryland 1974, Reed 1991). The timing of that change may have important implications for dispersal. Larvae of one bryozoan, *Mucropetraliella ellerii* settled almost exclusively in the dark (Klemke, personal communication), but over their

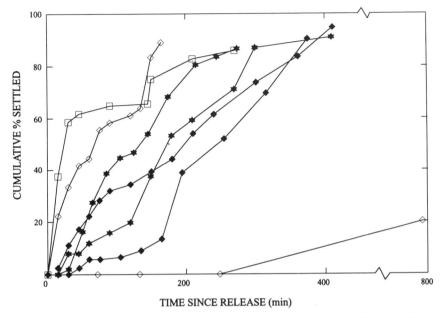

Figure 11.9.   Duration of the planktonic phase in the bryozoan *Bugula neritina*. The graph show cumulative frequency distributions of swimming time under laboratory conditions for larvae from seven different populations. Solid symbols represent Australian populations: Wollongong (NSW), Williamstown and Mornington (Vic), and Pt. Turton (SA). The open symbols represent U.S. populations. The two left curves, connected by solid lines represent two north Florida populations, and the right curve is a Santa Catalina Island (Cal) population. In each case, larvae were observed in small plastic containers, under uniform dim lighting at 22–25 °C. Sample sizes were generally 100.

planktonic period move between light and dark areas (Keough and Klemke 1991). Some larvae became negatively phototactic within one hour of release and remained that way. Others were positively phototactic for most of the time, becoming negative just before settlement. Other larvae changed taxis 3 times, moving into the dark, out into the light, then back into the dark again. A third alternative is that larvae may show behavioral responses that result in their remaining in various water bodies, such as thermoclines or fronts (Kingsford 1990).

Larval responses to settlement cues may also shorten the planktonic period substantially. These may be particularly important if larvae are swept away by currents, but turbulent motion brings them into contact with the substrate. If the cues present on the substrate are favorable, they may settle immediately. If cues are unfavorable, they may be resuspended in the water column. *Bugula neritina* larvae from Florida settle within an hour or so of release, but if they are provided with an appropriate cue, will settle within a few minutes (Keough 1986); those from Santa Catalina Island, in California, with a longer planktonic period, showed a similar effect of the presence of settlement substrates, although many still swam for 1–2 h (Keough 1989). In other species, such as *Mucropetraliella ellerii*, settlement cues may shorten the planktonic period less dramatically (Keough and Klemke 1991). It is important to note that, as for swimming time, all larvae do not respond identically to the presence of settlement cues. In addition to observed variation between individuals within a population, different populations of the same species may show different responses to a particular stimulus (Raimondi and Keough 1990). This variation in larval behavior between local populations could result in some local populations being largely self-seeding, while others may produce larvae that disperse widely.

We wish to make three important points. First, we emphasize that larval behavior is variable among individuals within a population. Whatever larval attributes are used to predict dispersal, one must recognize that it is both the average value of an attribute *and its frequency distribution* that are important. Second, the existence of interpopulation variation in some species means that existing literature, which is often drawn from studies in one or a few geographic localities, may not be a reliable indicator of populations within a local impact area. The extent of such interpopulation variation is poorly known for the vast majority of species (Raimondi and Keough 1990). Third, most of the data about larval biology, including planktonic duration and settlement behavior, are taken from laboratory studies. There are very few cases of this information being present for natural populations studied in situ. The uncertain relationship between laboratory and field data should be a cause for major caution.

In many ways, despite a long history of investigations into the biology of larvae, we remain ignorant about the behavior of larvae in the field (Keough 1988). Hydrodynamic models at spatial scales appropriate for EIA appear to have developed much more rapidly in the past few years than has knowledge of behavior.

The Relative Importance of Hydrodynamics and Biology

In situations such as those described for the Great Barrier Reef, hydrodynamic models with a relatively small amount of biological data may be adequate. The relatively high velocity of these environments may make it difficult for larvae to do anything but be swept along.

Vertical structure always exists in the currents. A near-bed boundary layer is always present and the shear stress increases with bed roughness. A surface boundary layer will also occur in windy conditions. Stratification due to salinity or temperature will amplify the effect of vertical shear. Thus, hydrodynamic complexity of a region will be a function of the intensity of the vertical and horizontal shear but some structure will always be present. As such, behavioral responses of larvae will always need to be considered if they alter their vertical level in the water column.

The most interesting, but least tractable, situations will be those with extensive vertical stratification. They require more complex models of circulation, but in these situations, restricted movements of larvae may allow them to move between strata and therefore greatly alter their transport. In these environments, complex models of circulation must be combined with detailed information about the vertical movements of larvae, as well as the duration of their planktonic period. The frequent reports of ontogenetic shifts in behavior and buoyancy, together with the variability of these characters, will require detailed data-gathering for individual species.

In all environments, there are situations in which larval behavior may still contribute substantially to dispersal. Even in high-flow environments, there will generally be a boundary layer of reduced flow, into which benthic animals and plants may release dispersive propagules. Negatively buoyant propagules or those showing the appropriate behavior may not be transported very far. Some larvae, however, may vary their behavior in response to the free-stream velocity. Larvae of the bivalve *Cerastoderma edule* in the laboratory swam in an upwards spiral in still water, while they drifted within 1 mm of the bottom at velocities above 0.05 m/s (Jonsson et al. 1991). Jonsson et al. noted that the larval behavior changed at what is a relatively low velocity in the field (see modeling discussion above). Pawlik et al. (1991) observed similar entrainment of polychaete larvae within near-bed flows, also in a laboratory flume.

Boundary layers may be enhanced considerably in topographically complex benthic habitats. This complexity may be provided by macrophytes, such as kelps and seagrass beds, or by large three-dimensional animals, such as reef-forming corals. Complexity may also come from broken, highly-reticulated reefs. McShane et al. (1988) suggested that on rocky reefs there is a zone of reduced water movement below the canopy of kelps, and a further reduction in water movement within crevices and beneath ledges. Similar reductions in water movement have been recorded for other habitats, such as seagrass meadows (Peterson et al. 1984, Eckman 1987, Eckman et al. 1989). These zones of reduced water movement may extend for meters, or even tens of meters above the seafloor. Larvae staying within the zone of reduced water movement would not disperse far, but those moving up beyond the kelp canopy could be dispersed far from their parental reefs. Their destination would depend on local hydrodynamics, but McShane et al. (1988) suggested that larvae that remained deep within areas of low water movement could be retained on local reefs for 3–7 days. Some of the reefs were hydrodynamically isolated from other larval sources.

In any situation with areas of reduced flow or vertically-stratified flow, additional information about the propagules will be necessary. For inactive propagules, such as spores or eggs, the initial buoyancy or sinking rates may be sufficient biological data. For some larvae, the behavior at the time of release may be sufficient. For example, the initial positive phototaxis shown by most bryozoan larvae would probably be sufficient to take them out of a seagrass bed or small macroalgal canopy. If larvae lack such an initial behavior, or are likely to move up and down in the water column during their planktonic period, additional biological data will be required. The behavior of the propagules must be understood to decide upon the likely contribution of behavior and hydrodynamic processes. This initial information may be used as a guide to the period of intensive data-gathering and modeling that will be at the core of the EIA activity.

Requirements for Individual Monitoring or EIA Programs

The desired output for any EIA will be a description of water circulation, and more importantly, a description of the likely dispersal profile of larvae. That description will, given the nature of larvae and the variability of winds, be a probability distribution. That distribution would itself be the product of a number of other probability distribution functions, each of which must be estimated separately.

Construction of Hydrodynamic Models

Usually a single comprehensive model code can be applied to a wide variety of regions. Different boundary conditions, bathymetry and coefficients, such as the bed friction, only need to be altered to change the behavior of the model. To create a model, three stages of development are normally needed:

(1) To be generally applicable, the model code must simulate most terms in the momentum balance, including the nonlinear terms, horizontal eddy viscosity, Coriolis force, wind stress, and pressure gradients. For shallow regions, flooding and drying of intertidal zones is also required. Headlands, islands, etc., also need to be resolved.

(2) The applicability of the model to the region must be tested against data. Inadequacies in the model, which should have been resolved in stage 1, and inadequacies in the input boundary conditions are often confused. To ensure that the model is adequately reflecting local dynamics, the boundary conditions need to contain sufficient information to ensure that all relevant processes are modeled. This is most easily achieved using actual measurements of sea level or current, wind, rainfall, evaporation, barometric pressure, river inflows, etc. The boundary conditions for a numerical model of a bay are more easily specified than for an open coast. The open coast is affected by remote phenomena such as ocean-scale circulation and coastal-trapped waves, which usually require actual measurements at a number of sites for their treatment. In Port Phillip Bay, measurements of sea levels at one site in the adjacent strait outside the entrance were sufficient to operate an accurate model, in real time if required (Black et al. 1992). Of course, measurements within the bay were a prerequisite to verify the model during its establishment. If these data are not available, then an assessment of the amount of total variance included in the (simplified) schematized condition

would be undertaken, so that our expectations of the model can be properly established.

(3) The model needs to be calibrated and verified at the scale of interest. Due to the inevitable model resolution problems, subgrid scale processes usually need to be specified and considered at this stage.

The model may then be applied to an environmental assessment, knowing the level of predictability being achieved. More than 90% of the variance in the sea levels and currents can be simulated when the inputs are well specified (e.g., Black et al. 1992). However, the predictions may be less well specified in other cases.

One of the most important decisions to be made is whether a two- or three-dimensional model is required. Two-dimensional models are simpler to construct, but the examples discussed above emphasize that it is difficult to determine *a priori* which model will be most appropriate. In the open coast environment, a simplified two-dimensional model was not useful, even though a number of studies have drawn on two-dimensional simulations for environmental assessments. The same result was obtained in an enclosed bay. On reefs of the Great Barrier Reef, three-dimensional simulations gave very similar predictions to the two-dimensional models (Black et al. 1991a). This latter result is explained by the strong topographic steering of the currents in the immediate vicinity of reefs. In more open waters in windy conditions, the results would be different. We emphasize that there is considerable potential for simple, two-dimensional models to provide a highly detailed and misleading specification of an overly simplified system. It is often tempting, because of limited time and money available for EIA, to use simplified models. These models may have some uses as a research tool to break down the complexity of the system, rather than as a practical aid in a management problem.

What Are the Data Requirements for a New Model Situation?   Sea levels, currents, temperature and density stratification, winds, barometric pressures, river inflows, evaporation, and wave climate would normally need to be measured prior to establishing a numerical model of a new region (Black et al. 1992). Observations of the sea bed to provide an independent check of the bed roughness would also be recommended (Black et al. 1993). Some of these parameters may not be needed in special locations. For example, waves may not be highly important inside an enclosed bay. The vertical eddy viscosity and diffusivity need to be specified for a three-dimensional dispersal simulation, and selected values confirmed by comparing model predictions with currents measured at a number of vertical levels. The relationship between the levels depends on the vertical eddy viscosity, which is often taken to be equal to the eddy diffusivity.

The technology to measure the above parameters is available, although sometimes the costs of field data collection are large.

Are Models Created for Other Purposes Usable for Prediction of Dispersal? The models of Bass Strait and the Great Barrier Reef have been used successfully to predict the dispersal of oil and to make ecological assessments. EIA studies of effluent discharge associated with tourist facilities on the Great Barrier Reef have also been undertaken (Black 1990).

The model calibration establishes the accuracy of the model and the confidence that can be placed on the results. For dissolved effluents, sediment transport or heat transport, the models are able to make very reasonable predictions.

The complexity and often the accuracy of the models discussed above are often a result of the history of the modeling exercise. The most sophisticated models are the result of a number of related projects, each of which may equal the effort put into a single EIA.

The continental shelf model was initially established to assess the likely dispersal of effluent from a proposed pulp mill, and was a prefeasibility study aimed at specifying site suitability. This was a government initiative to provide the basis to respond to proposals from potential developers.

The Port Phillip Bay model was developed to examine residence times in the bay to assess the impact of effluent and nutrient discharge from the sewage treatment plant on the bay's northern shoreline. Data analysis demonstrated that most of the variance in the circulation could be explained (Black et al. 1992; Black and Hatton 1992). The model was then viewed as a useful aid for environmental assessment and has been applied to studies of sediment dispersal, proposed port development and the impact of coastal construction. The Port Phillip Bay case study has shown that a well calibrated simulation, once established, remains a useful tool for a wide variety of applications. The continuing development of the model is being funded by a consortium of users. These are the Victorian Institute of Marine Sciences, the Environment Protection Authority of Victoria and the Melbourne Water Board. Initial support came also from the Water Police for search and rescue applications. The consortium arrangement broadens the value of the model and spreads the cost of development.

Biological Data

Exactly which biological data are needed will depend on the hydrodynamic conditions at a particular Impact site. In most cases, some estimates of the planktonic period will be necessary. The only exception is likely to be when the modeling exercise predicts retention periods greatly exceeding the larval duration of a whole taxonomic group. For example, most ascidians have very short-lived larvae, and a retention period of 10 days may be sufficient for a researcher to be confident that no larvae would be in the plankton for long enough to be transported from the area.

Duration of the Planktonic Period. The simplest data to obtain are the length of the precompetent phase and the physiological limit to the competent phase. These data are most often obtained from laboratory experiments. The values obtained should be viewed as upper limits and treated with some caution. The plasticity of the length of the precompetent phase can be dealt with by varying laboratory food supplies, temperature, etc. *In situ* larval culturing experiments (e.g., Olson 1987) can provide estimates of larval development rates under natural physical regimes. The major cause for caution is the ability of most species to delay metamorphosis. It is relatively simple to demonstrate in the laboratory that a particular stimulus shortens the larval period by inducing settlement. The methodology for these laboratory studies has been known for many years, and there is already a large volume of experimental data on invertebrates. However, there are relatively few studies that describe the effect of settlement cues under field conditions. These experiments are more difficult to do, although there are some notable examples (Raimondi 1988).

There is a further, logical problem in determining the extent to which settlement cues reduce the planktonic period. While the effect of individual stimuli can be framed as falsifiable hypotheses, the overall result of a range of such experiments is an inductive statement. We can never be sure that the stimuli we have chosen are those that produce the greatest reduction in swimming. It is always possible that some undiscovered cue may have a greater effect than those that we have already examined.

Buoyancy. Measurements of the buoyancy of propagules, i.e., rates at which they sink or rise to the top of a water column, are relatively simple to do. Because they are estimates of passive properties of propagules, they can be done with killed or anesthetized propagules, as long as care is taken not to damage appendages, mucous threads, or any other structure that could influence sinking rates. They can easily be done in the laboratory, as long as the temperature and salinity of the water match that of the water body under consideration. The major caution is that the diameter of the container in which the experiments are done should be large enough to preclude "wall" effects and problems caused by convection currents (Nowell and Jumars 1987).

For a given taxon, it is preferable to make measurements on propagules from a local population, in case there is geographic variation in, for example, larval size or the density of cilia, which would be expected to alter the drag on a propagule. There will be the usual requirement for adequate replication, to describe the frequency distribution of sinking rates. If the species in question undergoes planktonic development, it will be necessary to test for the existence of ontogenetic shifts in buoyancy. These data could be obtained from laboratory-reared propagules for species that can be reared easily, or from repeated sampling of the plankton during a reproductive season to obtain propagules at different developmental stages.

Larval Taxes and Swimming Speeds. The likely importance of boundary layer phenomena in temperate areas means that larval biology, beyond simple estimates of larval duration, will be necessary. This step is likely to be the messiest and most time-consuming. Variation in larval parameters, even as manifest in simple laboratory situations, occurs between species and between populations. Within populations, there is still considerable behavioral variability, so models must incorporate this variation as well. These data can only be obtained on an impact by impact basis, and the labor involved will be considerable. It may be that the work can only be done for "key" or indicator species, which may prove difficult in practice, as impacted species are only identified post hoc (Keough and Quinn 1991).

In vertically-stratified flow environments, the most important larval behavior is that involving movement up and down in the water column. It is important not only to document this behavior at a particular stage of larval development, but to describe the ontogeny of the behavior.

As with settlement cues, it is likely that most data will be obtained under laboratory conditions, and caution should be exercised in extrapolating from such data to field situations. One recent useful laboratory method is the use of laboratory flumes to describe larval behavior under varying flow regimes (e.g., Jonsson et al. 1991, Mullineaux and Butman 1991, Pawlik et al. 1991). These experiments have the advantage of increased realism and have already shown that larval behavior may be velocity dependent. Construction of flumes has been discussed extensively by Nowell and Jumars (1987). Other workers have developed ways of marking small larvae, making it easier to keep track of them. Fluorescent marking (Lindegarth et al. 1991) of larvae may even make it feasible to follow their movements in the field, either unconfined, or within *in situ* mesocosms.

Impacts on Fecundity

Effects of anthropogenic activities on fecundities of adult organisms are well understood. The methodology for estimating fecundity is described for many species, and the statistical designs necessary to detect changes in fecundity are described in many parts of the literature (Stewart-Oaten et al. 1986, Underwood 1991). We will not treat them in any detail here, but we do note that the output of the dispersal modeling will be necessary to identify the local populations that can contribute propagules into the plankton. Relatively restricted larval dispersal will mean that only a small range of populations need be sampled, while wide dispersal will require correspondingly widely distributed spatial sampling.

The Demographic Importance of Recruitment

We will not discuss the role of recruitment in population structure in detail here. A decade of intense debate about its importance has produced little but agreement that its importance varies on a case by case basis. There are no predictive models for which populations are recruitment limited and which are controlled by postrecruitment processes. Connell (1985) suggested that recruitment was not limiting when settlement rates were high, but such a suggestion hardly amounts to a major conceptual advance. Other recent attempts to produce syntheses (e.g., Menge and Sutherland 1987) are largely untested.

Compensatory effects are not well known, and the difficulties in dealing with such effects are addressed by Nisbet et al. (Chapter 13). We generally understand little of the general level of mortality experienced by larvae of most species (reviewed by Rumrill 1990). One important possibility is that the anthropogenic activity itself may affect not only the performance of adults, but may affect aspects of larval biology. For example, even a brief passage of abalone larvae through a produced-water outfall may reduce their ability to respond to the settlement-inducing GABA (Raimondi and Schmitt 1992, Raimondi and Reed, Chapter 10), and various lesions in larval fish have been associated with passage through sewage outfalls (Kingsford and Gray, Chapter 12).

There are general models about responses of marine benthic populations to disturbance that indicate when recovery comes from the plankton (Connell and Keough 1985, Sousa 1985). Although these models would appear to be useful for determining whether recruitment or interactions involving adults are more important, they are drawn mostly from studies of plants and sessile animals, with a conspicuous lack of data from mobile animals (reviewed by Lake 1990). These models do not include consideration of dispersal distance, and, some years after those reviews, there are still few observational, and fewer experimental studies of subtidal disturbances and the role of recruitment in population recovery.

Conclusions

There are as yet few combinations of numerical modeling and larval behavioral information and variability, but we suggest that both components are essential to an understanding of dispersal in an ecological (as opposed to a genetic) context. Behavior may reduce any relationship between larval duration and dispersal distance to an artifact (e.g., Keough 1988), but, correspondingly, knowing that larval behavior results in individuals moving out of boundary layers into faster-moving water is not sufficient to predict dispersal, unless we also understand where larvae will be carried. We have not tried to provide solutions, but to try and illustrate the potential of the numerical models for modeling dispersal, and to give an idea of the biological information that may be required, together with the knowledge gaps that restrict the use of larval dispersal in EIA.

In complex systems, there will be an additional need for information about larval behavior, and at the moment, there are few field data, and such data must be obtained on a species- and site-specific fashion.

The past few years have seen major advantages in our understanding of water movements. Importantly, models of circulation have been developed at intermediate spatial scales. These scales, hundreds to thousands of meters, correspond to the more common distances between Control and Impact sites in EIA studies. In hydrodynamically-simple systems, models are now at the stage where they can produce accurate, quantitative predictions of the dispersal of larvae and of toxicants.

There are now examples of successful prediction of dispersal of (mostly) invertebrate larvae (e.g., Black and Moran 1991, Farrell et al. 1991). Most of these examples have involved propagules that disperse only weakly, and thus situations in which larval behavior is unlikely to be important. There are at the moment relatively few examples of successful prediction of dispersal in topographically-complex environments or for larvae showing extensive behavior.

We consider that information about *field* behavior of larvae now lags considerably behind the development of models of transport. The extensive body of laboratory data suggests that biological data will need to be obtained in a species, or even population-specific fashion. This complexity is likely to make larval behavior a major impediment to successful prediction of dispersal for EIA.

Acknowledgments

Parts of this work were supported by grants from the Australian Research Council, the Victorian Institute of Marine Sciences, the Crown-of-Thorns Starfish Research Committee through the Great Barrier Reef Marine Park, and the National Pulp Mills Research Program. We appreciate the comments of A.J. Boxshall and the editors and reviewers.

References

Alcala, A. C., and G. R. Russ. 1990. A direct test of the effects of protective management on abundance and yield of tropical marine reserves. Journal du Conseil **47**:40–47.

Andrews, J. C., and M. J. Furnas. 1986. Subsurface intrusions of coral sea water into the central Great Barrier Reef - I. Structures and shelf-scale dynamics. Continental Shelf Research **6**:491–514.

Black, K. P. 1990. Effluent residence times and numerical simulations for application to engineering works on coral reefs. Pages 141–148 *in* Proc engineering in coral reef regions, Magnetic Island, Townsville. Institution of Engineers, Australia and the Great Barrier Reef Marine Park Authority, Townsville, Queensland.

Black, K.P. 1992. Evidence of the importance of deposition and winnowing of surficial sediments at a continental shelf scale. Journal of Coastal Research **8**:319–331.

Black, K.P.. 1993. The relative importance of local retention and inter-reef dispersal of neutrally buoyant material on coral reefs. Coral Reefs **12**:43–53.

Black, K. P., S. L. Gay, and J. C. Andrew. 1990. Residence times of neutrally-buoyant matter such as larvae, sewage or nutrients on coral reefs. Coral Reefs **9**:105–114.

Black, K. P., P. Greilach, and D. Hatton. 1991a. Calibration of a numerical hydrodynamic model of the central Great Barrier Reef on a 2.25 km grid. Victorian Institute of Marine Sciences, Melbourne, Australia.

Black, K. P., P. J. Moran, and L. S. Hammond. 1991b. Numerical models show coral reefs can be self-seeding. Marine Ecology Progress Series **74**:1–11.

Black, K. P., and D. N. Hatton. 1992. Hydrodynamic and sediment dynamic measurements in eastern Bass Strait. Volume 4. Two and three-dimensional numerical hydrodynamic and dispersal models. Victorian Institute of Marine Sciences, Melbourne, Australia (Technical Report No. 17).

Black, K. P., D. N. Hatton, and M. Rosenberg. 1993. Locally and externally-driven dynamics of a large semi-enclosed bay in southern Australia. Journal of Coastal Research **9**:509–538.

Black, K. P., T. R. Healy, and M. Hunter. 1989. Sediment dynamics in the lower section of a mixed sand and shell-lagged tidal estuary. Journal of Coastal Research 5:503–521.

Black, K. P., and P. J. Moran. 1991. Influence of hydrodynamics on the passive dispersal and initial recruitment of larvae of *Acanthaster planci* on the Great Barrier Reef. Marine Ecology Progress Series 69:55–65.

Black, K. P., M. Rosenberg, G. Symonds, R. Simons, C. Pattiaratchi, and P. Nielsen. 1994. Measurements of wave, current and sea level dynamics of an exposed coastal site. *In* C. Pattiaratchi (ed.) Mixing processes in estuaries and coasta seas. American Geophysical Union.

Burrage, D. M., J. A. Church, and C. R. Steinberg. 1991. Linear systems analysis of momentum on the continental shelf and slope of the central Great Barrier Reef. Journal of Geophysical Research 96:22169–22190.

Butman, C. A. 1987. Larval settlement of soft sediment invertebrates: the spatial scales of pattern explained by active habitat selection and the emerging role of hydrodynamic processes. Oceanography and Marine Biology Annual Review 25:113–165.

Chia, F.-S., J. Buckland-Nicks, and C. M. Young. 1984. Locomotion of marine invertebrate larvae: a review. Canadian Journal of Zoology 62:1205–1222.

Connell, J. H. 1985. The consequences of variation in initial settlement vs. post-settlement mortality in rocky intertidal communities. Journal of Experimental Marine Biology and Ecology 93:11–46.

Connell, J. H., and M. J. Keough. 1985. Disturbance and patch dynamics of subtidal marine animals on hard substrata. Pages 125–151 *in* S. T. A. Picket and P. S. White, editors. The ecology of natural disturbance and patch dynamics. Academic, New York, New York.

Crisp, D. J. 1976. Settlement responses in marine organisms. Pages 83–124 *in* R. C. Newell, editor. Adaptations to environment: essays on the physiology of marine animals. Butterworths, London, England.

Cronin, T. W., and R. B. J. Forward. 1986. Vertical migration cycles of crab larvae and their role in larval dispersal. Bulletin of Marine Science 39:192–201.

Davies, A. M. 1985. Application of a sigma coordinate sea model to the calculation of wind-induced currents. Continental Shelf Research 4:389–423.

Davis, A. R. 1987. Variation in recruitment of the subtidal colonial ascidian *Podoclavella cylindrica* (Quoy and Gaimard): the role of substratum choice and early survival. Journal of Experimental Marine Biology and Ecology 106:57–71.

Davis, A. R., and A. J. Butler. 1989. Direct observations of larval dispersal in the colonial ascidian *Podoclavella moluccensis* Sluiter: evidence for closed populations. Journal of Experimental Marine Biology and Ecology 127:189–203.

Eckman, J. E. 1987. The role of hydrodynamics in recruitment, growth, and survival of *Argopecten irradians* (L.) and *Anomia simplex* (D'Orbigny) within eelgrass meadows. Journal of Experimental Marine Biology and Ecology 106:165–192.

Eckman, J.E. 1990. A model of passive settlement by planktonic larvae onto bottoms of differing roughness. Limnology and Oceanography 35:887–901.

Eckman, J. E., and D. O. Duggins. 1991. Life and death beneath macrophyte canopies: effects of understory kelps on growth rates and survivorship of marine, benthic suspension feeders. Oecologia 87:473–487.

Eckman, J. E., C. H. Peterson, and J. A. Cahalan. 1989. Effects of flow speed, turbulence and orientation on growth of juvenile bay scallops *Argopecten irradians concertricus* (Say). Journal of Experimental Marine Biology and Ecology 132:123–140.

Fairweather, P. G. 1991. Implication of 'supply-side' ecology for environmental assessment and management. Trends in Ecology and Evolution 6:60–63.

Fandry, C. B., G. D. Hubbert, and P. C. McIntosh. 1985. Comparison of predictions of a numerical model and observations of tides in Bass Strait. Australian Journal of Marine and Freshwater Research 36:737–752.

Farrell, T. M., D. Bracher, and J. Roughgarden. 1991. Cross-shelf transport causes recruitment to intertidal populations in central California. Limnology and Oceanography 36:279–288.

Fogarty, M. J., M. P. Sissenwine, and E. B. Cohen. 1991. Recruitment variability and the dynamics of exploited marine populations. Trends in Ecology and Evolution 6:241–246.

Gaines, S. D., and J. Roughgarden. 1987. Fish in offshore kelp forests affect recruitment to intertidal barnacle populations. Science 235:479–481.

Gay, S. L., J. C. Andrews, and K. P. Black. 1991. Dispersal of neutrally buoyant material near John Brewer Reef. Pages 95–119 in R. Bradbury, editor. Acanthaster and the coral reef: a theoretical perspective. Springer-Verlag, Berlin, Germany.

Jenkins, G. P., and H. May. 1994. Variation in larval duration and settlement of King George whiting, Gillaginoides punctata, in Swan Bay, Victoria. Bulletin of Marine Science 54:281–296.

Jones, H. A., and P. J. Davies. 1983. Superficial sediments of the Tasmanian Continental Shelf and part of Bass Strait. Bureau of Mineral Resources, Geology and Geophysics Bulletin 218. Australian Government Publisher, ACT, 25 pp.

Jonsson, P. R., C. Andre, and M. Lindegarth. 1991. Swimming behaviour of marine bivalve larvae in a flame boundary-layer flow - evidence for near-bottom confinement. Marine Ecology Progress Series 79:67–76.

Keough, M. J., 1986. The distribution of the bryozoan Bugula neritina on seagrass blades: settlement growth and mortality. Ecology 67:846–857.

Keough, M.J., 1988. Benthic populations: is recruitment limiting or just fashionable? Proceedings of the 6th International Coral Reefs Symposium 1:141–148.

Keough, M.J., 1989. Dispersal of the bryozoan Bugula neritina and effects of adults on newly meta-morphosed juveniles. Marine Ecology Progress Series 57:163–171.

Keough, M. J., and J. Klemke. 1991. Larval dispersal and the scale of recruitment variation in benthic invertebrates. Pages 64–70 in D. A. Hancock, editor. Australian Society for Fish Biology Workshop. Recruitment Processes. Australian Government Publishing Service, Canberra, Australia.

Keough, M. J., and G. P. Quinn. 1991. Causality and the choice of measurements for detecting human impacts in marine environments. Australian Journal of Marine and Freshwater Research 42:539–554.

Kingsford, M. J. 1990. Linear oceanographic features: a focus for research on recruitment processes. Australian Journal of Ecology 15:27–37.

Kingsford, M. J., E. Wolanski, and J. H. Choat. 1991. Influence of tidally induced fronts and Langmuir circulations on distribution and movements of presettlement fishes around a coral reef. Marine Biology 109:167–180.

Knight-Jones, E. W. 1953. Decreased discrimination during settling after prolonged planktonic life in larvae of Spirorbis borealis (Serpulidae). Journal of the Marine Biology Association of the United Kingdom 32:337–345.

Lake, P. S. 1990. Disturbing hard and soft bottom communities: a comparison of marine and fresh-water environments. Australian Journal of Ecology 15:477–489.

Levin, L. A. 1990. A review of methods for labeling and tracking marine invertebrate larvae. Ophelia 32:115–144.

Lindegarth, M., P. R. Jonsson, and C. Andre. 1991. Fluorescent microparticles—a new way of visu-alizing sedimentation and larval settlement. Limnology and Oceanography 36:1471–1476.

Manahan, D. T. 1986. Energy resources for larval development. Bulletin of Marine Science.

McShane, P., K. P. Black, and M. G. Smith. 1988. Recruitment processes in Haliotis rubra (Mollusca: Gastropoda) and regional hydrodynamics in southeastern Australia imply localized dispersal of larvae. Journal of Experimental Marine Biology and Ecology 124:175–203.

Menge, B. A., and J. P. Sutherland. 1987. Community regulation—variation in disturbance, competi-tion, and predation in relation to environmental-stress recruitment. American Naturalist 30:730–757.

Middleton, J. F., and K. P. Black. 1994. The low frequency circulation in and around Bass Strait. Continental Shelf Research 14:1495–1521.

Middleton, J. F., and F. Viera. 1991. The forcing of low frequency motions within Bass Strait. Journal of Physical Oceanography 21:695–708.

Morse, D. E. 1990. Recent progress in larval settlement and metamorphosis: closing the gap between molecular biology and ecology. Bulletin of Marine Science 46:465–483.

Mullineaux, L. S., and C. A. Butman. 1991. Initial contact, exploration and attachment of barnacle (*Balanus amphitrite*) cyprids settling in flow. Marine Biology 110:93–103.

Munk, W., and E. R. Anderson. 1948. Notes on the theory of the thermocline. Journal of Marine Research 7:276–295.

Murdoch, W. W., R. C. Fay, and B. J. Mechalas. 1989. Final report of the Marine Review Committee to the California Coastal Commission. Marine Review Committee, Inc.

NRC (National Research Council). 1990a. Managing troubled waters: the role of marine environmental monitoring. National Academy Press, Washington, D.C.

NRC (National Research Council). 1990b. Monitoring Southern California's coastal waters. National Academy of Sciences, Washington, D.C.

Nowell, A. R. M., and P. A. Jumars. 1987. Flumes: theoretical and empirical considerations for simulation of benthic environments. Oceanography and Marine Biology Annual Review 25:91–112.

Olson, R. R. 1985. The consequences of short-distance larval dispersal in a sessile marine invertebrate. Ecology 66:30–39.

Olsen, R. R. 1987. In situ culturing as a test of the larval starvation hypothesis for the crown-of-thorns starfish, *Acanthaster planci*. Limnology and Oceanography 32:895–904.

Pawlik, J. R., C. A. Butman, and V. R. Starczak. 1991. Hydrodynamic facilitation of gregarious settlement in a reef-building tube worm. Science 251:421–424.

Pechenik, J. A. 1990. Delayed metamorphosis by larvae of benthic marine invertebrates: does it occur? is there a price to pay? Ophelia 32:63–94.

Peterson, C. H., H. C. Summerson, and P. B. Duncan. 1984. The influence of seagrass cover on population structure and individual growth rate of a suspension-feeding bivalve, *Mercenaria mercenaria*. Journal of Marine Research 42:614–631.

Possingham, H. P., and J. Roughgarden. 1990. Spatial population dynamics of a marine organism with a complex life cycle. Ecology 71:973–985.

Raimondi, P. T. 1988. Settlement cues and determination of the vertical limit of an intertidal barnacle. Ecology 69:400–407.

Raimondi, P. T., and M. J. Keough. 1990. Behavioural variability in marine larvae. Australian Journal of Ecology 15:427–437.

Raimondi, P. T., and R. J. Schmitt. 1992. Effects of produced water on settlement of larvae: field tests using red abalone. Pages 415–430 in J. P. Ray and F. R. Englehardt, editors. Produced water: technological/environmental issues and solutions. Plenum Press, New York, New York.

Reed, C. G. 1991. Bryozoa. Pages 86–246 in A. C. Giese, J. S. Pearse and V. B. Pearse, editors. Reproduction of marine invertebrates. Volume 6. Echinoderms and Lophophorates. Boxwood Press, Pacific Grove, California.

Rosenberg, M., R. Hodgkinson, K. Black, and R. Colman. 1992. Hydrodynamics of Port Phillip Bay. Volume 1: Field data collection and data reduction. Victorian Institute of Marine Sciences, Melbourne, Australia. (Working Paper No. 23).

Roughgarden, J., Y. Iwasa, and C. Baxter. 1985. Demographic theory for an open marine population with space-limited recruitment. Ecology 66:54–67.

Rumrill, S. S. 1990. Natural mortality of marine invertebrate larvae. Ophelia 32:163–198.

Russ, G. R., and A. C. Alcala. 1989. Effects of intense fishing pressure on an assemblage of coral reef fishes. Marine Ecology Progress Series 56:13–27.

Ryland, J. S. 1974. Behaviour, settlement and metamorphosis of bryozoan larvae: a review. Thalassia Jugoslavica 10:239–262.

Sammarco, P. W., and J. C. Andrews. 1988. Localized dispersal and recruitment in Great Barrier Reef corals: The Helix experiment. Science 239:1422–1424.

Scheltema, R. S. 1971. Larval dispersal as a means of genetic exchange between geographically separated populations of shallow-water benthic marine gastropods. Biological Bulletin 140:284–322.

Scheltema, R.S. 1974. Biological interactions determining larval settlement of marine invertebrates. Thalassia Jugoslavica 10:263–296.

Scheltema, R. S., and I. B. Williams. 1983. Long-distance dispersal of planktonic larvae and the bio-geography and evolution of some Polynesian and western Pacific mollusks. Bulletin of Marine Science 33:545–565.

Shanks, A. L. 1983. Surface slicks associated with tidally forced internal waves may transport larvae of benthic invertebrates and fishes shoreward. Marine Ecology Progress Series 13:311–315.

Sinclair, M. 1988. Marine populations. University of Washington Press, Seattle, Washington.

Smith, J. D., and T. S. Hopkins. 1972. Sediment transport on the continental shelf off Washington and Oregon in light of recent current measurements. Pages 143–180 in D. J. P. Swift et al., editor. Shelf sediment transport process and pattern. Dowden, Hutchison and Ross, Stroudsburg, Pennsylvania.

Sousa, W. P. 1985. Distrubance and patch dynamics on rocky intertidal shores. Pages 101–124 in S. T. A. Pickett and P. S. White, editors. The ecology of natural distrubance and patch dynamics. Academic Press, New York, New York.

Stewart-Oaten, A., W. W. Murdoch, and K. R. Parker. 1986. Environmental impact assessment: "psuedoreplication" in time? Ecology 67:929–940.

Thorson, G. 1950. Reproductive and larval ecology of marine bottom invertebrates. Biological Reviews 25:1–45.

Underwood, A. J. 1991. Beyond BACI: experimental designs for detecting human environmental impacts on temporal variations in natural populations. Australian Journal of Marine and Freshwater Research 42:569–587.

Underwood, A. J., and E. J. Denley. 1984. Paradigms, explanations and generalizations in models for the structure of intertidal communities on rocky shores. Pages 151–180 in D. R. Strong, D. S. Simberloff, L. G. Abele and A. Thistle, editors. Ecological communities, conceptual issues and the evidence. Princeton University Press, Princeton, New Jersey.

Viarengo, A., and L. Canesi. 1991. Mussels as biological indicators of pollution. Aquaculture 94:225–243.

Willis, B. L., and J. K. Oliver. 1988. Inter-reef dispersal of coral larvae following the annual mass spawning on the Great Barrier Reef. Proceedings of the 6th International Coral Reefs Symposium 2:667–672.

Young, C. M. 1986. Direct observations of field swimming behavior in larvae of the colonial ascid-ian Ecteinascidia turbinata. Bulletin of Marine Science 39:279–289.

Young, C.M. 1990. Larval ecology of marine invertebrates: a sesquicentennial history. Ophelia 32:1–48.

CHAPTER 12

INFLUENCE OF POLLUTANTS AND OCEANOGRAPHY ON ABUNDANCE AND DEFORMITIES OF WILD FISH LARVAE

Michael J. Kingsford and Charles A. Gray

A broad spectrum of pollutants including industrial waste, sewage, insecticides and storm water are released into coastal waters of the world through outfalls (e.g., Beder 1989). Concerns about the health of fish in the vicinity of outfalls include public health and the effects of pollutants on the demography of fish (especially species of commercial and recreational value), and the ecology of local marine assemblages. It is well recognized that pollutants have lethal (Mileikovsky 1970) and sublethal effects on marine larvae (Rosenthal and Alderdice 1976). Although historically the primary focus was on lethal doses (LD50s) on larvae, there is a rapidly expanding literature on sublethal effects of pollutants on fish eggs and larvae (reviewed by Westernhagen 1988, Weis and Weis 1989). Our perception of significant environmental degradation, however, still focuses on juvenile and adult phases in terms of "measurable effects" around ocean outfalls (e.g. McLean et al. 1991).

There is a common misconception that only juvenile and adult fish that dwell within a concentration-dependent "halo" around an outfall will be affected by pollutants. Unlike many juvenile and adult fish that are relatively site attached and are subjected to pollutants over long periods of time, larvae are generally ignored because they are thought to be exposed to pollutants for only a brief period (too short to elicit an effect) as they pass through a plume. Fish at an early stage of development are particularly vulnerable to pollutants over short periods of time (e.g., Westernhagen 1988) and may be sensitive to brief exposures. Furthermore, oceanographic features may constrain and concentrate fish larvae and pollutants in particular patches of water, thus enhancing exposure (e.g., Shanks 1987). Because the dispersal distances of many larvae may exceed tens of kilometers (e.g., Williams et al. 1984), longshore currents that sweep larvae

Detecting Ecological Impacts: Concepts and Applications in Coastal Habitats, edited by R. J. Schmitt and C. W. Osenberg

through plumes may be deleteriously affecting the young of fish from distant locations and thus mediate impacts on large spatial scales.

Thus, it is critical to examine the effects of pollutants on larvae and how these effects can be modified by oceanographic processes. Many studies have been done on the effects of pollutants on fish eggs and larvae in the laboratory (Westernhagen 1988, Hall 1991), but investigators have found it difficult to translate findings on LD50s and sublethal effects to field situations. Predicting effects under field conditions from the results of laboratory experiments are problematic because (i) we typically lack appropriate models for "scaling up" the results of laboratory studies to cohorts of wild larvae and populations (see Underwood and Peterson 1988); (ii) concentrations of chemicals used to measure LD50s of larvae in the laboratory are often "outrageously high" compared to estimations of concentrations measured in the field (Westernhagen 1988); (iii) it is difficult to determine concentrations of pollutants that larvae are subjected to, given variable conditions of oceanography in space and time and flow rates from outfalls; and (iv) larvae are small and mobile in a three-dimensional environment, and unlike conspicuous fish kills and deformities in adults the detection of effects on fish larvae resulting from a pollutant are difficult.

In this chapter we review how oceanographic features can concentrate fish larvae and pollutants and lead to exposure regimes of greater concentration and longer duration than expected based on simple diffusion models. Available evidence from the laboratory and field suggests pollutants can lead to decreased viability of larvae, resulting in part from morphological deformities. Field studies can provide a powerful way to study the interaction between oceanographic processes and biological effects.

Oceanographic Features and the Accumulation of Fish Larvae and Pollutants

It should not be assumed that pollutants simply disperse from a point source. Linear Oceanographic Features (LOFs) such as the fronts of plumes can facilitate the concentration of pollutants (Hardy et al. 1985, Cross et al. 1987, Kingsford 1990, Brown et al. 1991). For example, Tanabe et al. (1991) found that organochlorines concentrated in coastal fronts, while Shanks (1987) described the accumulation and transport of tar in the slicks of internal waves. Highest concentrations of metals have been found in surface particulate matter in some studies (e.g., Balls 1990) and may become more concentrated in fronts. Moreover, many pollutants are lipophilic (Södergren and Okla 1988) and would be expected to bind with the breakdown products of organisms and anthropogenic lipids that accumulate in LOFs. Concentrations of "free" and particle-bound pollutants in fronts as well as levels of pollutants in the prey of fish larvae are not well known (e.g., Rainbow 1989).

Fish larvae may not just drift through a plume, but may accumulate in fronts or haloclines. Accumulation may result from advective processes or from the behavior of larvae. For example, the concentration of zooplankton is often an order of magnitude greater in LOFs and thermoclines than in adjacent water (Haney 1988, Kingsford 1990). Thus fish larvae may be attracted to fronts and haloclines by high concentration of their food. As a result, larvae would be exposed to pollutants for longer periods than expected from simple diffusion and transport models.

Thermoclines are another category of oceanographic feature where planktonic organisms (Haney 1988) and pollutants may accumulate. In contrast to near surface outfalls, deep ocean outfalls (e.g., those operating off Sydney, at 60–80 m depth) release pollutants from diffusers from a depth. In the case of Sydney, the probability that sewage will rise to the surface depends on the presence of a thermocline (CDM Report 1989). Although some greases may rise to the surface, in the presence of a thermocline most pollutants should stay at or beneath it, creating a potential hazard for larvae at depth.

Sewage outfalls (which include industrial waste) along the coast of Sydney, Australia, demonstrate how pollution plumes can be significant intrusions into mainstream coastal currents. Until recently, sewage was released from outfalls at the base of Sydney's cliff line (most sewage is now released through deep water outfalls, 2–4 km from shore). Plots of the position of plumes, from the BEACH-WATCH helicopter, indicated that plumes (prior to deep water outfalls) extended up to 5 km from shore and greater distances along the coast (Figure 12.1). Fish larvae within 5 km of the outfall, therefore, were likely to encounter plumes. The orientation of the plumes varied according to the direction of mainstream currents and wind. Over a 173 day period, 46% of the time the North Head plume flowed in a N-NE direction, 24% East and 30% S-SE. In conditions under 15 knots of wind clearest frontal edges were observed on the incurrent side of the plume.

Plumes are clearly visible as turbid water and are usually of low salinity. A conductivity temperature depth device (CTD) towed slowly from the plume and transversely across the front to clear coastal waters clearly showed the position of the front (Figure 12.2). Although variation in salinity of the plume was probably due to upwelling of uncontaminated seawater from below, some isolated peaks were probably the result of foreign bodies getting caught around the sensors of the CTD. The magnitude of salinity change at the front, depth of the plume edge and the interaction of the plume with coastal currents would alter the intensity of convergence in this area (Wolanski and Hamner 1988). A marked change in salinity was generally observed at the front, but this was not always the case (e.g., Figure 12.2, 12 Dec.).

The buoyancy of the plume suggested that larvae in the neuston to a few meters below the surface would be most vulnerable to pollutants in the plume. The depth of low-salinity water (70–150 m from the outfall) in the plume varied between 2 and 8 m among times (Figure 12.3). The plume, therefore floated as a

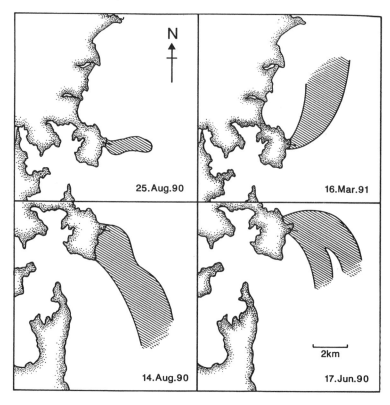

Figure 12.1. The orientation of cliff-face sewage plumes at 4 times off the North Head sewage outfall, Sydney, Australia. Typical flow = 325 megaliters per day (dry); 460 megaliters per day (wet).

low salinity wedge. Depth of the plume would vary according to distance from the outfall, volume of sewage released, and the effects of wind and currents (also see Washburn et al. 1992).

Abundance Patterns of Fish Larvae in Plumes

Investigators have compared polluted and unpolluted waters for major differences in the abundance of fish larvae and other zooplankters. The purpose of this approach has been to look for dramatic reductions in the ichthyoplankton fauna, as found for the benthos near point sources of pollution (Green 1979); such a pattern would be expected if pollutants caused acute mortality of plankters over short periods of time. However, no consistent patterns of larval fish and zooplankter abundance have been found in the vicinity of polluted plumes. For example, Karas et al. (1991) found low abundances of perch larvae in an area affected by effluent from a pulp mill, whereas Scarrett (1969) found that effluent from a mill had no effect on the abundance patterns of lobster larvae.

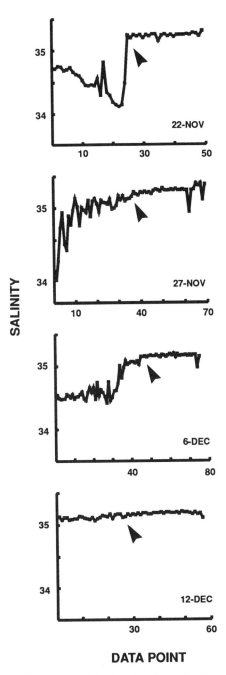

Figure 12.2. Horizontal CTD profiles of the frontal region of the North Head plume at 4 times. The CTD was towed at 0.2–0.4 m depth over a distance of 240 –320 m; 50–70 CTD data points were collected over that distance. Arrows indicate the position of fronts.

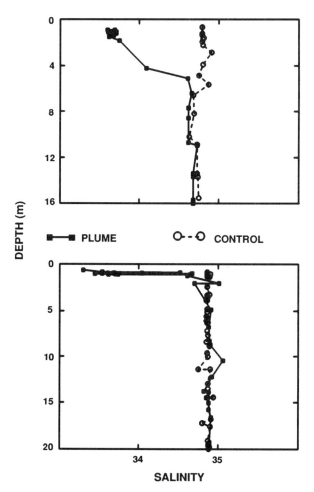

Figure 12.3.   Vertical CTD profiles of the plume and adjacent clear waters (Control) on 2 days, showing shallow plumes (2–6 m) of low salinity water.

Gray et al. (1992) sampled fish larvae in and out (> 8 km) of three sewage plumes in the Sydney area at the surface and at 20 m (deep). Patterns of abundance varied among taxa (Figure 12.4), but no clear relationship was found between the presence of an outfall and abundance of fish. This is not surprising as rates of immigration and emigration from plumes are probably high and may vary according to supply of larvae (from upstream), larval behavior and oceanography. Moreover, even if larvae die as a result of pollutants in the plume, time delays for lethal effects may result in death occurring outside the plume. Many species of fish were captured in highest abundance at the 20 m depth which

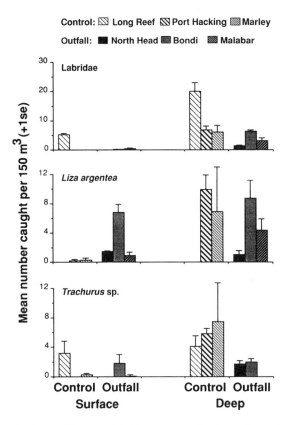

Control: ▧ Long Reef ▨ Port Hacking ▩ Marley

Outfall: ■ North Head ▦ Bondi ▨ Malabar

Figure 12.4.   Abundance of 3 groups of fish larvae captured at the surface and at 20 m (Deep) in April –May 1990 in and out of three outfalls in the vicinity of Sydney, Australia; Control sites were 5 –10 km from plumes. Samples were collected with a ring net, area of mouth 0.64 m², filtration efficiency 1:5.6, mesh 500 µm, towed at 2–4 knots for 15 min., volume of water filtered was measured using a flow meter. We use the term "fish larvae" to mean those developmental stages between egg hatching and metamorphosis (after Ahlstrom; see Kingsford 1988).

suggests that some groups may be more vulnerable to pollutants coming from the substrate and accumulating at the thermocline (as for deep ocean outfalls) than surface plumes and fronts.

Exposure might also vary depending on the horizontal distribution of larvae. Groups of fish that remain close to shore during their early development would be more vulnerable to the deleterious effects of plumes than fish found at a range of distances from shore. Although fish larvae are found both nearshore and offshore in different parts of the world, patterns of abundance vary between taxonomic groups (reviewed by Kingsford 1988, Leis 1991). For example, tripterygiids are usually most abundant in nearshore waters (e.g., Kingsford and Choat 1989, Kingsford et al. 1991).

Although estuarine plumes and associated fronts have been demonstrated to have major effects on the abundance patterns of meroplankters (e.g., Govoni et al. 1989), investigators have rarely attempted to stratify sampling according to oceanography in studies of pollution. In a study of fish larvae in the vicinity of a sewage plume (that includes industrial waste) on the coast of Sydney, Australia, samples were taken in the plume, front and a control area, which was approximately 2 km outside the plume in pristine waters (Figure 12.5). Because abundance of larval fish varies with time, sampling was done on six occasions. There was no evidence for consistently low numbers of fish larvae in the plume or front. Large numbers of fish larvae are, therefore, bathed in pollutants of the plume.

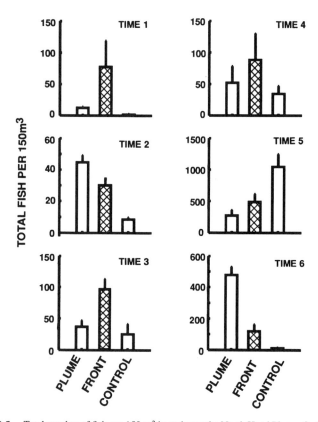

Figure 12.5.   Total number of fish per 150 m³ in and near the North Head Plume, Sydney, Australia (Mean ± SE). Samples were collected with a 0.67×0.67 m neuston net, filtration efficiency 1:12, mesh 500 µm, towed at 2 –4 knots for 3 min., volume of water filtered was measured using a flow meter, $n = 3$ tows. Two-way ANOVA, ln $(x + 1)$ transformation, F = fixed factor, R = Random factor, MS = mean square, C = Cochran's C-test for homogeneity of variances, $*P < 0.05$, $**P < 0.01$, $***P < 0.001$, NS = not significant; C = 0.19 NS; (R) Time (5,36) MS = 13.9***; (F) Hydrology (2,10) MS = 7.6 NS; T×H (10,36), MS = 2.89***; Residual (36) MS = 0.378 (from Kingsford et al. 1991).

Highest numbers of larvae were caught at the front at Times 1, 3 and 4. In general, however, the rank abundance of sites varied among times (ANOVA significant Time × Hydrology interaction; Figure 12.5). The intensity of convergence will vary according to the interaction of the plume with coastal currents (which vary in strength) and attributes of the plume. When convergence is intense, more larvae may be advected into the front and/or they may be attracted to high concentrations of prey that typically accumulate in LOFs (Kingsford 1990).

Vulnerability of Fish Larvae to Pollutants

"The developing fish embryo or larva is generally considered the most sensitive stage in the life history of a teleost, being particularly sensitive to all kinds of low-level environmental changes to which it might be exposed" (Westernhagen 1988). Most evidence on the vulnerability of fish larvae to pollutants is from laboratory studies, which indicate that metals, insecticides, acidification and chlorine can cause fatal embryological changes to eggs as well as high mortality rates and chronic deformities to fish larvae (Table 12.1; also see Westernhagen 1988, Hall 1991). Importantly, many of these pollutants can cause major deleterious effects to larvae in under 24 hr (see Raimondi and Schmitt 1992 for data on abalone larvae). Thus a protracted immersion in pollutants is not required to elicit acute or chronic effects in larvae. This has important implications if larvae are drifting into plumes in mainstream currents; they do not need to be subjected to pollutants for long before deleterious effects manifest themselves. It has also been noted by some investigators that the susceptibility of larvae to pollutants varies among species (Hoss et al. 1974, Blaxter 1977). Susceptibility within a taxonomic group may also vary, possibly depending on the lipid content of larvae (e.g., Södergren and Okla 1988, Lassiter and Hallam 1990) and their age (Mehrle et al. 1987). Accordingly, consequences of exposure to pollutants may be lethal, sublethal, or zero depending on the taxonomic group of fish larvae.

Many laboratory studies have only looked at simple effects of pollutants on larvae. It is clear that many synergistic effects of multiple pollutants (typical of municipal wastewaters) are possible and have been found in some larvae. For example, Weis and Weis (1982) found that the ability of larval *Fundulus heteroclitus* to resist methylmercury decreased as their resistance to lead increased. Hall et al. (1981) found that *Morone saxatilis* larvae were more susceptible to ozone in the presence of freshwater. Mehrle et al. (1987) concluded in a study of bass larvae that age of the larvae, concentration of contaminants and salinity of the environment must be considered in evaluating the influence of environmental contaminants on fish larvae. Hence normal loads of freshwater in plumes and increased loads during storms may not only cause changes in salinity, but result in synergistic combinations of chemicals and freshwater that are more toxic to fish larvae.

Table 12.1.   Examples of Pollutants Demonstrated, or Argued to, Affect Fish Larvae

Pollutants	Effect	Conditions	Source
Cd	M, PH	FW(L)	Woodworth and Pascoe 1982
	D	SW(L)	Westernhagen and Dethlesen 1975
	B,M	FW-E(L)	Nakagawa and Ishio 1988
Hg	V	FW-SW(L)	Servizi and Martens 1978
Hg, Pb	M	E(L)	Weis and Weis 1982
Pb, Zn, NH_4	M	FW(L)	Hall et al. 1989
Chlordane			Hall et al. 1989
Cu	G, M, B	SW(L)	Blaxter 1977
Al, pH	M	FW(L)	Buckler et al. 1987
Ag	M, PH	SW-E(L)	Klein-MacPhee et al. 1984
Ozone	M	E(L)	Hall et al. 1981
Hydrocarbon	D, M, G	SW(F)	Cross et al. 1987
Organochlorine			Cross et al. 1987
DDT	D, M, V	SW(L)	Smith and Cole 1973
PCB	LH	SW(L)	Black et al 1988
	V	SW(L)	Westernhagen et al. 1989
Industrial waste	G	FW(L)	Barron and Adelman 1984
Chlorine	M	FW(L)	Brooks and Bartos 1984
Bacteria metabolites	D	FW(L)0	Neilson et al. 1984
pH	M	E(F)	Hall 1984
Temp	D, M	SW(F)	Purcell et al. 1990
Oil	G, M, D, L, H	SW(L)	Carls and Rice 1989
Oil dispersant	M	SW(L)	Singer et al. 1990
Benzene	D, M, P	SW(L)	Struhsaker et al. 1974

Note: B = behavior, D = deformities, G = growth, LH = length at hatching, M = mortality, PH = premature hatch, P = physiological functions (e.g., heartbeat rate), V = viability of eggs. Conditions: FW = freshwater, SW = saltwater, E = estuarine, (L) laboratory study, (F) field study.

In addition to lethal acute effects of pollutants on larvae, larvae may suffer acute or chronic sublethal effects. We use the term sublethal in the following context: sublethal effects may lead to reduced survival potential during ontogeny (adapted from Rosenthal and Alderdice 1976). Primarily laboratory studies have described abnormal physiology, behavior, growth and deformities that may influence feeding, the ability to avoid predators and other factors influencing survivorship of larvae (Table 12.1). For example, larvae that have been immersed in pollutants display erratic swimming, depression of heart rate, and a loss of equilibrium (Nakagawa and Ishio 1988, Weis et al. 1989). The growth rate of fish may also be influenced by pollutants (Barron and Adelman 1984), as for sea urchins (Krause et al. 1992). It is generally accepted that fish larvae with zero or low rates of growth suffer higher rates of mortality than those with high rates of growth (Anderson 1988): i.e., reduced growth will decrease chances of survival to a given size.

Many pollution-related effects manifest themselves as deformities in fish larvae (Table 12.2). Types of deformities that have been found in many studies include those to the notochord, eyes, mouth, and yolk. A variety of conditions within these major categories have been identified. Deformities on many fish larvae are so severe that the larvae are unlikely to survive for more than a few days. Although larvae may not be killed immediately in plumes, they may suffer irreversible damage.

The possible fates of fish that may encounter pollutants during their early life history are summarized in Figure 12.6. Pollutants may induce genetic change (Stegeman, personal communication) and fatal developmental change to eggs and larvae (e.g., Rosenthal and Alderdice 1976, Swedmark and Granmo 1981, Nakagawa and Ishio 1988). If effects are not acute, larvae may suffer chronic effects that lead to reduced survival potential or death. Some larvae may survive pollutants and retain these xenobiotics as juveniles or adults. Given the small

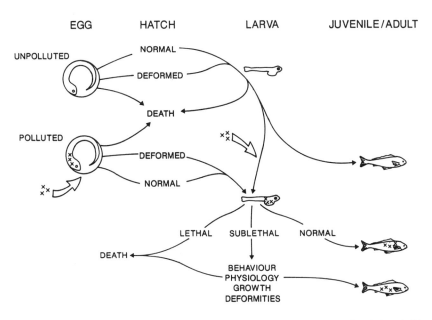

Figure 12.6. A schematic diagram of the effects pollutants may have on fish during their early life history. Pollutants may affect unfertilized eggs in the adults or fertilized eggs. Crosses indicate the entry of pollutants or the accumulation of pollutants in tissue. Death for unpolluted larvae may result from birth defects, predation, starvation or physical factors (e.g., temperature).

Table 12.2. Types of Deformities That Have Been Found in Wild and Laboratory Reared Fish Larvae

(1) Notochord	(i)	Scoliosis
	(ii)	Lordosis (or multiple dorsal kinks)
	(iii)	Sharp kink
	(iv)	Lateral curl
	(v)	Not defined
(2) Skeletal defects	(i)	Cranial anomalies
	(ii)	Unspecified
(3) Otic capsule	(i)	Missing otoliths
	(ii)	Absent capsules
(4) Eyes	(i)	Deformed retina
	(ii)	Bubble eye
	(iii)	Incomplete pigmentation
	(iv)	Opaque eye
	(v)	Monophathalmia
	(vi)	Size
	(vii)	Disorganized retinal tissue
(5) Hemorrhaging	(i)	Of myomeres
	(ii)	Of vertebrae junctions
	(iii)	Not defined
(6) Abnormally formed mouth	(i)	Missing jaw
	(ii)	Incomplete jaw
	(iii)	Delayed formation
	(iv)	Misaligned mandibles
(7) Fin defects	(i)	Reduced finfold diameter
	(ii)	Crooked rays
	(iii)	Smaller fins
(8) Yolk	(i)	Deformed
	(ii)	Bloat
	(iii)	Necrotic tissue
	(iv)	Incomplete circulation
	(v)	Underutilized

(9) Intestine

(10) Altered pigmentation

Note: Codes given with each source correspond to those on the list of deformities. Smith and Cole (1973:2,5ii); Struhsaker et al. (1974: 1v,6ii); Westernhagen and Dethlesen (1975: 1v); Rosenthal and Alderdice (1976: 1iii,v;3i,ii,4vi,vii); Weis and Weis (1982: 2i,ii;8iv); Woodworth and Pascoe (1982: 1v;5iii); Neilson et al. (1984: 1v;5i); Birchfield (1987: 2i;6i,iv); Cross et al. (1987: 1ii;6;7i;8v); Carls and Rice (1989: 1iv;8i,ii,6i;10); Weis et al. (1989: 2); Westernhagen et al. (1989: 4v); Purcell et al. (1990: 6ii;7iii); Karas et al. (1991: 1iii); Kingsford et al. (1991: 1i,ii,iv;4ii,iii,iv;7ii;9).

amount of tissue and lipid in larvae it is unlikely that accumulated pollutants greatly impair the physiological function and behavior of large juveniles and adults. Clearly, however, deformities and cell necrosis (to vital organs) accrued during early life history may greatly impair the normal function of juveniles and adults.

Deformities in Wild Larvae from Plumes

Deformities have been found in larval fish from many laboratory studies (Table 12.1). Few investigators, however, have attempted to examine larval deformities from field collections. Karas et al. (1991) found that larvae from the demersal eggs of perch had deformities only when they hatched in polluted waters. Ten percent of the embryos had sharp bends in the posterior part of the spinal cord. Purcell et al. (1990) described mass abundances of abnormal Pacific herring. Larvae had contortions to the body, reduction in the size of the jaws and pectoral fins and reduced size. The authors argued that deformities were caused by high temperatures when demersal eggs were exposed to the air (at very low tides) or warm shallow water. Although this study does not deal with pollutants, fish subjected to thermal plumes may suffer a similar fate.

Kingsford et al. (1991) found a variety of deformities in larvae in or near a sewage plume off Sydney (Table 12.2). Most abnormal fish had deformed notochords (lordosis) and/or deformed eyes. Deformed fish were found in the plume, front and control (Figure 12.7). With the exception of Time 6, highest percentages of fish with serious deformities were found in the plume and front. Variation in rank percentage of deformed fish among water masses (plume, front, ocean) resulted in a significant Time × Hydrology interaction (ANOVA, Figure 12.7). Change in rank percentage of deformed fish may be due to the different exposure times that larvae experienced in each water mass on each sampling occasion. Although low sample sizes can distort percentages, there was no relationship between abundance of larvae and number of deformed fish ($df = 1,53$; $r^2 = 0.02$; NS). Only 17 fish larvae were found with deformed or infected eyes: "bubble eye" (a pigmented bubble pinched off the perimeter of the eye), protruding eyes, opaque eyes or incomplete pigmentation on one half of the eyes. No clear patterns were found among water masses for these deformities (number, percentage of total); Plume (2, 0.08), Front (7, 0.21), Control (7, 0.2), possibly due to the greater times taken for these conditions to manifest themselves.

Caveats to Quantifying Deformities in Wild Fish Larvae and Other Approaches

Small numbers of deformed larvae in samples should not be considered as major evidence that sources of pollution have little effect. Although they may make up a small percentage of total fish in samples, this is not unexpected. Many

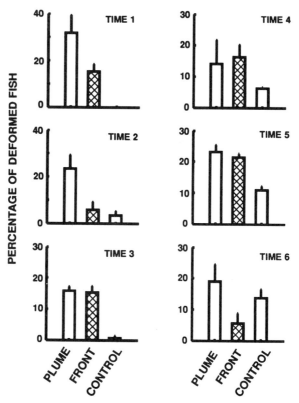

Figure 12.7. Percentage of fish per 150 m³ with "serious" categories of notochord deformities (lordosis) and deformities to the eyes; net as for Figure 12.5. Two-way ANOVA, arcsine square root transformation of proportions (abbreviations: see Figure 12.5): C = 0.34*; (R) Time (5,36), MS = 0.029 NS; (F) Hydrology (2,10), MS = 0.32*; T×H (10,36), MS = 0.054***; Residual (36), MS = 0.013 (from Kingsford et al. 1991).

pollutants act quickly on larvae and there may be only a small window of time in which larvae are deformed before they die and are lost from the water column or are eaten by predators.

The occurrence of deformed fish in unpolluted areas may be due to natural deformities, or because larvae affected by plumes have emigrated to clear waters. With the exception of data gathered by Purcell et al. (1990) little information is available on the occurrence of natural deformities in wild fish larvae. Different patterns of distribution of fish with deformities may be found according to the spatial scale of studies and oceanography. On scales <10 km it is possible with current reversal, eddies, and the behavior of larvae, that larvae affected by pollutants may be found at considerable distances from point sources of pollution

(Raimondi and Reed, Chapter 10, Keough and Black, Chapter 11). Multiple Controls (see "Beyond BACI" designs: Underwood 1991, Chapter 9) at different distances from plumes, and information on oceanography should give better resolution of this problem.

Classification of fish with deformities is a problem. Many authors do not describe what they mean by notochord deformity. Fish larvae are so delicate that effects of preservatives and other material in tows may influence the frequency of deformities. Such effects have been demonstrated for deformities to the head (Birchfield 1987). Birchfield (1987) found that the frequency of cranial deformities in larvae depended on the concentration of preservatives, temperature and the degree of agitation. Weis et al. (1989) classified newly hatched flounder according to a skeletal index as (i) normal, (ii) mild, and (iii) moderate and severe. However, this classification may be confounded for preserved fish larvae, as their category (ii) is a form of lateral curl and could be caused by preservation or fibrous material.

Kingsford et al. (1991) tested if preservation technique and the presence of fibrous material affected the quantification of larval deformities. The fibrous material we used was toilet paper, which is common in many plumes in Australia, but can be thought of as mimicking the effect of many types of particulates that might occur in the water column and alter the shape of larvae (e.g., high concentrations of *Oscillatoria* and other plankters in LOFs may lead to similar effects). When larvae were subjected to four treatments of preservation the presence of fibrous material in samples was shown to influence the number of fish with "lateral curl" (i.e., Weis et al. 1989, skeletal index ii–iii), but not fish with serious deformities such as lordosis (see Figure 12.8). The percentage of fish with deformities varied among treatments and times for serious deformities, indicating no consistent pattern (significant Time × Treatment interaction). In contrast, there was a significant treatment effect for fish with lateral curl and no interaction. The highest percentage of deformed fish was found for paper with formalin. The high mean value for MS-222 was due to some samples with very low numbers of fish (< 5). The two greatest values from the data set were in this treatment (62 and 50%; mean recalculated without these values = 11 ± 4). We conclude, therefore, that fibrous material can increase the frequency of larvae with lateral curl during preservation.

Data on the abundance of larvae in tows (design as for Figure 12.8) suggested that tows with paper caught more very small larvae (< 3 mm SL, 8–16% representation in treatments) than in treatments without fibrous material in nets, presumably by altering the functional mesh size of the net (mean number of fish larvae: Formalin = 16.5, Ethanol = 20.5, MS-222 = 15, Paper = 29; Time MS = 10.4***, Treatment MS = 0.72*, Time × Treatment MS = 0.2, Residual MS = 0.314). This indicated that the presence of fibrous material may influence the number of small larvae caught, but does not negate the fact that large numbers of larvae are found in plumes and fronts and are therefore subjected to pollutants.

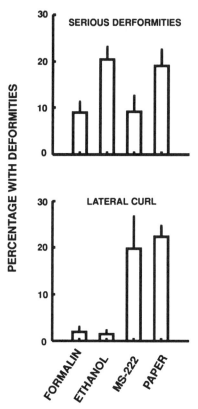

Figure 12.8. Deformities in fish from the preservation study, expressed as percentage of fish per 150 m³ (Mean ± SE). Treatments are: (i) 2–5% "Formalin"; (ii) "Ethanol"; (iii) "MS-222" and 2–5% formalin (it was thought that a fish anesthetic would relax fish prior to preservation e.g., Westernhagen 1988); (iv) toilet "paper" and 2% formalin. The fourth treatment was included because fibrous material is often caught in nets towed through sewage plumes. It was possible, therefore, that larvae were deformed as they died and were compressed by fibrous material. Large quantities of toilet paper were thrown into the net as it was towed. Following the tow the paper had disintegrated and appeared very similar to that collected from the sewage plumes. Two replicates of each treatment were collected at 5 times; net as for Figure 12.5. Two-way ANOVAs were based on arcsine square root transformation of proportions (abbreviations: see Figure 12.5): Serious deformities: C = 0.18 NS, (R) Time (4,20), MS = 0.018 NS; (F) Treatment (3,12), MS = 0.131 NS, Ti×Tr (12,20), MS = 2.56*; Residual(20), MS = 0.02; Lateral Curl: C = 0.57*, (R) Time (4,20), MS = 0.058 NS; (F) Treatment (3,12), MS = 0.46***, Ti×Tr (12,20), MS = 0.036 NS; Residual (20), MS = 0.027 (from Kingsford et al. 1991).

It is difficult to detect larval deformities in field samples. Hence, a combination of field and laboratory based techniques may be valuable. For example, controlled experiments for testing the effects of pollutants in different water masses on larvae may include larval assays done by immersing larvae in water from polluted water masses (Cross et al. 1987, Weis et al. 1989) as well as immersion of larvae *in situ* by using floating enclosures (Hall 1991). Such enclosures could be deployed in different water masses, alleviating the problems associated with choosing the concentration and composition of pollutants in traditional laboratory studies.

Discussion

It is clear that fish eggs and larvae are more sensitive to pollutants than later life-history stages. Although a considerable amount of toxicity testing has been done in laboratories, the translation of findings to field situations has proved difficult. This is partly because oceanography has not been considered in relation to concentrations of pollutants and the distribution patterns of larvae. Historically, investigators have assumed pollutants simply diffuse from point sources and have not considered accumulation of pollutants in oceanographic features such as fronts (e.g., Tanabe et al. 1991). Fish eggs, larvae and prey are generally an order of magnitude (or more) higher in LOFs (reviewed by Kingsford 1990) and other features such as thermoclines, where pollutants also accumulate. Plumes may affect larvae spawned near to and at great distances from outfalls, due to longshore currents. Moreover, reversing currents and or multiple plumes may result in multiple exposure to pollutants thus increasing the chances of deleterious effects to larvae. For fish and invertebrates, relatively brief exposures (24–96 h) to pollution plumes have resulted in reduced viability of larvae, reduced growth and increased mortality, which in some cases were manifested first as deformities (Westernhagen 1988, Krause et al. 1992, Raimondi and Schmitt 1992). More investigation is required on the effects of multiple short pulses of pollutants on fish larvae. Because the effects of a pollution plume may be spread over a wide area (also see Raimondi and Reed, Chapter 10), it is plausible that the size of adult fish stocks will be affected even with different forms of "compensation" by the stock for loss of young, as argued for the larvae entering the intakes of nuclear power plants (Nisbet et al., Chapter 13). Ultimately, the location of plume point sources, oceanography, concentration and type of pollutant, the distribution patterns of larvae, duration of the larval phase and the resistance of larvae to pollution will determine taxa that are most vulnerable and the potential influence to adult stocks (see Keough and Black, Chapter 11).

Fish often die or are deformed when they are raised in polluted water in the laboratory. Deformed larvae have been found in the field, but natural levels of deformities (e.g., Purcell et al. 1990) are poorly described and methods of preservation and fibrous material caught in plumes may artificially influence the quantification of certain types of deformities. It is unlikely that high numbers of larvae with some deformities will be captured in the field as they may quickly die or be targeted by predators. Deleterious effects of pollutants on larvae may arise without conspicuous deformities. For example, chronic damage to vital organs such as the kidney, liver, and brain (Gardner and La Roche 1973) may result in early death. Because the organs of larvae are small, the consequences of cell necrosis for the health of individuals may be easier to elucidate than for older animals (see Underwood and Peterson 1988).

We conclude that the historical view that organisms in the water column are unlikely to be affected by pollution plumes is incorrect. Moreover, oceanographic processes have a major influence on the distribution of plankters and concentrations of pollutants and thus provide an important link between pollutant discharge and dispersion and larval exposure. Increased effort should be given to field studies that focus on the oceanography, biology and chemistry of pollution plumes on small and large spatial scales.

Acknowledgments

Financial support for this work was provided by the State Pollution Control Commission (SPCC) of New South Wales and an ARC Grant to Kingsford. Thanks to I. Suthers, K. Tricklebank and P. Rendell for discussion on pollution and fish larvae, and especially to C. Osenberg, P. Raimondi, and R. Schmitt for their critical evaluation of the manuscript and J. Jeffreys for assistance with figures.

References

Anderson, J. T. 1988. A review of size dependent survival during pre-recruit stages of fishes in relation to recruitment. Journal of Northwest Atlantic Fishery Science 8:55–66.

Balls, P. W. 1990. Distribution and composition of suspended particulate material in the Clyde Estuary and associated Sea Lochs. Estuarine, Coastal and Shelf Science 30:475–487.

Barron, M. G., and I. R. Adelman. 1984. Nucleic acid, protein content, and growth of larval fish sublethally exposed to various toxicants. Canadian Journal of Fisheries and Aquatic Sciences 41:141–1150.

Beder, S. 1989. Toxic fish and sewer surfing. Allen & Unwin, Sydney, Australia.

Birchfield, L. J. 1987. Inducement of cranial anomalies in freshwater larval fish during collection and fixation. American Fisheries Society Symposium 2:170–173.

Black, D. E., D. K. Phelps, and R. L. Lapan. 1988. The effect of inherited contamination on egg and larval winter flounder, *Pseudopleuronectes americanus*. Marine Environmental Research 25:45–62.

Blaxter, J. H. S. 1977. The effect of copper on the eggs and larvae of plaice and herring. 57:849–858.

Brooks, A. S., and J. M. Bartos. 1984. Effects of free and combined chlorine and exposure duration on rainbow trout, channel catfish and emerald shiners. Transactions of the American Fisheries Society 113:786–793.

Brown, J., W. R. Turrell, and J. H. Simpson. 1991. Aerial surveys of axial convergent fronts in UK Estuaries and the implications for pollution. Pollution Bulletin 22:397–400.

Buckler, D. R., P. M. Mehrle, L. Cleveland, and F. J. Dwyer. 1987. Influence of pH on the toxicity of aluminium and other inorganic contaminants to East Coast striped bass. Water, Air, and Soil Pollution 35:97–106.

Carls, M. G., and S. Rice. 1989. Abnormal development and growth reductions of pollock *Theragra chalcogramma* embryos exposed to water-soluble fractions of oil. Fisheries Bulletin 88:29–37.

CDM Report. 1989. Camp, Dresser, McKee Review of Sydney's Beach Protection Programme. *Report prepared for the Ministry of the Environment New South Wales Government.*

Cross, J. N., J. T. Hardy, J. E. Hose, G. P. Hershelman, L. D. Antrim, R. W. Gossett, and A. E. Creelius. 1987. Contaminant concentrations and toxicity of sea-surface microlayer near Los Angeles, California. Marine Environmental Research 23:307–323.

Gardner, G. R., and G. La Roche. 1973. Copper induced lesions in estuarine teleosts. Canadian Journal of Fisheries and Aquatic Sciences 30:363–368.

Govoni, J. J., D. E. Hoss, and D. R. Colby. 1989. The spatial distribution of larval fishes about the Mississippi River Plume. Limnology and Oceanography 34:178–187.

Gray, C. A., N. M. Otway, F. A. Laurenson, A. D. Miskiewicz, and Pethebridge. 1992. Distribution and abundance of ichthyoplankton in relation to effluent plumes from sewage outfalls and depth of water. Marine Biology 113:549–559.

Green, R. H. 1979. Sampling design and statistical methods for environmental biologists. Wiley and Sons, New York, New York.

Hall, L. W., Jr. 1987. Acidification effects on larval striped bass, *Morone saxatilis* in Chesapeake Bay tributaries: a review. Water, Air, and Soil Pollution 35:87–96.

Hall, L. W., Jr. 1991. A synthesis of water quality and contaminants data on early life stage of striped bass, *Morone saxatilis*. Reviews in Aquatic Sciences 4:87–96.

Hall, L. W., Jr., D. T. Burton, and L. B. Richardson. 1981. Comparison of ozone and chlorine toxicity to the developmental stages of striped bass, *Morone saxatilis*. Canadian Journal of Fisheries and Aquatic Sciences 38:752–757.

Hall, L. W., Jr., M. C. Ziegenfuss, S. J. Bushong, M. A. Unger, and R. L. Herman. 1989. Studies of contaminant and water quality effects on striped bass prolarvae and yearlings in the Potomac River and Upper Chesapeake Bay in 1988. Transactions of the American Fisheries Society 118:619–629.

Haney, J. F. 1988. Diel patterns of zooplankton behavior. Bulletin of Marine Science 43:583–603.

Hardy, J. T., C. W. Apts, E. A. Creelius, and N. S. Bloom. 1985. Sea-surface microlayer metals enrichments in an urban and rural bay. Estuarine, Coastal and Shelf Science 20:299–312.

Hoss, D. E., L. C. Coston, and W. E. Schaaf. 1974. Effects of seawater extracts of sediments from Charleston Harbor, S.C., on larval estuarine fishes. Estuarine and Coastal Marine Science 2:323–328.

Karas, P., E. Neuman, and O. Sandstrom. 1991. Effects of a pulp mill effluent on the population dynamics of perch, *Perca fluviatilis*. Canadian Journal of Fisheries and Aquatic Sciences 48:28–34.

Kingsford, M. J. 1988. The early life history of fish in coastal waters of northern New Zealand: a review. New Zealand Journal of Marine and Freshwater Research 22:463–479.

Kingsford, M. J. 1990. Linear oceanographic features: a focus for research on recruitment processes. Australian Journal of Ecology 15:27–37.

Kingsford, M. J., and J. H. Choat. 1989. Horizontal distribution patterns of presettlement reef fish: are they influenced by the proximity of reefs? Marine Biology 101:285–297.

Kingsford, M. J., B. E. Druce, I. M. Suthers, B. M. Gillanders, and K. A. Tricklebank. 1991. Abundance and deformities of larval fish near the North Head sewerage outfall: A precommissioning study on the influence of preservatives on deformities in fish. Report to the State Pollution Control Commission of NSW.

Klein-MacPhee, G., J. A. Cardin, and W. J. Berry. 1984. Effects of silver on eggs and larvae of the Winter Flounder. Transactions of the American Fisheries Society 113:247–251.

Krause, P. R., C. W. Osenberg, and R. J. Schmitt. 1992. Effects of produced water on early life stages of a sea urchin: state-specific responses and related expression. Pages 431–444 *in* J. P. Ray and F. R. Englehardt, editors. Produced water: technological/environmental issues and solutions. Plenum Press, New York, New York.

Lassiter, R. R., and T. G. Hallam. 1990. Survival of the fattest: implications for acute effects of lipophilic chemicals on aquatic populations. Environmental Toxicology and Chemistry 9:585–595.

Leis, J. M. 1991. The pelagic stage of reef fishes: the larval biology of coral reef fishes. Pages 183–230 *in* P. F. Sale, editor. The ecology of fishes on coral reefs. Academic Press, New York, New York.

McLean, C., A. G. Miskiewcz, and E. A. Roberts. 1991. Effect of 3 primary treatment sewage outfalls on metal concentrations in the fish *Cheilodactylus fuscus* collected along the coast of Sydney, Australia. Marine Pollution Bulletin 22:134–140.

Mehrle, P. M., L. Cleveland, and D. R. Buckler. 1987. Chronic toxicity of an environmental contaminant mixture to young (or larval) striped bass. Water, Air, and Soil Pollution 35:107–118.

Mileikovsky, S. A. 1970. The influence of pollution on pelagic larvae of bottom invertebrates in marine nearshore and estuarine waters. Marine Biology 6:350–356.

Nakagawa, H., and S. Ishio. 1988. Toxicity of cadmium and its accumulation on the egg and larva of medaka *Oryzias latipes*. Nippon Suisan Gakkaishi 54:2153–2158.

Neilson, A. H., A. Allard, S. Reiland, M. Remberger, A. Tarnholm, T. Viktor, and L. Landner. 1984. Tri- and tetra- chloroveratrole, metabolites produced by bacterial O-methylation of tri- and tetra-chloroguaiacol: an assessment of their bioconcentration potential and their effects on fish reproduction. Canadian Journal of Fisheries and Aquatic Sciences 41:1502–1512.

Purcell, J. E., D. Grosse, and J. J. Gover. 1990. Mass abundance of abnormal Pacific herring larvae at a spawning ground in British Columbia. Transactions of the American Fisheries Society 119:463–469.

Raimondi, P. T., and R. J. Schmitt. 1992. Effects of produced water on settlement of larvae: field tests using red abalone. Pages 415–430 *in* J. P. Ray and F. R. Englehardt, editors. Produced water: technological/environmental issues and solutions. Plenum Press, New York, New York.

Rainbow, P. S. 1989. Copper, cadmium and zinc concentration in oceanic amphipod and euphausid crustaceans, as a source of heavy metals to pelagic birds. Marine Biology 103:513–518.

Rosenthal, H., and D. F. Alderdice. 1976. Sublethal effects of environmental stressors, natural and pollutional, on marine fish eggs and larvae. Canadian Journal of Fisheries and Aquatic Sciences 33:2047–2065.

Scarrett, D. J. 1969. Lobster larvae off Pictou, Nova Scotia, not affected by bleached kraft mill effluent. Canadian Journal of Fisheries and Aquatic Sciences 26:1931–1934.

Servizi, J. A., and D. W. Martens. 1978. Effects of selected heavy metals on early life of sockeye and pink salmon. International Pacific Salmon Fisheries Commission Progress Report 39:1–26.

Shanks, A. L. 1987. The onshore transport of an oil spill by internal waves. Science 235:1198–1200.

Singer, M. M., D. L. Smalheer, R. S. Tjeerdema, and M. Martin. 1990. Toxicity of an oil dispersant to the early life stages of four California marine species. Environmental Toxicology and Chemistry 9:1387–1898.

Smith, R. M., and C. F. Cole. 1973. Effects of egg concentrations of DDT and dieldrin on development in winter flounder (*Pseudopleuronectes americanus*). Journal of Fisheries Research Board of Canada 30:1894–1898.

Södergren, A., and L. Okla. 1988. Simulation of interface mechanisms with dialysis membranes to study uptake and elimination of persistent pollutants in aquatic organisms. Verh. International Verem. Limnolology 23:1633–1638.

Struhsaker, J. W., M. B. Eldridge, and T. Echeveria. 1974. Effects of benzene (a water soluble component of crude oil) on eggs and larvae of Pacific herring and northern anchovy. Pages 245–284 *in* F. J. Vernberg and W. B. Vernberg, editors. Pollution and physiology of marine organisms. Academic Press, New York, New York.

Swedmark, M., and A. Granmo. 1981. Effects of mixtures of heavy metals and a surfactant on the development of cod (*Gadus morhua L.*) Journal du conseil. Conseil international pour l'exploration de la mer. Rapports et procès-verbaux des rèunions 178:95–103.

Tanabe, S., A. Nishimura, S. Hanaoka, T. Yanagi, H. Takeoka, and R. Tatsukawa. 1991. Persistent organochlorines in coastal fronts. Marine Pollution Bulletin 22:344–353.

Underwood, A. J. 1991. Beyond BACI: experimental designs for detecting human environmental impacts on temporal variations in natural populations. Australian Journal of Marine and Freshwater Research 42:569–587.

Underwood, A. J., and C. H. Peterson. 1988. Towards an ecological framework for investigating pollution. Marine Ecology Progress Series 46:227–234.

Washburn, L., B. H. Jones, A. Bratkovich, T. D. Dickey, and M. S. Chen. 1992. Mixing, dispersion, and resuspension in the vicinity of an ocean wastewater plume. Journal of Hydraulic Engineering 118:38–58.

Weis, J. S., and P. Weis. 1989. Effects of environmental pollutants on early fish development. Reviews in Aquatic Sciences 1:46–73.

Weis, P., and J. S. Weis. 1982. Toxicity of methylmercury, mercuric chloride and lead in killifish (Fundulus heteroclitus) from Southampton, New York. Environmental Research 28:364–374.

Weis, P., J. S. Weis, and A. Greenberg. 1989. Treated municipal wastewaters: effects on development and growth of fishes. Marine Environmental Research 28:527–532.

Westernhagen, H. von. 1988. Sublethal effects of pollutants on fish eggs and larvae. Pages 253-346 in W. S. Hoar and D. J. Randall, editors. Fish physiology XIA. Academic Press, New York, New York.

Westernhagen, H. von, and V. Dethlesen. 1975. Combined effects of cadmium and salinity on development and survival of flounder eggs. Journal of the Marine Biology Association of the United Kingdom 55:945–957.

Westernhagen, H. von, P. Cameron, V. Dethlesen, and D. Janssen. 1989. Chlorinated hydrocarbons in North Sea whiting (Merlangus merlangus L.), and effects on reproduction. I. Tissue burden and hatching success. Helgoländer Meersuntersuchungen 43:45–60.

Williams, D. M., E. Wolanski, and J. C. Andrews. 1984. Transport mechanisms and the potential movement of planktonic larvae in the central region of the Great Barrier Reef. Coral Reefs 3:229–236.

Wolanski, E., and W. M. Hamner. 1988. Topography controlled fronts in the ocean and their biological influence. Science 241:177–181.

Woodworth, J., and D. Pascoe. 1982. Cadmium toxicity to rainbow trout, Salmo gairdneri Richardson: a study of eggs and alevins. Journal of Fish Biology 21:47–57

CHAPTER 13

CONSEQUENCES FOR ADULT FISH STOCKS OF HUMAN-INDUCED MORTALITY ON IMMATURES

Roger M. Nisbet, William W. Murdoch,
and Allan Stewart-Oaten

This chapter looks at a difficult and unsolved question in fisheries biology: how, and to what extent, can populations of marine fish "compensate" for additional negative impacts imposed upon them? It was stimulated by a pressing practical problem. A coastal nuclear power plant in Southern California was known to kill in its intakes each year several billion eggs, larvae and juvenile stages of marine fish. A regulatory body needed to decide whether or not to require mitigation of these effects and, if so, how much. Other chapters in this collection (Osenberg et al., Chapter 6, Stewart-Oaten, Chapter 7, Bence et al., Chapter 8, Underwood, Chapter 9) discuss methods of *statistically evaluating* environmental impacts in the local area focused around the polluting or degrading activity. We are concerned with a different problem, namely *deducing* an impact where it is suspected but cannot be detected because it is diffused over a large region.

In the particular example motivating our analyses it was calculated that the effect of the additional mortality caused by the power plant is probably spread at least throughout the Southern California Bight, both because immature fish are carried large distances by longshore currents and because adults can move in their lifetime over equivalent distances. Calculations suggested that these deaths might represent an increment of 1–10% in the mortality of immature stages of the Bight-wide population of one or more fish species (Parker and DeMartini 1990). But even if it were feasible to detect a consequent change in density in these populations after the power plant became operational, there is no reasonable Control population to which the Impact population could be compared (Stewart-Oaten et al. 1986). Yet if we were to conclude that no impact is present because it cannot be detected, there would never be grounds for controlling successive incremental activities that could lead cumulatively to massive reductions in adult stock, and even to extinction. This problem is but one example of the generic difficulty of determining the effects of local activities on a spatially distributed population.

A perfect answer would translate annual, local, immature losses for each species into a change in the mean abundance (and biomass) of its total adult population. As noted by Barnthouse and Van Winkle (1988), many researchers have indeed tried, and failed, to translate additional immature deaths into effects on standing stock in detail and with precision. The calculations need to include compensatory mechanisms (Savidge et al. 1988) since in their absence the additional mortality would cause extinction. The capacity to compensate is, however, necessarily limited indeed, if enough additional sources of mortality are imposed, any population will eventually be driven extinct. Sardines, white seabass, barracuda, yellowtail, Pacific mackerel, and pelagic sharks are all California marine fish species that have declined drastically in the face of additional human-imposed mortality, in spite of any compensation (CalCOFI 1983).

Thus a precise answer requires detailed, quantitative information on compensation mechanisms. This is commonly unavailable; this was, for example, the situation for the species that motivated our work. These are wholly marine populations with dynamics and potential compensatory mechanisms virtually unstudied. Yet, in spite of the difficulties, the calculation of long-term changes in abundance needs to be attempted, since, as already noted, many human actions will lead to effects which are detectable at one level (deaths of immatures), but have ultimate effects at the population level that are potentially large though undetectable because of measurement difficulties. This in turn requires population models which include *explicit assumptions* about compensatory mechanisms that cannot be tested directly. While somewhat unsatisfactory, this approach is preferable to one which simply assumes that there is an equilibrium, without identifying the mechanism (e.g., Schaaf et al. 1987). Confidence in the predictions of any particular model then depends on the extent to which its conclusions are insensitive to the specific assumptions; conversely assumptions which turn out to be critical can be identified as natural targets for further research. The notorious Hudson River controversy was exacerbated by lack of consensus among the protagonists on the role (and limitations) of models in environmental impact evaluations (e.g., Levin 1979, Limburg et al. 1986, Barnthouse et al. 1988).

In this chapter we use simple models to clarify the critical issues involved in determining the impacts of human activities on populations of marine fish. We do this in part because we could not defend the details of more complex models, but also because the issues are general. First, can *any* of the various compensation mechanisms that may exist in marine fish prevent additional immature losses from causing a reduction in the average abundance of adult fish? Second, what is the likely relationship between the fractional increase in immature mortality and the fractional change in adult equilibrium density? Third, is it likely that additional immature losses will lead to a change in stability? To sharpen the presentation, we discuss these issues in the context of our original problem—the impact of fish loss due to entrainment by a power plant—but the results are potentially much more broadly applicable.

Ideally, we would like to write detailed models of the dynamics of the fish populations in a particular locality—e.g., the Southern California Bight—based upon extensive information about the real populations. The information would include, for example, fecundity, development rate, and survival of each age class, all in relation to the density of various age classes, and also annual variation in these rates. Such information is unavailable for most species, and believable detailed models are out of the question. We therefore use the simplest models that are consistent with fish life history. They do *not* aim to be realistic portrayals of the detailed dynamics of any particular fish populations, nor are they intended to provide a precise measure of potential changes in stock size. However, the models provide a more rigorous guide than mere intuition to the likely consequences, for adult stocks, of various possible forms of compensation. By making assumptions explicit they also allow the reader to judge whether they are acceptable.

Modeling Compensatory Processes in Fish Populations

Compensation Mechanisms Modeled

We define *compensation* to be the effect on population size of density-dependent factors operating on per capita birth and death rates. It occurs because the individual fish responds to its environment, which includes other individuals like it: a reduction in the density of like individuals may make the environment more hospitable for those that remain. For example, reducing the density of larvae might cause the remainder to survive or grow better. Density-dependent factors may also result in rate processes associated with one life stage being influenced by the population density in another stage; we then use the terminology "*density-dependent response via* process x *to* stage y" to mean that the rate of some process (development, survival or reproduction) x is affected by the density of individuals in stage y.

Adults can respond via increased survival, growth or fecundity. As we focus our discussion on numbers of individuals, not biomass, we do not discuss growth *per se*: compensation by increased size becomes an implicit part of our discussion of compensation via increased survival or fecundity. We further simplify our model by assuming that adult compensation will operate in response to the density of adults only, i.e., that immature individuals do not compete with adults or prey on them.

Immatures can respond via increased survival or enhanced growth. The latter may influence population dynamics through reduced time to maturation or enhanced fecundity subsequent to maturation, adequate modeling of which involves many complexities and much species-specific detail: in particular knowledge of whether growth is size- stage-, or age-dependent, and whether fecundity is size- or age-dependent. Gurney and Nisbet (1985) explore some of the subtleties of such models in a continuous time context, but the only work of which we are aware that attempts a similar analysis for a discrete time model is that of Cushing and Li (1992) who modeled a semelparous species with a two-year juvenile development period followed by a single burst of reproduction. Having decided not to model adult biomass dynamics, we do not incorporate such mechanisms in our present model; instead, we argue that the essence of all mechanisms other than that involving enhanced fecundity is a decrease in immature mortality in response to decreased density of immatures. This we model.

In some situations, the focus of concern may be not only on the impacted fish species, but on some *predator* on that species. This was the case in our power plant study where the economic importance of some of the impacted species (e.g., queenfish) is that they are food for sport or commercial species. Thus later in the chapter we investigate the "knock-on" effects on predator populations.

A "Single-Species" Model and Its General Properties

In our investigation of most forms of compensation, we make extensive use of one particular "single-species" model, defined in Table 13.1. The model includes as special cases many of the "stock-recruitment" based models in the fisheries literature (see e.g., Rothschild 1986 and references therein) and in particular closely resembles one developed by Mittelbach and Chesson (1987) for freshwater fish; however, our analyses address different issues from those considered by these authors. Although fish development is a complex process, the model represents explicitly only two life stages: adults and immatures. We assume that reproduction occurs in a short period each year and produces a distinct cohort of immatures. Compensation occurs through functions f, g and h which describe respectively density dependence of adult fecundity, immature survival and adult survival. We assume these are nonincreasing functions, with at least one of them being strictly decreasing, so a power plant-induced decrease in adults or juveniles will lead to increased (or, at least, not decreased) fertility or survival.

Depending upon the mechanisms we wish to investigate, the "immatures" can include both the planktonic and the juvenile stages, or only the latter. When the immatures represent only juveniles, the survival of the planktonic stages is included in the adult fecundity (the parameter b and the function f in Table 13.1). The parameter S and the function g describe survival of juveniles only. Otherwise S and g describe survival from egg to first year adult.

The population dynamics are described by two equations in Table 13.1. Note that the left hand side of the first of these equations is I_t, not I_{t+1}. For example, the birth process (including hatching and survival through the planktonic stage) could take from April 1 to May 1; the juvenile stage could last until the following

Table 13.1.   The Single-Species Fish Population Model

		Variables
t	=	time in years
I_t	=	density of immatures at time t
A_t	=	density of adults at time t

		Population dynamics
I_t	=	$bf(A_t)A_t$
A_{t+1}	=	$Sg(I_t)I_t + S_A h(A_t)A_t$

		Parameters
b	=	maximum number of "births" per adult, where a "birth" is an egg that hatches and survives through the planktonic stage to become an immature
S	=	maximum survival of immatures at low immature density
S_A	=	maximum year to year survival of adult fish

		Functions
$f(A_t)$	=	ratio of per capita birth rate when the adult density is At to the maximum rate
$g(I_t)$	=	ratio of immature survival when immature density is It to maximum immature survival
$h(A_t)$	=	ratio of adult survival when adult density is At to maximum adult survival

March 31, when surviving juveniles would become full adults, and contribute to the next set of births. A_t would be the number of adults on April 1, and I_t the number of juveniles on May 1, of year t.

We assume that the power plant directly reduces the model parameters b and/or S which represent the *density-independent* components of adult fecundity and immature survival (Table 13.1).

Equilibrium and Stability

I_t can be eliminated from the model equations. The numbers of adults in successive years are thus related by the single difference equation

$$A_{t+1} = F(A_t,b,S)A_t \tag{1}$$

where

$$F(A,b,S) = bSf(A)g(bAf(A)) + S_A h(A). \tag{2}$$

The equilibrium adult population satisfies

$$F(A^*,b,S) = 1. \tag{3}$$

Sufficient conditions for Equation 3 to have at least one solution are:

(i) $bS + S_A > 1$,

AND

(ii) EITHER
      (a) at least one of the functions $f(A)$ or $g(I)$ approaches zero at sufficiently large populations (adult or immature respectively),

      OR
      (b) $f = g = 1$, $h(A) \to 0$ for sufficiently large A, and $bS < 1$.

Condition (i), demanding a positive intrinsic growth rate at sufficiently small populations, is also necessary. The conditions in (ii) define "sufficiently strong" compensation to guarantee the existence of an equilibrium, but are not necessary. The second part (b) of the second condition (ii), referring to $h(A)$, indicates that a population cannot be regulated by adult survival alone if sufficient offspring are produced after one year to yield a positive intrinsic growth rate irrespective of subsequent adult mortality.

As noted, the power plant is assumed to reduce the values of b and/or S. It is obvious from condition (i) above that if the value of this product drops below $1 - S_A$, then extinction is inevitable. Restated, this implies that a fractional decrease in the product bS of $1 - (1 - S_A)/bS$ will make extinction inevitable, *irrespective of the form of compensation*.

An equilibrium is of little relevance unless it is *locally stable*, i.e. following a small perturbation, the stock returns to its previous level. If $a_t\ (= A_t - A^*)$ is a small deviation from equilibrium in year t, then (to first order of approximation)

$$a_{t+1} \approx z\, a_t \tag{4}$$

with $z = A^*[dF/dA]^* + 1$,

(Note: * means that a derivative is evaluated at (A^*, b, S), i.e., at the unimpacted equilibrium and values of b and S.) The "eigenvalue," z, determines whether, after a small perturbation, the system oscillates ($z < 0$) or changes monotonically ($z > 0$) and also whether it returns to equilibrium ($|z| < 1$) or not ($|z| > 1$). With a little algebra, we can show that

$$z = (1 - S_A h(A^*))PQ + S_A h(A^*)R \tag{5}$$

where

$$P = 1 + A^* f'(A^*)/f(A^*) \tag{6}$$
$$Q = 1 + I^* g'(I^*)/g(I^*) \tag{7}$$
$$R = 1 + A^* h'(A^*)/h(A^*). \tag{8}$$

With a little further algebra, it can be proved that if the derivatives f', g' and h' are all nonpositive, then $z \leq 1$. The inequality is strict if at least one of the derivatives f', g' and h' is strictly negative. Thus, *provided there are no depensatory mechanisms* (i.e., provided f, g and h are all nonincreasing, so that increased populations cannot have increased fecundity or survival), *monotonic divergence from equilibrium is not possible in this model*. However, it *is* possible to have $z < -1$ and hence oscillatory instability. Here the cause of instability is "overcompensation", i.e., compensation which is so strong that after a small perturbation, A_{t+1} is further above (or below) the equilibrium than A_t was below (or above) it. With most plausible forms for the functions f, g, and h, these oscillations do not grow in amplitude forever, rather the population exhibits sustained oscillations or chaotic fluctuations (e.g., May 1976, May and Oster 1976).

Sensitivity of Adult Equilibria

As already noted, the operation of the power plant is assumed to reduce S and/or b. We define the *sensitivity* of the equilibrium value, A^*, to power plant-induced changes in S (or b) to be the fractional change in A^* relative to a small fractional change in S (or b) that causes it, other parameters being unchanged. More precisely, if a change of δS in S (or of δb in b) causes a change δA^* in A^*, then the sensitivities σ_S and σ_b are defined by

$$\sigma_S = [\delta A^*/A^*]/[\delta S/S] = [S/A^*] \, \partial A^*/\partial S; \tag{9a}$$

$$\sigma_b = [\delta A^*/A^*]/[\delta b/b] = [b/A^*] \, \partial A^*/\partial b . \tag{9b}$$

A sensitivity index greater than one implies (for a stable population) that the fractional change in adult stock due to the power plant will exceed in magnitude the fractional change in b or S.
With a little algebra, we can relate the sensitivity indices to previously-defined quantities. Thus,

$$\sigma_S = \{1 - S_A h(A^*)\}/(1 - z); \tag{10a}$$

$$\sigma_b = Q \, \sigma_S \tag{10b}$$

with Q defined by Equation 7. Since $z \leq 1$, Equation 10 shows that in this model σ_S is always positive; thus, *a reduction in immature survival inevitably leads to a reduction in the equilibrium adult population*. If $Q > 0$, the same holds for the fertility, i.e., $\sigma_b > 0$ so a reduction in b leads to a reduction in A^*. However, it is possible that $Q < 0$, so the number of immatures surviving to adulthood actually increases when the number of immatures produced decreases; a decrease in b could then lead to an increase in A^*. This is discussed in Model 3 in the next section.

Sensitivity and Stability

Equation 10 shows that the questions of the impact on equilibrium densities and on stability of the additional mortality due to the power plant are not independent. If two populations have the same adult mortality at equilibrium, i.e., the same value of $1 - S_A h(A^*)$ but different values of z, then the population with the smaller value of z will be less sensitive to a change in S, i.e., its equilibrium value will change less. If there is no density dependence in adult survival, this implies a simple relationship between σ_S and z which is illustrated in Figure 13.1. Of particular interest is the fact that situations which lead to a small value of the sensitivity σ_S also lead to a high likelihood of instability ($z < -1$). Equation 10 shows that a similar result holds for the sensitivity index σ_b in the absence of immature compensation ($Q = 1$ in Equation 7). With immature compensation, the interrelation between sensitivity and stability is more complex since as already noted σ_b can be negative. However the general insight that small (now possibly negative) sensitivities are associated with an enhanced likelihood of instability appears to remain valid.

In the following section, we look at some special cases of the model, with particular emphasis on cases where only one of f, g and h is not constant.

Predicted Consequences of Increased Immature Mortality

The results in the previous section are useful, but too general to permit focused comment on the questions of interest for environmental impact studies. To the critical question of whether the sensitivity indices will be less than one (implying that compensation acts to ameliorate impacts), we can only answer that it depends on functional forms. In order to sharpen our understanding we now consider three models, each a special case of the more general model

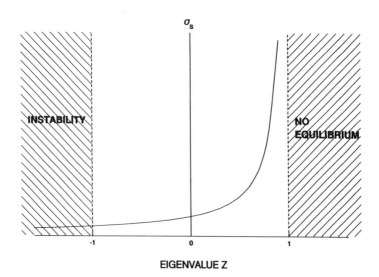

Figure 13.1. Variation of equilibrium sensitivity σ_S with eigenvalue z for the situation where $h(A) = 1$ for all A, i.e., no compensation via adult mortality.

presented above, but incorporating only one of the possible compensatory mechanisms. Thus, each model is obtained by assuming two of the three functions f, g, h to be identically equal to one. The details of the models are in Table 13.2, which also contains expressions for the sensitivity indices, eigenvalue, and the fractional loss required for extinction. Equipped with the results of these analyses, we conclude the section with a brief discussion of the situation where more than one compensation mechanism operates.

Model 1: Density-Dependent Fecundity

In this model, "birth" means recruitment to the juvenile population, so that fecundity includes the survival of eggs and planktonic larvae. We assume there is no density dependence in either juvenile or adult survival, so the population compensates only via $f(A_t)$, i.e., through an increase in fecundity and/or planktonic survival in response to a decrease in adult density. The dynamic consequences are the same, regardless of whether it is adult fecundity or planktonic survival that is density dependent. The adult equilibrium is obtained from Equation 3 which simplifies to

$$f(A^*) = (1 - S_A)/bS. \tag{11}$$

The effect of the power plant is to reduce b and/or S. Since in this case both sensitivity indices are positive and equal ($Q = 1$; see previous section), this implies a reduced adult population. From Table 13.2, the sensitivity indices are less than 1 only if, at equilibrium, $d[Af(A)]/dA < 0$; thus, the fractional reduction in the equilibrium density is less than the fractional reduction in bS only if fecundity is very strongly density dependent. This can be made more precise by noting that the slope of the function $bSAf(A)$, which represents the total annual recruitment to the adult population, is $bS[f(A^*) + A^*f'(A^*)]$ and is thus positive if (and only if) the sensitivity indices are greater than one. Thus, in this model *the magnitude of the sensitivity index may be inferred from the slope of the stock-recruitment relationship at equilibrium.*

Table 13.2. Summary of Results for the Three Models

Response via To	Model 1 Fecundity Adult density	Model 2 Adult survival Adult density	Model 3 Immature survival Immature density
z	$1 + bSA^*f'(A^*)$	$1 + S_A A^* h'(A^*)$	$1 + b^2 SA^* g'(bA^*)$
σ_b	$-f(A^*)/[A^*f'(A^*)]$	$-bS/[S_A A^* h'(A^*)]$	$-1-g(bA^*)/[bA^*g'(bA^*)]$
σ_S	$-f(A^*)/[A^*f'(A^*)]$	$-bS/[S_A A^* h'(A^*)]$	$-g(bA^*)/[bA^*g'(bA^*)]$

Note: z denotes the eigenvalue characterizing the stability of the population, and σ_b and σ_S are sensitivity indices, defined in Equation 9.

An instructive example that illustrates these points is obtained by generalizing the well-known Beverton-Holt stock-recruitment relation and assuming

$$f(A) = 1/[1 + (A/A_0)^n].$$ (12)

where A_0 is a parameter interpretable as the adult density at which fecundity is reduced by 50%. If $n > 0$, per capita birth rates decrease as A increases; if $0 < n < 1$, total recruitment increases without limit as A increases; $n = 1$ is the standard Beverton-Holt model; if $n > 1$, total recruitment peaks when $A = A_0/(n - 1)^{1/n}$, and then declines, so that high stock levels lead to low recruitment (see Figure 13.2). This choice of f gives

$$A^* = A_0[bS/(1-S_A) - 1]^{1/n},$$ (13)

the eigenvalue is

$$z = 1 - n(1-S_A)[1 - (1-S_A)/bS],$$ (14)

and the sensitivity indices are given by

$$\sigma_S = \sigma_b = n^{-1}bS/[bS - (1-S_A)].$$ (15)

Equation 14 shows that with this (rather broad) family of generalized Beverton-Holt functions *a power plant-induced reduction in the product bS will never destabilize the system* (by decreasing z to a value less than -1). Equation 15 tells us that *the equilibrium sensitivity indices are always greater than one unless $n > 1$*, i.e., the percentage reduction in population is greater than the percentage reduction in bS unless compensation is stronger than the standard Beverton-Holt form. In the case $n > 1$, no general comment is possible, as the sensitivity indices may take values greater or less than one depending on the precise values taken by b, S, and S_A. Note that Equations 14 and 15 combine to give $z = 1 - (1 - S_A)/\sigma$, where $\sigma = \sigma_S$ or σ_b. Thus increasing n to reduce σ will eventually lead to $z < -1$ and instability.

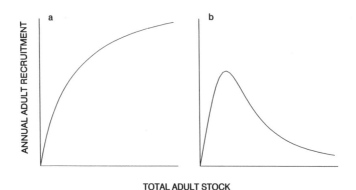

Figure 13.2.   Stock recruitment relations with the generalized Beverton-Holt form. (a) $n = 1$, (b) $n = 3$.

More generally, if A is large and $f(A)$ small, $f(A)$ might behave like cA^{-n} for some constants c and n; then σ_S is approximately $1/n$. The fractional decrease in A^* is smaller than the corresponding decrease in bS if n is large (strong compensation) and larger if n is small. The borderline case, $n = 1$, occurs when total recruitment is approximately constant, independent of the size of the adult stock. However there is again a problem with the "optimistic" case $n > 1$: the possibility of instability. Specifically the system is unstable whenever $n > 2/(1 - S_A)$.

It is possible to develop similar arguments with countless families of functions, and furthermore to generalize the above results to the situation where the developmental time is greater than one year (using e.g., models in Bergh and Getz 1988, Getz and Haight 1989), but we are aware of no plausible examples that contradict our main conclusions for this model, namely that *unless there is overcompensation, the power plant will produce a reduction in adult stock at least as large as the reduction in bS and will never destabilize the system.*

Model 2: Density-Dependent Adult Survival

In this case, $h(A)$ is the only function not identically equal to one. The equilibrium is given by

$$h(A^*) = (1 - bS)/S_A, \tag{16}$$

and exists only if $bS < 1 < bS + S_A$. The first of these inequalities reflects the fact that adult mortality (i.e., mortality after the year of growth from egg to adult) cannot control the population if each adult produces more than one replacement in its first year; the population will then grow without limit. The second inequality is simply the requirement that at low adult densities (where $h(A)$ is approximately one) the population should be viable: on average, the new population each year will consist of a fraction S_A of the adults of the old population plus bS new adults for each adult in the old population. Where there is an equilibrium, the eigenvalue and the equilibrium sensitivity indices are given by the expressions exhibited in Table 13.2.

Further progress requires assumptions about the function $h(A)$. If this function is convex, it turns out that the eigenvalue must be positive, implying exponential stability (i.e., a steady return to equilibrium), and that

$$\sigma_S = \sigma_b > bS/S_A[1-h(A^*)] = bS/(bS + S_A - 1) > 1, \tag{17}$$

implying that *the percentage decrease in adult stock will be more than that in b and/or S.*

If h is sigmoidal, as might happen if density dependence is weak until the population grows large enough to strain resources, it is possible that $-A^*h'(A^*) > 1 - h(A^*)$, so that σ_S and σ_b are smaller than 1, implying a smaller fractional change in A^* than in bS. But if $-A^*h'(A^*) > 2/S_A$, the equilibrium will be unstable because of overcompensation.

Model 3: Density Dependence within Immature Stage

Here, $f = h = 1$, so compensation acts through immature survivorship, $g(I_t) = g(bA_t)$. The equilibrium is given by

$$g(bA^*) = (1 - S_A)/bS. \tag{18}$$

The mathematical expressions for the eigenvalue, sensitivity indices, and the fractional loss that produces extinction are again given in Table 13.2, but particular care in interpretation is necessary. It is useful to distinguish three situations characterized by whether density dependence operates on immatures (i) younger than, (ii) the same age as, or (iii) older than those impacted by the power plant. Case (i) is equivalent to Model 1 provided (as in previous section) we define fecundity to include survival through the earliest life stages, and determine the sensitivity from σ_S. We now discuss the other two cases in more detail.

Case (ii): Density Dependence in Stages Impacted by the Power Plant.   Here the power plant is assumed to affect only S. The parameter b now refers strictly to the (fixed) number of eggs produced per adult. Since b is fixed, σ_S is the relevant sensitivity index (see Table 13.2). Much of the discussion of Model 1 carries over with minor reinterpretation. In particular, a reduction in S implies an increase in $g(bA^*)$ (see Equation 18); this implies a decrease in A^*, since g is a decreasing function and b is fixed. Thus, *even though immature survival increases in response to power plant effects on the immatures, fewer individuals recruit to the adult population, and hence average adult population is reduced.* Also, the sensitivity is greater than unity (so the fractional reduction in A^* will be greater than that in S), unless under preimpact conditions a decrease in new immatures would have led to an increase in new adults at equilibrium.

Case (iii): Density-Dependent Survival of Late Immatures.   We assume that the late immatures (i.e., juveniles) suffer no mortality from the power plant, as an extreme case of very low mortality on this age group. We further assume, however, that the power plant does increase mortality on previous immature stages. The variable I_t now represents the juveniles, and the parameter affected by the plant is b, which in turn represents the number of "births" per adult into the *juvenile* age class (I_t). It is affected via mortality on all immatures up to the late juvenile stage. S, the maximum survival of juveniles at low juvenile density, is assumed unaffected because this stage is assumed to have negligible power plant-induced mortality. Compensation again acts through $g(I_t)$. Thus, σ_S is ignored, and σ_b is the relevant sensitivity index. From Table 13.2, it is clear that if compensation is strong enough (i.e., if the magnitude of g' is large enough) σ_b can be negative, implying that a small decrease in b can lead to an actual increase in the equilibrium adult population.

The generalized Beverton-Holt model, (Equation 12 with "g", "I", and "I_0" replacing "f", "A", and "A_0"), is also plausible here. We write $E = (bS + S_A - 1)/bS$; when the population is low (no compensation); this is the "excess" fraction of new adults beyond those needed to replace the adults who died. We then get $z = 1 - n(1 - S_A)E$ and $\sigma_b = -1 + 1/En$. Thus, the sensitivity index is less than 1 if $nE > 0.5$ and is negative if $nE > 1$; but the system is unstable if $nE > 2/(1 - S_A)$.

Combinations of Compensatory Mechanisms

There is very likely to be more than one compensatory mechanism operating, and it is of interest to know how they interact. The full model with all mechanisms operating is very complex, its properties being summarized in Equations 1–10. While we have made no attempt to interpret these general formulae, it is of interest to study the combined effects of two of the most likely of the mechanisms studied.

We assume that there is density-dependent adult fecundity and density-dependent immature survival, but no density dependence in adult mortality. We also assume that the equilibrium population occurs in a region of the curves $f(A)$ and $g(I)$ where

$$f(A) = cA^{-n}, \; g(I) = dI^{-m}. \tag{19}$$

The sensitivity indices are now

$$\sigma_S = 1/[1-(1-n)(1-m)] \tag{20}$$

and

$$\sigma_b = (1-m)/[1-(1-n)(1-m)]. \tag{21}$$

The sensitivity index to changes in immature survival is greater than one unless $(1-m)(1-n) < 0$, which requires overcompensation in one of the processes (i.e., $m > 1$ or $n > 1$). The case of double overcompensation ($m > 1$ and $n > 1$) is very complex (Tschumy 1982; Rodriguez 1988; Nisbet and Onyiah 1994), since multiple equilibria can occur; being aware of no evidence to suggest that this mechanism for multiple equilibria is of importance in marine fish, we do not explore this point further. The expression for sensitivity to changes in b is of interest since it includes the only case we studied (Model 3, Case (ii)) where total compensation was plausible. We see that the results of that section appear quite robust: perfect compensation occurs with $m = 1$ irrespective of the other compensation via adult fecundity.

Impact on Predators

We now consider the possibility that the fish species that suffers additional immature mortality is a major food item for a predator species. Here, our standard model is not suitable, since the life cycles of fodder and predator species are probably different. We therefore model this situation in a continuous-time framework.

First, we assume that the compensatory mechanisms for the fodder fish operate in response to adult density only. We also assume that the predatory fish species depends entirely upon the fodder fish for food.

A standard type of model is

$$dA/dt = AF(A,b) - Pk(A) \qquad (22)$$
$$dP/dt = cPk(A) - mP, \qquad (23)$$

where A and P are respectively the densities of adult fodder fish and of predators. The term $k(A)$ is the "functional response" giving the number of prey killed per predator per unit time. These prey are converted into new predators with efficiency c. The term $F(A,b)$ represents the *per capita* rate of change of the adult stock, as a function of the stock size, but in the absence of predation; b is the recruitment rate, which may be affected by the power plant. Thus, F plays a role similar to that in Equation 1. The effect of the power plant is to reduce F by reducing b, which combines the roles previously played by b and S. We assume the predator suffers a constant death rate, m.

We assume that, for a given b, a reduction in A leads to an increase in F; thus, compensation will act if A is reduced. But from Equation 23, the equilibrium fodder fish population is given by $k(A^*) = m/c$. Thus *this population is completely unaffected by the reduction in F caused by the power plant, so no compensatory mechanism is brought into operation.*

However, the equilibrium population for the *predator* species is now given by Equation 22:

$$P^* = A^*F(A^*,b)/k(A^*). \qquad (24)$$

A reduction in b will thus not affect A^* but will cause a reduction in the predator species. How large the reduction is depends on the role of b in F. For example, if $F(A^*,b) = bF_1(A^*) - dF_2(A^*)$, i.e., growth rate = birth rate − death rate (from causes other than predation), then the fractional change in P^* is greater than that in b.

There are *a priori* reasons to doubt the realism of this simple, Lotka-Volterra-type model as it excludes many factors, including some emphasized in the single species models. Furthermore, it turns out that with many forms for the function F (or for F_1 or F_2), the equilibrium is likely to be very unstable, with large amplitude cycles in both prey and predator. However the result that the fodder fish density is set by attributes of the *predator* is rather general and will hold provided there is no direct dependence on *predator* density of the predator death rate, the efficiency of converting prey into new predators, and/or the functional response. It does however require that the predator is a specialist, regulated exclusively by the availability of one particular prey species.

Modification of the model to cover either of the above weaknesses is beyond the scope of this chapter. However, it would seem plausible that relaxation of the model assumptions will prevent the complete transfer of losses from the fodder to the predator fish (cf. McCauley et al. 1988, and Kretzchmar et al. 1993 for related studies). The overall conclusions are thus rather weak; the model predicts a "greater than one" sensitivity index implying that fractional loss of predators will be greater than fractional loss of immature fodder fish, but plausible modifications may reverse this. There appears to be little experimental evidence either way.

Discussion

The main message to emerge from our analysis is unsurprising: most, but not all, forms of compensation fail to prevent a decline in adult stock in response to enhanced mortality of immatures due to power plant operation. Only with very strong compensation, as might occur with a "humped" stock-recruitment relation, is the fractional decline in adult stock likely to be significantly smaller than the fractional increase in immature mortality. The most important exception to these generalizations is the situation where there is sufficient compensation in the *late* immature stages (i.e., after impact but before maturation) to prevent an appreciable decline in adult stock. Thus, while the present suite of "strategic models" does not provide an unambiguous pointer to the likely magnitude of the decline in adult density caused by death of a specified fraction of the immatures, it does significantly reduce the field of candidate mechanisms that might lead to an "optimistic" outcome (fractional decline in adult stock that is significantly less than the fraction of immatures killed).

There are disturbingly few data to help us assess the likelihood that these mechanisms may be important in real marine fish populations. Research by Ziljstra and Witte (1985) and by Van der Veer (1987) suggests density-dependent mortality (related to predation) of immature plaice. A review (Saila et al. 1987) of evidence concerning compensation in fish considered 13 species of which only three spend all of their lives (including the egg and larval stages) entirely in the ocean: Pacific herring, northern anchovy and Atlantic cod. The authors argued, on general ecological grounds, that compensation must be common in both the adult and immature stages, though they noted that there was only weak evidence for density dependence in herring, perhaps occurring via reproduction, and that herring populations are very sensitive to increased negative impacts. They found evidence for density-dependent growth of adults in anchovy and cod, and in some herring populations but not others. Density-dependent adult growth can be expected to be accompanied by density-dependent fecundity, which has been observed in anchovy but has not been well established in cod, except via increased growth.

There is no evidence for increased adult survival at lower stock sizes in the three species reviewed by Saila et al. (1987). The relationship between adult survival and adult density is notoriously difficult to estimate, however, and it is possible that there might be such a response via increased food intake. On the other hand, McCall (personal communication) suggests that at least one component of adult mortality is likely to be depensatory: marine birds and mammals that feed on adult fish are such information-rich feeders that they are likely to take a constant amount of food over a very wide range of fish densities. McCall suggests that in this sense these predators are analogous to modern information-rich fishing that uses technology such as sonar and aerial surveys to achieve similar constant yields regardless of fish density.

Direct evidence on compensation in immature stages is sparse for marine fish, though compensation via immature growth and survival is well established for freshwater fish. Reductions in the density of immatures may lead to compensation via: (1) increases in their food intake and hence in their growth rate and survival (e.g., Shepherd and Cushing 1980), and (2) decreases in the rates of predation and cannibalism.

Process (1) could arise in two ways: (i) the per head rate of intake of food could depend *directly* on the number of individuals searching for the food; i.e., a consumer-dependent functional response, or (ii) lower densities of immature fish could lead to higher densities of their food, and thus to higher feeding rates. Hypothesis (i) is presently contentious (e.g., the debate on "ratio dependence" reviewed by Hanski 1991, see also Ruxton et al. 1992, for discussion of possible mechanisms) and we know of no evidence for any species of marine fish that the functional response of individual immatures in nature depends on the density of other immatures in the population. Hypothesis (ii) can be valid only if food density had previously been suppressed by the immature stages of the species of concern. It is possible for this condition to be satisfied. But for the species that motivated our study, and for those studied by Saila et al. (1987), food particles come from such a variety of sources and are eaten by such a variety of species that the condition is unlikely to hold.

Process (2), density-dependent predation or cannibalism on immatures, is also controversial. There is extensive evidence that, in general, the short-term effects of predation (i.e., the response of individual predators) are typically either density independent or even inversely density dependent (Murdoch and Bence 1987). The predatory behavior of individual cannibals (larger immatures) feeding on their own species presumably obeys similar rules to predation: the functional response is typically type 2 and hence depensatory in the short term.

There is no good empirical evidence for compensation via response to immature density specific to the three marine species considered by Saila et al. (1987). We were not able to find reliable evidence for density-dependent predation or cannibalism. Saila et al. (1987) found evidence for *spatially* density-dependent predation in anchovy, but this is as likely to be destabilizing as stabilizing (Murdoch and Stewart-Oaten 1989, Murdoch et al. 1992). At least one piece of evidence presented on anchovy shows predation to be depensatory (Saila et al. 1987, p. 23) and the remaining evidence (Saila et al. 1987, p. 24) does not specify whether the *number* or the *fraction* of larvae eaten increased with their density. There was no evidence for density-dependent predation in the other species.

Finally we consider the mechanism which led to exceptions to our general conclusions: compensation which acts in a late stage in response to a change in an earlier stage. The hypothesis here is that there is a bottleneck at the late juvenile stage, so that, over a wide range of larval densities, a more or less constant number passes through the late juvenile stage to adulthood, regardless of the number entering the stage. This could occur, for example, if there were a fixed number of refuges or territories for late juveniles. There are well-documented examples of this phenomenon in freshwater species, with support from detailed individual-based models (e.g., DeAngelis et al. 1991 show how size-related winter survival of juvenile small-mouth bass can lead to overcompensation for earlier low survival). However, again we know of no evidence for this type of compensation in exclusively marine species.

To summarize the evidence on compensation mechanisms, there seems to be adequate evidence in wholly marine species for density-dependent adult fecundity. The fraction of immatures cannibalized may also increase with increasing adult density, though there does not seem to be good evidence for this. The evidence is at best weak that immature stages (larval or juvenile) compensate for reductions in immature density, for example by faster growth or higher survival. Thus, we are forced to the conclusion that optimistic outcomes all appear to demand mechanisms which have not been proved *in any* marine fish *anywhere*. Still, compensation holding the percentage loss in adult stock to about the same as the percentage of immatures killed seems plausible. It is more plausible if several compensatory mechanisms operate simultaneously, and still more if one of these is adult survival.

A second important property of our models is that the adult population is unlikely to be *destabilized* by the action of the power plant. This is because with most forms of compensation, decreases in the model parameters b and/or S cause an increase in the eigenvalue z. Levin (1979), in testimony to the U.S. Environmental Protection Agency concerning power plants on the Hudson River, emphasized that calculations of shifts in equilibrium population densities were meaningless unless these equilibria were stable. We agree. However, our results suggest that calculations based on equilibrium shifts are robust unless the population of interest was unstable *prior* to impact.

This argument might be challenged on the grounds that our results are fragile as they rest on a particular form for the assumed fish life cycle; i.e., reproduction occurs after exactly one year. Certainly, the main mechanism producing instability in this and other single species fish models is negative feedback occurring after a time lag which is comparable in duration with the natural response time of the population. However in most simple discrete or continuous time models with time delays, reduction in the intrinsic growth rate of the population at low densities has a stabilizing effect (e.g., Nisbet and Gurney (1982) Chapters 2 and 8 for general discussion or Levin and Goodyear (1980) for a model similar to our Model 1 which was applied to the Hudson River striped bass stock). The effect of adding time delays to prey-predator models, such as that used in the section above, can be more complex (Hastings 1984, Nunney 1985a, 1985b, Murdoch et al. 1987) but we know of no studies involving biologically plausible functions and parameters where reduction in intrinsic prey growth rate causes instability.

We conclude with some comments on the limitations of our modeling. While we hope our conclusions will prove robust against changes in model detail, there are some aspects that merit careful further investigation. First, we have neglected two aspects of size structure: intersize-class interactions within a cohort, and intercohort interactions. The former is important if cannibalism of planktonic stages by juveniles is of importance. We have conducted a preliminary study of this mechanism using a continuous-time model and (to date) find equilibrium shifts consistent with our broad conclusion that in the absence of the "late juvenile bottleneck", sensitivity indices are likely to be around unity or larger. However the results are numerical, and further analysis is required before these conclusions can be accepted as being more than a consequence of an arbitrary selection of parameter values. Intercohort interactions are likely to be of greatest concern in contexts where immature stages last more than one year and older immatures influence the survival of new recruits. Cushing and Li (1989) and Nisbet and Onyiah (1994) demonstrate that such interactions can lead to instability and multiyear cycles in populations of semelparous organisms; a similar study in the context of the iteroparous life cycle assumed here would be valuable. Second, our somewhat cavalier treatment of compensation via variation in time to maturity is a weakness in the present work. We do not believe this aspect to be of critical importance, as we do not anticipate major changes from those obtained using our model of immature survival.

The most important limitation of our modeling arises when we are concerned with situations where compensation operates in response to the *same* stage as suffers the impacts (e.g., our Model 3, Case (ii)). We represented compensation in terms of functions whose arguments are stage populations *on one date of the year*. This is a convenient, phenomenological approach, which can be defended on the basis that although mortality throughout the year is the result of some complex interactions which should be modeled in continuous time (probably using differential equations), the final survival will be some function of initial density. (This argument is often used to justify the traditional fisheries models - e.g., Rothschild 1986) However, a change in the density-independent component of mortality will in general change the *functional dependence* of survival on initial density and not just the density-independent component of some predetermined function as assumed here. This point has motivated much research on models which include a very detailed description of the youngest year classes. A particularly instructive example of this type of model is presented by Christensen et al. (1977). These authors presented a very detailed model of the dynamics of young striped bass, which was coupled with a simple description of year-to-year survival of all other year classes. Their primary interest was in a measure of equilibrium sensitivity equivalent to that used in this chapter. Interestingly, they were able to establish for that model a relationship between that sensitivity index and the slope of the stock-recruitment relation at equilibrium that is equivalent to our Equation 10. More recent work (e.g., Van Winkle et al. 1993) uses an individual-based description of the first year of life to determine the size of the cohort recruiting to the next year of life. Future research using individual-based descriptions of *within-year* dynamics for critical life stages and simplified descriptions of the remaining year classes would appear particularly appropriate.

Acknowledgments

We thank Craig Osenberg, Russ Schmitt and two anonymous reviewers for detailed criticism of an earlier draft of this chapter. This research was supported in part by Marine Review Committee, Inc., and in part by grants to AS-O from the Minerals Management Service, U.S. Department of the Interior under MMS Agreement No. 14-35-0001-30471 (The Southern California Educational Initiative) and to WWM from the U.S. Department of Energy (Grant DE-FG03-89ER60885). None of the above organizations necessarily agrees with the results or conclusions.

References

Barnthouse, L. W., and W. VanWinkle. 1988. Analysis of impingement impacts on Hudson River fish populations. Pages 182–190 *in* L. W. Barnthouse, R. J. Klauda, D. S. Vaughan and R. L. Kendall, editors. Science, law, and Hudson River power plants: a case study in environmental impact assessment. American Fisheries Society, Bethesda, Maryland.

Barnthouse, L. W., R. J. Klauda, D. S. Vaughan, and R. L. Kendall, editors. 1988. Science, law, and Hudson River power plants: a case study in environmental impact assessment. American Fisheries Society, Bethesda, Maryland.

Bergh, M. O., and W. M. Getz. 1988. Stability of discrete age-structured and aggregated delay-difference population models. Journal of Mathematical Biology 26:551–581.

CalCOFI. 1983. The larger pelagic fishes of the California current. Symposium. A. McCall convener. Report Vol. 24 (introduction).

Christensen, S. W., D. L. DeAngelis, and A. G. Clark. 1977. Development of a stock-progeny model for assessing power plant effects on fish populations. Pages 196–226 *in* W. Van Winkle, editor. Assessing the effects of power-plant-induced mortality on fish populations. Pergamon Press, New York, New York.

Cushing, J. M., and J. Li. 1989. On Ebenman's model for the dynamics of a population with competing juveniles and adults. Bulletin of Mathematical Biology 51:687–713.

Cushing, J. M., and J. Li. 1992. Intra-specific competition and density-dependent juvenile growth. Bulletin of Mathematical Biology 54:503–519.

DeAngelis, D. L., L. Godbout, and B. L. Shuter. 1991. An individual-based approach to predicting density-dependent dynamics in smallmouth bass populations. Ecological Modeling 57:503–519.

Getz, W. M., and R. Haight. 1989. Population harvesting: demography models of fish, forest and animal resources. Princeton University Press, Princeton, New Jersey.

Gurney, W. S. C., and R. M. Nisbet. 1985. Generation separation, fluctuation periodicity and the expression of larval competition. Theoretical Population Biology 28:150–180.

Hanski, I. 1991. The functional response of predators: worries about scale. Trends in Ecology and Evolution 6:141–142.

Hasting, A. 1984. Delays in recruitment at different trophic levels: effects on stability. Journal of Mathematical Biology 21:35–44.

Kretzchmar, M., R. M. Nisbet, and E. McCauley. 1993. A predator-prey model for zooplankton grazing on competing algal populations. Theoretical Population Biology 44:32–66.

Levin, S. A. 1979. The concept of compensatory mortality in relation to impacts of power plants on fish populations. Testimony prepared for the U.S. Environmental Protection Agency.

Levin, S. A., and C. P. Goodyear. 1980. Analysis of an age-structured fishery model. Journal of Mathematical Biology 9:245–274.

Limburg, K. E., S. A. Levin, and C. C. Harwell. 1986. Ecology and estuarine impact assessment:

lessons learned from the Hudson River (U.S.A.) and other estuarine experiences. Journal of Environmental Management 22:255–280.

May, R. M. 1976. Simple mathematical models with very complicated dynamics. Nature 261:459–467.

May, R. M., and G. F. Oster. 1976. Bifurcations and dynamic complexity in simple ecological models. American Naturalist 110:573–599.

McCauley, E., W. W. Murdoch, and S. Watson. 1988. Simple models and variation in plankton densities among lakes. American Naturalist 110:573–599.

Mittelbach, G. G., and P. L. Chesson. 1987. Predation risk: indirect effects on fish populations. Pages 315–332 in W. C. Kerfoot and A. Sih, editors. Predation: direct and indirect effects on aquatic communities. New England University Press, Hannover, New Hampshire.

Murdoch, W. W., and J. R. Bence. 1987. General predators and unstable prey populations. Pages 17–30 in W. C. Kerfoot and A. Sih, editors. Predation: direct and indirect impacts on aquatic communities. University Press of New England, Hanover, New Hampshire.

Murdoch, W. W., and A. Stewart-Oaten. 1989. Aggregation by parasitoids and predators: effects on equilibrium and stability. American Naturalist 134:288–310.

Murdoch, W. W., R. M. Nisbet, S. P. Blythe, W. S. C. Gurney, and J. D. Reeve. 1987. An invulnerable age class and stability in delay-differential parasitoid-host models. American Naturalist 129:263–282.

Murdoch, W. W., C. J. Briggs, R. M. Nisbet, W. S. C. Gurney, and A. Stewart-Oaten. 1992. Aggregation and stability in metapopulation models. American Naturalist 140:41–-58.

Nisbet, R. M., and W. S. C. Gurney. 1982. Modelling fluctuating populations. John Wiley and Sons, Chichester, United Kingdom.

Nisbet, R. M., and L. C. Onyiah. 1994. Population dynamic consequences on competition within and between age classes. Journal of Mathematical Biology 32:329–344.

Nunney, L. 1985a. The effects of long time delays in prey-predator systems. Theoretical Population Biology 27:202–221.

Nunney, L. 1985b. Short time delays in population models: a role in enhancing stability. Ecology 66:1849–1858.

Parker, K. R., and E. DeMartini. 1989. Technical report to the California Coastal Commission. D. Adult-equivalent loss. Marine Review Committee, Inc.

Rodriguez, D. J. 1988. Models of growth with density regulation in more than one stage. Theoretical Population Biology 34:93–119.

Rothschild, B. J. 1986. Dynamics of marine fish populations. Harvard University Press, Cambridge, Massachusetts.

Ruxton, G., W. S. C. Gurney, and A. M. DeRoos. 1992. Interference and generation cycles. Theoretical Population Biology 42:235–253.

Saila, S. B., X. Chen, K. Erzini, and B. Martin. 1987. Compensatory mechanisms in fish populations: Literature Reviews. Volume 1: Critical evaluation of case histories of fish populations experiencing chronic exploitation or impact. Report prepared for Electric Power Research Institute, EA-5200, Volume 1. Research Project 1633-6.

Savidge, I. R., J. B. Gladden, P. Campbell, and J. S. Ziesenis. 1988. Development and sensitivity analysis of impact assessment equations based on stock-recruitment theory. American Fisheries Society Monograph 4:191–203.

Schaff, W. E., D. S. Peters, D. S. Vaughan, L. Coston-Clements, and C. W. Krouse. 1987. Fish population responses to chronic and acute pollution: the influence of life history strategies. Estuaries 10:267–275.

Shepherd, J. G., and D. H. Cushing. 1980. A mechanism for density-dependent survival of larval fish as the basis for a stock-recruitment relationship. Journal du Conseil 39:160–167.

Stewart-Oaten, A., W. W. Murdoch, and K. R. Parker. 1986. Environmental impact assessment: "pseudoreplication" in time? Ecology 67:929–940.

Tschumy, W. O. 1982. Competition between juveniles and adults in age-structured populations. Theoretical Population Biology 21:255–268.

Van der Veer, H. W. 1986. Immigration, settlement, and density-dependent mortality of a larval and early post-larval 0-group plaice (*Pleuronectes platessa*) population in the western Wadden Sea. Marine Ecology Progress Series **29**:223–236.

Van Winkle, W., K. A. Rose, K. O. Winemiller, D. L. DeAngelis, S. W. Christensen, R. G. Otto, and B. J. Shuter. 1993. Linking life history theory, environmental setting, and individual-based modeling to compare responses of different fish species to environmental change. Transactions of the American Fisheries Society **122**:459–466.

Ziljstra, J. J., and J. J. Witte. 1985. On the recruitment of 0-group plaice in the North Sea. Netherlands Journal of Zoology **35**:360–376.

THE LINK BETWEEN ADMINISTRATIVE ENVIRONMENTAL IMPACT STUDIES AND WELL-DESIGNED FIELD ASSESSMENTS

CHAPTER 14

THE ART AND SCIENCE OF ADMINISTRATIVE ENVIRONMENTAL IMPACT ASSESSMENT

Russell J. Schmitt, Craig W. Osenberg, William J. Douros, and Jean Chesson

A fundamental goal of "environmental protection" in the administrative review of planned developments is to ensure that features of the natural world considered important by the public are not unduly degraded. Detecting ecological impacts after they have occurred is not a desired mechanism for achieving such protection. The ideal, of course, is to accurately forecast adverse effects of a proposed intervention before they occur, and to modify the project design at the planning stage to avoid or minimize anticipated impacts. These principles are central to the administrative environmental review process, which is a legal mandate for proposed development projects in most industrialized countries. The presumption is that comprehensive planning involving appraisal of potential impacts ("environmental impact assessment" or EIA) will result in developments that are consistent with policy designed to protect the public's interest in the natural environment (for a recent discussion, see Hildebrand and Cannon 1993). Because the process is designed to weigh the often diametric concerns of various interested parties, an intended outcome is to reduce conflict and engender consensus (Lester, Chapter 17). Hence, the process is conceived as a vehicle to achieve balanced decisions that promote expeditious development of environmentally sensible projects, and prohibit ventures that are perceived to be deficient in this regard.

There can be little dispute that administrative environmental review generally has served its intended purposes since its inception in the late 1960s (but see Wenner 1989). What can be disputed is the degree to which the process has protected the environment from substantial impact (e.g., Ambrose et al., Chapter 18), and, as a corollary, its success in reducing conflict and facilitating expeditious development (e.g., Lester, Chapter 17). Often, public response to a proposed development is shaped by perceptions of attendant environmental risks (Ruckelshaus 1983, Lester Chapter 17). A fundamental issue, then, is how well administrative environmental review—and more specifically legislated environmental impact assessment—reduces uncertainty in our understanding of potential effects of development. Surprisingly, there have been few comprehensive evaluations of the EIA process that address this issue (see Ambrose et al., Chapter 18). This situation exists in part because of structural features that hinder our ability to verify administrative findings through subsequent scientific evaluation (e.g., Piltz, Chapter 16; Ambrose et al., Chapter 18). As a consequence of limited feedback between administrative and scientific findings, it is difficult to determine in other than the most general sense whether the planning and review process should be made more effective, and if so, how.

Here we explore the interface between the "art" of administrative environmental review and the "science" of impact assessment research. The intent is to highlight some of the more prominent issues that limit the contribution of science to the review process, as well as to point out some of the consequences of this limitation. We describe the administrative environmental review process and explore some of the types of scientific information and analyses typically collected for such paper assessments. Subsequent chapters in this volume (Carney, Chapter 15, Piltz, Chapter 16, Lester, Chapter 17, Ambrose et al., Chapter 18) explore in detail related aspects, all of which are drawn from such contentious developments as offshore oil and gas activities and nuclear power generating stations.

Insights arising from examination of these controversial developments are far-reaching. For reasons that are both scientific and socio-political, administrative environmental review often fails to decrease uncertainty to a level acceptable to the public. Failure of environmental review policy is most evident for contentious developments where prolonged conflict not only obviates expeditious decisions, but can lead to adjudication by entities—the judiciary—poorly equipped to assess technical information (Lester, Chapter 17). Much of the cause for such dissension is rooted in the imperfect nature of the "factual" evidence upon which decisions are based (see Ruckelshaus 1983). It perhaps is less well appreciated that these imperfections exist to some extent in the deliberations of all proposed developments, irrespective of public response. This again underscores the need to more fully evaluate a basic premise of administrative environmental review, that the process protects the environment from meaningful impact.

Administrative Environmental Review

The Policy, Legislative and Legal Setting

Administrative review, including environmental impact assessment (EIA), is mandated for most proposed developments by legislation that sets out the prescription for the process. In the United States, developments carried out or approved by the Federal government fall under the National Environmental Policy Act (NEPA). Public and private projects developed under the purview of individual states are governed by comparable state legislation; examples include the California Environmental Quality Act (CEQA), and the State of Washington's Environmental Policy Act. Administrative environmental review in other industrialized countries also is governed by similar legislation, such as the Environmental Planning and Assessment Act for Australia (Fairweather 1989; for a listing by country, see Westman 1985).

The National Environmental Protection Act (NEPA) and equivalent state laws (e.g., CEQA) have two main purposes. The first is to specify a uniform framework within which decision makers are provided information and facts about the environmental impacts of a proposed development. The second is to provide full public disclosure of the information and reasons leading to decisions, as well as obligations for mitigation by the applicant and enforcement of permit conditions by relevant agencies. Thus, laws such as NEPA stipulate *how* environmental review is to be conducted, and *what* information must be considered in that deliberation.

NEPA sets policy aimed at achieving broad environmental objectives, such as allowing beneficial uses of the environment that do not result in degradation, preserving the quality of renewable resources, and protecting the environment for enjoyment by future generations. However, NEPA and comparable state laws do *not* specify standards as to what does or does not constitute an acceptable level of environmental perturbation from a proposed development. How, then, are the broad environmental objectives of NEPA achieved? NEPA simply requires that the policies and programs of the Federal government be consistent with NEPA's broad environmental aims. Conformity rests on the regulatory system, also predicated in Federal law, that provides specific policies, standards, limitations and restrictions with which developments must comply. Hence, such Federal legislation as the Clean Water Act, Clean Air Act, and Endangered Species Act not only sets out environmental objectives that are more specific than—but consistent with—those of NEPA, it also provides the foundation for specific regulatory policy and the regulatory machinery.

Environmental review conducted under NEPA or equivalent state laws produces an informational record (document) for a proposed development that applicable regulatory authorities use to determine whether to issue discretionary permits. As the name implies, a discretionary permit is one where the decision to issue hinges on the judgment of the relevant agency whether the action conforms with stated policy. (By contrast, an administerial permit is issued based solely on whether an action will meet explicit requirements, such as discharge standards for an ocean outfall.) While a single agency typically has primary responsibility for conducting the environmental review, approval of a proposed project can require separate consent from a number of different permitting authorities (see below). It should be noted that permitting agencies have latitude in their interpretation of policy, and that environmental policies often must be balanced against other, potentially conflicting mandates of an agency (e.g., Lester, Chapter 17). Possible environmental impact is but one of a number of issues to be considered in making a decision about a proposed development. Consequently, it is quite possible for various authorities to reach different permitting decisions for the same proposed development.

The following example illustrates the jurisdiction of various agencies and the regulatory filters for a proposed development of moderately high complexity. In the mid-1980s, Atlantic Richfield Company (ARCO) submitted a proposal to develop oil and gas leases in State waters (≤ 3 miles offshore) off Santa Barbara County, California. The proposed development, entitled the "Coal Oil Point Project," was to consist of 3 new offshore production platforms in State waters, new onshore processing and storage facilities, and new onshore and offshore pipelines for oil and gas transport. Because the State of California has jurisdiction over projects in its onshore environs and in "tideland" waters (to 3 miles offshore), the State Lands Commission of California conducted the administrative environmental review as required by the California Environmental Quality Act (CEQA). However, the proposed project included the placement of permanent obstructions in navigable waters of the United States (i.e., the platforms), and approval for such actions rests with the U.S. Army Corps of Engineers under the Federal Rivers and Harbors Act. Hence, the Coal Oil Point Project also was subjected to administrative review under the National Environmental Policy Act (NEPA), in this case conducted by the Army Corps of Engineers.

ARCO's proposed Coal Oil Point Project highlights the number and diversity of regulatory agencies that can be involved in the approval process of a project. ARCO needed no less than 27 permits or approvals from 15 Federal, State and County agencies (Table 14.1). The NEPA and CEQA environmental reviews for the Coal Oil Point Project, therefore, had to address the information requirements of these 15 permitting authorities. As "Lead Agencies," the State Lands Commission and Army Corps of Engineers were responsible for ensuring that the environmental documents were prepared in accordance with regulations of each applicable agency. Ultimately, the California State Lands Commission set aside ARCO's application (a *de facto* denial) on the grounds that the State first needed to formulate a master policy plan for development of California's tidelands. ARCO litigated this action (*ARCO vs California State Lands Commission and County of Santa Barbara*), and the Court sided with the State. ARCO subsequently dropped its appeal of the lower court's decision.

The judiciary plays several important roles, not the least of which is refining the administrative process of environmental review (Westman 1985, Hildebrand and Cannon 1993). As illustrated by the Coal Oil Point example above, parties affected by an agency's decision have recourse through the legal system. However, it is difficult to challenge a particular decision when it is generally consistent with stated policy. For example, NEPA and CEQA do not require that the most environmentally preferable project alternative be adopted, only that it be identified (quite often the "No Project" alternative will be so identified). Hence,

Table 14.1. Responsible Agencies and Permit Approvals for an Oil and Gas Development Proposed by ARCO in Santa Barbara County, California

Responsible agency	Permit or approval
Federal agencies	
U.S. Army Corps of Engineers	Section 10 of the Rivers and Harbors Act for Activities and Structures in Navigable Waters
	Section 404 of the Clean Water Act for Fill in Waters of the United States
Environmental Protection Agency	Spill Prevention Control and Countermeasure Plan
U.S. Coast Guard	Oil Spill Contingency Plan
	Approval of Aids to Navigation
Federal Aviation Administration	Approval of Heliport Operation Communication License
Federal Communication Commission	Communication License
State agencies	
State Lands Commission	Development Plan
	Land Lease for Pipeline Rights of Way
California Coastal Commission	Coastal Development Permit
	Coastal Plan Amendment
Regional Water Quality Control Board	NPDES Permit for Waste Water Discharge
Department of Fish and Game	Sections 1601—1603 Fish & Game Code: Stream or Lake Crossing Agreement
Department of Oil and Gas	Safety System Permit
	Permit to Conduct Well Operations
Department of Transportation	Right of Way Permit
	Encroachment on State Highway Permit
	Encroachment on State Highway Permit
Santa Barbara County agencies	
Resource Management Department	Development Plan
	Coastal Development Permit
	Conditional Use Permit
Environmental Health Division	Water Well Permits
	Water System Permits
	Hazardous Materials and Waste
Air Pollution Control District	Authority to Construct
	Permit to Operate
Department of Public Works	Permits for Onshore Facilities
	Excavation Permits

Note: Adopted from Environmental Impact Report / Statement Volume I, ARCO Coal Oil Point Project, State Lands Commission et al. 1986.

an agency's decision not to choose the most environmentally preferred alternative in and of itself provides no grounds for judicial relief. By contrast, failure to adhere to NEPA or CEQA *procedures* is subject to challenge in the courts. Included in this context is the adequacy of the environmental document in forecasting environmental impacts of project alternatives. Indeed, it is not an uncommon practice to litigate on the grounds that available information was insufficient to support the decision made (for some consequences, see Lester, Chapter 17). To the extent that scientific uncertainty contributes to this circumstance, the application of science can have a fundamental role in enhancing the effectiveness of administrative environmental review.

The Process of Administrative Environmental Review

Hildebrand and Cannon (1993) recently have discussed the National Environmental Policy Act (NEPA) in detail. Here the process of administrative environmental review is described in general, with specifics taken from the California Environmental Quality Act. (Except for details, the CEQA process is identical to that prescribed by NEPA. For example, NEPA requires the preparation of an Environmental Impact Statement (EIS), whereas the same document prepared under state legislation is termed an Environmental Impact Report (EIR)).

The process of administrative environmental review under CEQA is stipulated in enabling legislation and subsequent amending bills, which state:

> The purpose of an environmental impact report is to identify the significant effects of a project on the environment, to identify alternatives to the project, and to indicate the manner in which those significant effects can be mitigated or avoided.

Thus, CEQA involves the preparation of documents that provide information about the impacts on the environment of a particular project, and how those impacts can be avoided or reduced. The process consists of mandated steps to generate the information needed for applicable agencies and the public at large, all of which must be completed—and a decision reached—in a specified period of time (e.g., 1 year for CEQA). While such time limitation is designed to protect the interest of the applicant, it greatly constrains the types of new scientific information that can be gathered (see below).

The first step in the process is to determine if a particular development is subject to environmental review as a number of statutory and categorical exemptions from review have been granted (e.g., construction, replacement or demolition of single family residences or existing schools, emergency repairs to existing freeways; for NEPA, actions deemed in the interest of national security). Once it is determined that an EIA review applies to a particular proposal, the agency in charge then determines (within 30 days) if the application is complete (i.e., it contains all information needed to begin review of potential impacts of the proposed development). Once deemed complete, the time "clock" for reaching a decision begins.

The agency then determines whether an environmental impact report (or statement) (EIR/S) will need to be prepared, a decision which hinges on whether the proposed project may have any significant environmental effects. This is accomplished by cursory comparison of the project against criteria established by the agency for such "issue areas" as air quality, biological resources, ground-water resources and so forth. One possible outcome is that the proposed project may be judged not to have any significant adverse environmental effects. In this case, an informational document, termed a Negative Declaration, will be prepared; if adopted, the necessary permits will be issued and the project can proceed without further environmental review. Alternatively, the preparation of an EIR/S may be deemed necessary if significant impacts appear possible. In this case, a Notice of Preparation is made public and sent to other permitting agencies to initiate the process of defining the scope of the impending review. The "scoping" process serves to identify relevant issues, general concerns, and regulatory requirements that need to be addressed in the EIR/S.

The next phase is the preparation of the Environmental Impact Report (or Statement), which is released in draft form to the public and applicable agencies for comment. The EIR/S typically includes a description of the project, a summary of the environmental "setting" in which the project is to be developed, and some description of predevelopment—or "baseline"—environmental conditions. This information provides the local context for assessing potential impacts of the proposed development. The core of the EIR/S document is the "analysis" of potential impacts that may arise from the construction, operation and decommissioning of the proposed project. In essence, this is an exercise in predictive risk assessment, although evaluation of effects almost always is qualitative and without explicit estimates of uncertainty. Project- and site-specific information is coupled with data gathered previously for other situations to theorize possible impacts. In addition to the applicant's "Preferred Project" design, other feasible designs ("Project Alternatives") also are considered to determine if they reduce or eliminate potentially significant impacts, even if they do not allow the applicant to meet the full objectives of the project. In addition to effects that might arise expressly from the proposed project ("project-specific impacts"), "cumulative impacts" also are considered. Cumulative impacts are the combined environmental effects that arise from all developments and activities in a region, and to which the proposed project will contribute.

Once possible project-related and cumulative effects have been identified, their importance is determined by comparing effects against "significance" criteria. For legally protected habitats (e.g., wetlands) or organisms (e.g., endangered species; marine mammals), any adverse effect may be judged a substantial impact regardless of its magnitude or duration. Except for such mandated situations, assessment of an effect's significance is an opinion derived from "best professional judgment." In general, environmental effects will be classified into four categories: substantial adverse impact that cannot be mitigated; substantial adverse impacts that can be mitigated to insignificance; insignificant impacts (no mitigation needed); and positive (beneficial) effects. Other than in the broadest sense, there are no set operational criteria for making these evaluations, which therefore tends to be an *ad hoc* exercise done for each proposed project. In general, consideration is given to the species and habitats likely to be affected, the magnitude of possible impact ("effect size;" Osenberg et al. Chapter 6), and the spatial extent and duration of impact (Raimondi and Reed, Chapter 10). Because the imprecise nature of administrative impact assessment is recognized explicitly, aspects of the process make the evaluation more environmentally conservative. In particular, plausible "worst-case" scenarios are analyzed even if the probability of such events occurring is remote (but nonzero). Thus, a proposed project is evaluated on the basis of its worst possible effects on the environment, and not just its most likely or average impacts. Analyzing projects against this more stringent standard is an attempt to minimize possible environmental consequences of uncertainty in predicting impacts.

Part of evaluating whether potential project effects will be substantial is determining whether and how identified impacts could be reduced to "insignificant." This paper exercise involves exploring feasible mitigation alternatives. Since the preferred resolution is to avoid an impact altogether, methods are considered that might have this desired outcome. It often is the case that adverse effects could be minimized but not totally avoided, and additional measures may need to be sought to reduce impacts to an acceptable level. Nonetheless, there will be impacts for which no mitigation measures are known to be effective. In cases where restoration of natural resources cannot be accomplished, "out-of-kind" compensation might be contemplated.

The final phase of the administrative review is certification of the EIR/S document. This involves solicitation of comments on the Draft EIR/S from agencies and other interested parties, which can result in modification of the document. The EIR/S is certified after all requirements have been satisfied and the document is deemed adequate. By this action, it is accepted that the predicted impacts are those (and only those) the project will actually have, and that the stated mitigation actions will actually work. The "best guesses" become facts upon which decisions to issue discretionary permits are made.

Scientific Data Collection and Analyses in Environmental Impact Reports

The collection of new scientific data to assess likely impacts is a sanctioned activity in the administrative review of a proposed development. It seems inherently reasonable that acquisition of new information would reduce uncertainty for decision-makers regarding potential impacts. The question we ask is, are new data often gathered for the preparation of an EIR/S, and if so, how useful are they in assessing likely impacts? Here we examine recent environmental impact statements to explore the extent to which new field data were collected and analyzed to help predict effects as part of the EIR/S process. Collection of new, "site-specific" data on resources at risk is the most common activity in this regard, and the general rationale given is to provide a baseline of preimpact conditions which could then be compared with future measurements. We examined 18 recent EISs for major proposed coastal developments in Australia, where the administrative environmental review process parallels that in the United States (see Fairweather 1989, Lincoln Smith 1991). While the 18 assessments are not a random selection, they were chosen without prior knowledge of the type of data and data analysis they contained. Therefore, insights provided by examining these EISs likely apply to administrative environmental assessments in general. Each EIS assessment was categorized according to: (i) the level of the biological data collected to establish baseline conditions (pre-existing data only, data from a new quantitative survey, comparative study (quantitative surveys at > 1 site or time), or experimental study), (ii) the type of "statistical" analysis performed on these data (qualitative only (e.g., species list), raw data presentation, descriptive statistics, exploratory statistics or hypothesis testing), and (iii) the nature of suggested proposals for future monitoring of effects (none, general description, some detail, specific structure).

All 18 Environmental Impact Statements included data on biological resources that were newly collected as part of the assessment, although the extent varied from mere lists of species present to detailed field surveys. None included any investigation that might be classified as an experimental study. Ten (56%) included data gathered from two or more sites or two or more time periods; however, data collections rarely extended beyond a single year, and usually involved a single time period. Because temporal variability was not (or, at best, poorly) estimated, in no case was the data collection sufficient for subsequent statistical exploration of project impacts using a Before-After-Control-Impact Paired Series (BACIPS) type of design (for general requirements of BACIPS, see Stewart-Oaten et al. 1986, Stewart-Oaten, Chapter 7; for sample size considerations, see Osenberg et al., Chapter 6).

With respect to data analysis, six (33%) EISs presented results as species lists, descriptive accounts or in other non-quantitative forms. The remaining 12 (67%) provided some descriptive statistics ranging from simple summary tables to more sophisticated measures such as diversity indices (for discussions of the utility of diversity measures for impact assessment, see Carney 1987, Chapter 15). Three EISs used a standard statistical test (analysis of variance) to identify differences between locations or habitats. This hypothesis test is not particularly relevant to impact assessment as differences between locations or habitats always are expected (if not always detected); demonstrating statistically that sites differ provides no insight into the potential impact of a project. The most sophisticated data analysis presented in the 18 EISs was associated with a proposed power station; various classification techniques, diversity indices and a mathematical model of the effect of water temperature were provided, and model predictions were compared with survey data.

With respect to proposals for monitoring of subsequent impacts, 5 (28%) EISs made no mention of monitoring either during or after development was to take place (EIR/Ss in the United States must describe mandated compliance monitoring requirements, although many of these are inadequately designed: Osenberg and Schmitt, Chapter 1). Four (22%) mentioned monitoring in general terms, 5 (28%) gave some indication of what might be monitored, and 4 (22%) provided some insight on the overall design of a monitoring program. None contained a detailed proposal for monitoring, and only two hinted at such crucial design issues as the number of samples and effect of variability that would need to be considered. None included any explicit or implicit reference to a BACIPS-type sampling design.

This brief analysis of environmental impact statements indicates that the collection of new site-specific data is a common practice in the preparation of EIR/Ss. It is not uncommon for the environmental assessment to be the first biological survey of an area, and thus the data collected are an important first step. However, as practiced, the information is most useful in describing the habitats and species at risk to a proposed project. Despite the commonly stated aim of providing a baseline for subsequent measurement of impacts, data collected specifically as part of an EIR/S appear rarely to be useful for a subsequent BACIPS-type exploration of actual impacts. Further, statistical design considerations do not play a major role in the review process, and statistical analysis of even the most rudimentary kind was noticeably absent from the majority of assessments. Where statistical tests were applied, the hypotheses tested were irrelevant to impact assessment. Further, there appears to be widespread failure to appreciate that traditional hypothesis testing approaches are not well suited to environmental impact assessment where yes/no answers have limited applicability compared with estimating the magnitude of effects (see Stewart-Oaten, Chapter 2).

At least two aspects account for the paucity in EIR/Ss of "before impact" data useful for rigorous measurement of subsequent impacts. First, the nature of most individual development proposals makes it impractical to devote the time and resources necessary to conduct the type of field surveys needed for rigorous statistical exploration of impacts. For many proposed developments, the cost to gather adequate data is far out of proportion to the value of the proposed development or its likely environmental impact. Second, the mandated time limitation for completion of the administrative review, once initiated, greatly constrains the type of new scientific inquiry that can be conducted. This mismatch in the time needed for proper scientific inquiry and that available is perhaps secondary to another structural problem: scientific efforts that could improve the process as a whole have limited benefit to the individual proponent, and typically require an investment of resources that are unrealistic for most proposed developments. In few cases where many similar (and major) projects are anticipated—for example offshore oil and gas developments—ongoing programs to conduct comprehensive environmental research have been established to provide generic information useful in the administrative review of specific proposals. The Environmental Studies Program of the Minerals Management Service, U.S. Department of the Interior is one such example (see Carney, Chapter 15; Piltz, Chapter 16; Lester, Chapter 17). In other cases involving large developments, *ad hoc* research programs have been established that operate in parallel but independently of the EIR/S process (for an example, see Ambrose et al., Chapter 18). In general, however, there is little capacity in the EIR/S process for the provision of "before impact" monitoring data that are adequate for subsequent estimation of a project's environmental effects.

The Need for Better Scientific Feedback in the EIA Process

A legitimate question is whether better mechanisms to obtain adequate "before impact" data are needed in the administrative review of proposed developments. After all, the goal is to predict a project's impacts and devise means to ameliorate them. Scientific advances in "ecological risk assessment" (*sensu* Suter 1993a) show great promise for improving capabilities in this regard. While estimating environmental impacts through extensive monitoring programs will continue to be unwarranted for many proposed developments, there remain compelling reasons for more frequent monitoring of impacts as part of the administrative review process that would benefit the process as a whole. First, there have been virtually no rigorous checks on the extent to which the EIR/S process has successfully predicted the environmental effects of a development. Even when quantitative "ecological risk assessment" techniques become used more widely to predict effects, there is a crucial need to establish how well impacts are being forecast. This will not be possible unless such rigorous sampling designs such as BACIPS are used to estimate actual impacts. Results of perhaps the best quantitative audit of EIR/S predictions done to date are not comforting (Ambrose et al., Chapter 18); not only were the predicted environmental impacts highly inaccurate, real effects were not detected by the mandated compliance monitoring regime. Because such quantitative audits of environmental predictions in EIR/Ss are extremely rare, no feedback loop exists between administrative findings and scientific verification (Larkin 1984). Consequently, it is not possible to assess the extent to which current EIR/S practices protect the environment from meaningful impact.

Second, monitoring programs to estimate effects are useful in attempts to mitigate the environmental impacts of a development. They can provide quantitative estimates of the amounts of natural resources that are damaged or lost, and therefore provide a quantitative target for the levels of resources that need to be replaced. Perhaps more importantly, BACIPS-type monitoring programs can also be applied to the mitigation project to evaluate the efficacy of the mitigation actions. Currently, mandated mitigation actions are assumed to have the desired result, and scientific confirmation is virtually nonexistent. Permitting agencies are reluctant to approve mitigation measures that are not well proven, yet there is no systematic mechanism to establish the success of either currently used or promising techniques.

Finally, there is the general issue of improving the information base to facilitate and improve future decisions. While every proposed project has novel elements (e.g., specific location and design features), most are quite similar to other projects that have been undertaken numerous times in the past. For example, a large fraction of proposed projects are extensions or modifications of existing facilities (e.g., 8 of the 18 Australian assessments examined here), or essentially are "replicates" of others but in a new location (e.g., offshore oil production platforms) (for several examples, see Fairweather and Lincoln Smith 1993). Had the impacts from these previous projects been well-studied, there would have been improved potential to accurately predict the likely impacts of the proposed projects. Despite the opportunity for empirical study, there is a rather poor understanding of the general impacts that arise from any sort of development. This largely reflects a common failure to rigorously measure actual impacts and provide case studies; attempts to do so are infrequent (see Piltz, Chapter 16), and when done, the results often are equivocal because of inadequate or misguided sampling designs (Carney 1987, Osenberg and Schmitt, Chapter 1). Because "retrospective" assessment techniques (i.e., those which lack Before data and are applied only *after* impacts have occurred) provide weak evidence at best (Lincoln Smith 1991, Suter 1993b, Osenberg and Schmitt, Chapter 1), the consistent faiure to obtain adequate Before and After impact data through comprehensive monitoring programs is a major contributor to the problem.

Conclusions

Administrative environmental impact assessment (EIA) is a public process through which decision-makers are provided with information about the possible environmental effects of a proposed development. The information record (EIR/S) is used to support the decision made. While it is inevitable that such decisions will be made on incomplete scientific information (Ruckelshaus 1983), there are structural aspects of the process that constrain the potential contributions of science in improving its execution. One notable weakness is the lack of regular feedback between administrative findings and scientific corroboration. The system for collecting (or collating) information not only should support the present decision, but also should facilitate assessment of whether the decision was justified, and should support future decisions concerning similar developments. The challenge is to integrate administrative assessments with rigorous impact monitoring programs to maximize their benefits to one another and to environmental management in general.

Acknowledgments

We thank S.J. Holbrook and C. St. Mary for comments, I. Lamb, K. Tzafaris, M.L. Morris and N. Harvey for assistance in gaining access to Australian environmental impact statements, and the UC Coastal Toxicology Program and the US Minerals Management Service (under the Southern California Educational Initiative, MMS contract no. 14-35-001-3071) for assistance in preparation of this chapter. The views and conclusions in this chapter are solely those of the authors, and should not be interpreted as necessarily representing the official policies, either expressed or implied, of the Federal, State and County governments of the United States or Australia.

References

Carney, R. S. 1987. A review of study designs for the detection of long-term environmental effects of offshore petroleum activities. Pages 651–696 in D. F. Boesch and N. N. Rabalais, editors. Long-term environmental effects of offshore oil and gas development. Elsevier, New York, New York.

Fairweather, P. G. 1989. Environmental impact assessment - where is the science in EIA? Search 20:141–144.

Fairweather, P. G., and M. P. Lincoln Smith. 1993. The difficulty of assessing environmental impacts before they have occurred: a perspective from Australian consultants. Pages 121–130 in C. N. Battershill, D. R. Schiel, G. P. Jones, R. G. Creese and A. B. MacDiarmid, editors. Proceedings of the second international temperate reef symposium. NIWA Marine, Wellington, New Zealand.

Hildebrand, S. G., and J. B. Cannon. 1993. Environmental analysis: the NEPA experience. Lewis Publishers, Boca Raton, Florida.

Larkin, P. A. 1984. A commentary on environmental impact assessment for large projects affecting lakes and streams. Canadian Journal of Fisheries and Aquatic Sciences 41:1121–1127.

Lincoln Smith, M. P. 1991. Environmental impact assessment: the roles of predicting and monitoring the extent of impacts. Australian Journal of Marine and Freshwater Research 42:603–614.

Ruckelshaus, W. D. 1983. Science, risk, and public policy. Science 221:1026–1028.

State Lands Commission, County of Santa Barbara, and U.S. Army Corps of Engineers. 1986. Environmental impact report / statement: proposed ARCO Coal Oil Point Project. Vol. I. SLC No. EIR-401.

Stewart-Oaten, A., W. W. Murdoch, and K. R. Parker. 1986. Environmental impact assessment: "pseudoreplication" in time? Ecology 67:929–940.

Suter, G. W., II, editor. 1993a. Ecological risk assessment. Lewis Publishers, Boca Raton, Florida.

Suter, G. W. 1993b. Retrospective risk assessment. Pages 311–ß364 in G. W. Suter II, editor. Ecological risk assessment. Lewis Publishers, Baco Raton, Florida.

Wenner, L. 1989. The courts and environmental policy. Pages 238–260 in J. P. Lester, editor. Environmental politics and policy: theories and evidence. Duke University Press, Durham, North Carolina.

Westman, W. E. 1985. Ecology, impact assessment, and environmental planning. John Wiley and Sons, New York, New York.

CHAPTER 15

ON THE ADEQUACY AND IMPROVEMENT OF MARINE BENTHIC PRE-IMPACT SURVEYS

Examples from the Gulf of Mexico
Outer Continental Shelf

Robert S. Carney

Ecologists are called upon by environmental managers to provide scientific information answering two questions about impacts. The first asks whether an ongoing human activity is having an unacceptable detrimental impact upon some component of the biota and must be regulated. The second asks whether a proposed human activity will have an unacceptably detrimental impact and must be forbidden or permitted only with restriction. This review is intended as a critical and constructive examination of this second type of question, impact prediction, considering how effectively studies are designed, conducted, synthesized, and used. This critique is necessitated by a widening contrast in the quality of scientific information produced by studies testing for versus studies predicting impacts. Although proper design and effective execution of the former remain difficult, they are generally viewed as being conceptually tractable. In the context of hypothesis testing, impact can be envisioned as a specific change caused by a specific anthropogenic perturbation. This simplification required for hypothesis testing may ignore important aspects of perturbed natural systems, but it has been quite seminal. The resulting scientific dialogue has refined the conceptual base, improved design, identified mensurable impacts (responses), and identified mensurable perturbations (doses). As a consequence, decisions to regulate ongoing activities can be based upon increasingly rigorous scientific evidence. Unfortunately, decisions to permit new activities do not benefit from information of comparable quality. A theme that will be developed herein is that predictive environmental impact studies suffer from a lack of well-defined links between management's needs, scientific question, and study design. Lacking this critical linkage, designs are adopted and surveys executed according to traditional rather than scientifically appropriate standards. It shall be proposed that improvement could be effected through development of a simple conceptual base which might combine an understanding of processes and statistical description.

This review is restricted to soft-bottom faunal surveys undertaken as part of Environmental Impact Studies (EIS) of the outer continental shelf. While the contrast in scientific rigor between impact testing and impact predicting is generic to all but the simplest of systems, it is especially important with respect to the management of offshore lands between 3 and 200 nautical miles seaward, legally termed the Outer Continental Shelf (OCS). This is a poorly understood environment in the ecological and oceanographic sense, where management decisions must resolve real and imagined conflicts among proponents for oil and gas development and other concerned parties. Since offshore lands are uninhabited and not available for private ownership, the policy issues facing management are relatively straight forward and the role of science as the primary source of information well defined. Scientific information and policy merge during the competitive leasing process. These national lands are leased for development from the government in blocks (3 by 3 nautical miles) through a process of bidding. Under the requirements of the National Environmental Policy Act and the Outer Continental Lands Act, the Minerals Management Service (MMS) of the Department of Interior manages the OCS and prepares EIS prior to leasing. The EIS prepared must anticipate impacts and propose scenarios which protect the overall resource value (biological, commercial, cultural, socioeconomic, etc.) of the region. These EISs, the reports of the studies undertaken to develop them, and the data bases developed during those studies are the focus of this review.

MMS studies were selected, not as a criticism of a specific government agency, but as examples of some of the most extensive EIS surveys ever supported by a management agency. The MMS Environmental Studies Program's method of developing, conducting, and critiquing studies is perhaps unique in the sense that scientific and policy issues are easily discernible allowing criticism to be appropriately directed. MMS procures information using the best available technology via contracts and subcontracts with academic and private sector ecologists. The level of academic community involvement is exemplary, employing formal advisory groups at all phases of program development and execution. MMS makes frequent use of National Research Council reviews (NRC 1992) for both comprehensive retrospection and target issue evaluation. Thus, the science in these studies closely reflects the contemporary strengths and weaknesses of the disciplines conducting the research to an usually high degree.

Structure of the Review

This review of efforts to anticipate impacts in EIS produced prior to oil and gas leasing of OCS lands closely parallels a previous review of impact test studies around offshore platforms (Carney 1987). It has the same geographic focus upon the Gulf of Mexico, and makes similar use of government documents to track down the links between program needs, questions, planing, design, execution, and final utilization. This review departs from the previous, in that data archives were examined. This use of the archives served as a test of their utility, the quality of their content, and a means of determining the basic properties of the data.

MMS has supported three major surveys in the Gulf of Mexico specifically to establish a baseline and predict impacts associated with development. Geographically, these include portions of the Texas, Mississippi, Alabama, and Florida coasts. The Louisiana coast, dominated by the Mississippi-Atchafalaya delta complex, has not received similar baseline study. However, a platform effects study with sampling across the continental shelf did contain elements of a baseline survey and can be included. Thus, four surveys in the Gulf of Mexico were reviewed. Three were BLM/MMS baseline surveys: the 1975–1978 South Texas Baseline study (STBS, Flint and Rabalais 1981), the 1975–1978 Mississippi, Alabama, Florida Study (MAFL, Dames and Moore 1979), and the 1980–1982 Southwest Florida Study (SWFL, Environmental Science and Engineering, Inc. et al. 1987). The fourth was a BLM/MMS impact study, the 1978–1979 Central Gulf Platform study (CGP, Bedinger 1981). Although the newest of these studies was completed more than a decade ago, they still comprise a good basis for review. They are regionally adjacent, they represent the best opportunity to develop a larger-scale synthesis, and the same general design and sampling philosophy was still being followed in the most recent MMS large area survey, the 1986–1992 California Monitoring Program (Piltz, Chapter 16).

Five environmental impact statements were reviewed to determine how data from OCS surveys were used. A broad geographic range was included so that generic aspects rather than regional styles could be seen. The reviews included: California (MMS 1983a), North Atlantic (MMS 1983b), Mid Atlantic (MMS 1984a), Alaska (MMS 1984b), and Gulf of Mexico (MMS 1992).

Review Findings

Use of Survey Results in Environmental Impact Statements

The Environmental Impact Statements (EIS) reviewed were lengthy, very broad in the environmental and socioeconomic concerns addressed, and sometimes very broad in geographic application. The primary use of survey data was limited to presentation of a species inventory and delineation of faunally distinct habitats. Each EIS provided a summary of species composition and distribution as part of a general description of the environment subject to impact. Detail ranged from presentation of comprehensive data tables to minimal text citing previous EISs. Estimates of the type and magnitude of impacting activities emphasized oil spills, burial by drill cuttings, discharge of drilling fluids, and discharge of produced water. Using a working definition of impact as any decrease in population, spills were consistently discounted as having no impact on subtidal benthos, and the other activities were considered to have localized detrimental effects extending no more than 1000 m from an oil or gas platform.

In some instances the ecological scope of a particular EIS is so great as to force consideration of only a limited number of especially sensitive habitat types. The most recent Gulf of Mexico EIS is a good case in point (MMS 1992). The actions of Sale 142 and 143 may result in environmental impact on the bathyal, continental shelf, near shore, estuarine benthos, and coastal wetlands. Within these larger habitats there are chemosynthetic communities, deep carbonate banks, shallower carbonate banks, true coral reefs, fishing grounds, and seagrass meadows. Even though succinctly written and with supporting citations, the 18 pages devoted to a description of those special habitats were insufficient for careful anticipation of impacts.

Description of Data Provided by Benthic Surveys

The four studies (Table 15.1) did provide some useful information, but the omission of important results was conspicuous. All conducted a species-level inventory and developed an archive, but then omitted the descriptive species by species summary statistics necessary for the planning of future impact studies. All provided some type of habitat mapping in the form of geographic patterns and associations, but none considered the ecological relevance of the results or the appropriateness of the clustering or ordination analyses performed. All provided some discussion of patterns of diversity, usually using Shannon-Weaver H' and related indices (Shannon and Weaver 1949, Pielou 1975), but none discussed the meaning of these results. All discussed the possible environmental factors behind data variation, but none actually partitioned that variation into the relatively few components allowed by the sampling design (typically, position cross shelf, position along shelf, and time). The impression given by these common omissions is that the studies were preoccupied with providing geographic and temporal coverage with minimal consideration of how the data might be used to actually anticipate impacts or design subsequent studies.

Taxonomic Quality

Even though all studies were intended to provide species-level inventory, actual taxonomic resolution was low and interstudy comparisons reveal either unexpected regional endemism or inconsistent identification. Collectively, 2774 taxa were recorded in at least one of the Gulf of Mexico studies, but only 1667 (60%) were identified to species. Of all taxa, 134 were common to the four studies, 267 limited to three, 609 limited to two, and 1768 limited to one. Of the 1667 species recorded, only 72 were common to all studies, 158 to three, 365 to two, and 1070 to one. Consistent with the apparently high local endemism, interstudy faunal similarity (expressed simply as percent of species shared) is low and geographic gradients are not pronounced. The possibility of serious taxonomic error must be a major source of concern.

Table 15.1. Comparison of Gulf of Mexico OCS Surveys Reviewed

Study	Primary design macrofauna sampling	Equipment	Analyses	Data array recovered: samples × taxa
STBS	4 cross-shelf transects	0.1 m^2	H 'diversity	1671×799
	25 stations	Smith-McIntyre	Canberra metric	
	4–6 replicates	Grab	Similarity	
	3 seasons: W, S, & F		Flexible sorting	
CGB	4 oil platforms	0.1 m^2	H 'diversity	776×736
	3 control sites	Smith-McIntyre	Multiple clustering	
	Stations 500 and 2000m	Grab	Correlations	
	N, E, S, & W			
	5 replicates			
MAFL	8 cross-shelf transects	0.65 m^2	H 'diversity multiple	2235×1691
	Scattered stations	Box cores	Clustering	
	Replication inconsistent	? Smith-McIntyre	Linear regression	
SWFL	Cross shelf transects	0.057 m^2	H' diversity	297×1081
	19 stations	Box cores	Bray Curtis similarity	
	5 replicates		Flexible sorting	
	Data pooled		Discriminant analysis	

The finding that only 72 species occur in all four studies out of 1667 taxa identified to species is problematic. There may actually be a high level of regional endemism around the northern Gulf of Mexico associated with dramatic changes in the coastal environments. Or, there may be a serious problem in taxonomic inconsistency from study to study. Neither of these alternatives could be resolved from the archived data. However, a third possibility that endemism was simply an artifact of the log-normal distribution of abundance was explored and rejected. If the observed regional range of species is an artifact of abundance, then species occurring in four studies should be more abundant than those in three and so on with those in a single region tending to be the least abundant. This was not found in the data. The 72 Gulf-wide species were scattered across the abundance distribution from very abundant to relatively rare. Furthermore there was some positional consistency in ranges. The most common of the 72 Gulf wide species at both western studies, STBS and CGP, was *Paraprionospio pinnata,* and at both eastern studies, MAFL and SWFL, was *Prionospio cristata.*

While relatively common in all studies, neither of these polychaete worms was the most abundant species in any study.

General Properties of Samples and Taxa

Delineation of habitats, examination of spatial and temporal trends, and testing hypotheses about place to place and time to time differences all are based upon a sample-wise analysis of data. The basic attributes of a sample which may be examined are the number of specimens, the species, and the allocation of specimen to species. An alternate mode of examining data is taxon-wise with the basic attributes of a taxon being the samples (times and places) in which it occurs and the abundance of these occurrences. Currently, taxon-wise examination of survey data is restricted to compilation of species inventories and discussions of overall species diversity. However, with increased emphasis upon maintenance of diversity as a management objective, taxon-wise approaches must receive greater attention. Sample-wise or taxon-wise, the properties of both sets of attributes set limits upon their use and must be known.

Properties of Samples

When analyzing sample data, a decision must be made about the level of taxonomic resolution to be used. Although species-level data are collected, even the most abundant species may be missing in so many samples as to make comparisons at that level undesirable. This was the case in the four Gulf of Mexico studies. The most abundant species in each were absent from many samples (percent dominant species absence: STBS 81%, CGP 19%, MAFL 70%, SWFL 41%). Therefore, all four studies used total faunal abundance in a sample, summing all taxa, when exploring gross temporal and spatial patterns of variation among samples. This pooled variable was found to be statistically well behaved in the sense of approximating a log-normal distribution (as seen in ranked abundance plots, Figure 15.1). In each study, the number of samples having a very large number of specimens is proportionately low and roughly equal to the number of samples having a very few number of specimens. The largest portion of samples in each study show a relatively narrow range of faunal abundance. SWFL shows the narrowest range of total faunal abundance since pooling has the effect of combining high value replicates with low value replicates to produce intermediate values. These ubiquitous log-normal distributions suggest that variation in total faunal abundance is geometric and that the $\log_e$ (pooled counts) transformation produces the additive model required for most linear model analyses (Clarke and Green 1988).

All studies made some effort to examine patterns of taxa found in samples. Simple species richness, the number of species in a sample, and the conceptually more complex H' diversity were most often used. Due to the simplicity of species richness, it was emphasized in the examination of restructured data sets. Just as

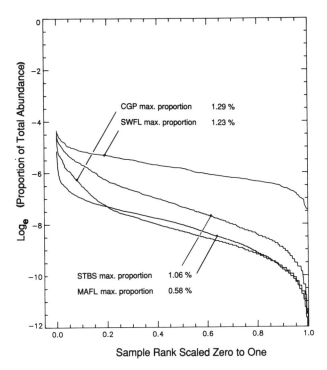

Figure 15.1. The rank abundance of total species abundance in all samples for the STBS, CGP, MAFL, and SWFL studies. Common scales are produced by plotting the proportional rank against the proportional abundance on a log scale.

with total abundance in samples, species richness in samples approximated a log-normal distribution with roughly equal numbers of samples having very many or very few species. Again, variance in species richness appears to be geometric and the $\log_e$(species richness) transformation required when analyses require additive variance. The log-normal distribution for species richness simply reflects a positive correlation between sample size and species number. It should not be taken as a model for diversity in the environment. It is to be expected that any index of diversity correlated to either species richness and/or abundance will show a similar distribution.

Properties of Taxa

The most conspicuous property of the taxa in a cross-shelf survey is that very few taxa are abundant and commonly encountered in samples, while a much greater number are far less common and less abundant. This asymmetry and great contrast between common and rare is more than a nuisance filling the data array with many small counts and many more zeros. It is a pervasive property of this type of data, largely unaltered by transformations, which manifests itself in virtually all analyses and when ignored leads to misinterpretation.

The distribution of species abundance is similar for all studies (Figure 15.2) showing the log-series or truncated log-normal series form common to large faunal surveys (May 1975). A useful consequence of the general similarity is that informative comparisons can be made which use only the low and high end of the curves. The low end in these distributions are the singletons, species with only a single specimen encountered. The percent of all taxa which are singletons fall in a relatively narrow range (STBS 16.6%, CGP 13.7%, MAFL 17.4%, and SWFL 19.1%). The high end is the dominant taxon, which show a broader range. The percent of the total specimens collected belonging to the dominant taxon were: STBS 10.6%, CGP 17.7%, MAFL 3.3%, and SWFL 11.2%. Rather than make inappropriate comparisons between studies, these values can be best used to predict limits; in cross-shelf studies in the Gulf of Mexico, 15% or fewer of all animals collected will belong to a single species, and roughly the same percent of the number of species encountered will be singletons. The value of these limits will be developed in the discussion section.

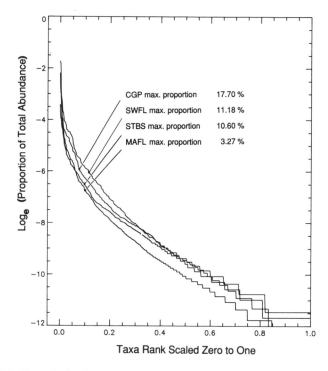

Figure 15.2. The rank abundance pooled over all samples of each species for the STBS, CCP, MAFLA, and SWFL studies. Common scales are produced by plotting the proportional rank against the proportional abundance on a log scale.

Variance Covariance Structure

Since the primary task of impact testing is separation of impact from natural variation, impact prediction studies should identify the primary sources and magnitudes of variation in the pre-impact environment. Analyses of variance, regression, ordination procedures, and cluster analysis which employ variance-like similarity are greatly influenced by the variance and covariance of the data. Optimally, estimates of this structure are used in design planning prior to a study. Minimally, this structure is estimated during a study and its effects considered in analysis and interpretation.

The most striking weakness of all the studies is the omission of those aspects of variance included in the design: estimates of variance, covariance, partitioning of variance into the components. All (except CGP) employed transect-station field designs which allow for the partitioning of variance into positional, temporal, station, replicate, and residual components. More ecologically-relevant parameters such as depth and sediment type were measured, but the sampling design did not specifically seek their contribution to faunal variation. Partitioning variance in the archived data sets proved impractical due to replicate pooling, temporally inconsistent sampling of stations, major addition or deletion of transects, and a lack of linkage between faunal and geological archives. Only the grossest statement is possible, and that is that no more than 40% of total variance of transformed counts in any study is attributable to cross-shelf position.

Examination of the data sets showed that more than half the variance in raw counts was due to only 1% of the taxa. Variance in $\log_e$(counts + 1) transformed data was dominated by no more than 10% of the taxa. Variance of even transformed counts is high relative to the mean. This is the common pattern in nature due to the relationship between variance and the mean (Taylor 1961, Venzia 1988). Those taxa with greatest variance are also those with the greatest overall abundance. The dependence of variance upon the mean is not completely eliminated by transformation since even the most abundant species may be absent (zero count) from a high percentage of the samples.

Surprisingly, covariance among species is overwhelmingly positive in all four studies, in spite of the traditional view that because nearshore species decrease offshore and offshore species decrease onshore, negative covariance would be expected. The preponderance of positive covariance when negative is anticipated has three interpretations. An ecological explanation is that few species actually show the onshore-offshore trend. A sampling explanation is that the trend is masked by a sample size effect causing on and offshore species to be more abundant in large samples. Finally, the numerous zeros in the data set may be the cause. Basically, a large number of zeros has the effect of distorting the meaning of covariance. For example, if there are two species found together interacting antagonistically or responding to the environment in opposite ways, their counts should have a negative covariance. The negative sign of the covariance reflects that when one species is above its mean value the other is below. There is a positive covariance when both species fall on the same side of their respective means in a sample. In a large area survey, many samples will lie beyond the range of these two species and will contribute zero counts to the data. These zeros enter into computation for the mean of each species, shifting the means closer to zero. If enough zeros are included, the mean for both species will actually lie below the range of nonzero counts for both. When covariance across the entire data set is calculated, the two species' non-zero counts will always fall above (and zero counts below) their respective means. Thus, species which covary negatively where mutually present on the small scale become positively covariant on a larger scale that introduces many zero counts into the data. Positive covariance is to be expected in faunal data with many zeros, and is largely uninformative.

Discussion

The surveys reviewed here, and probably many others, are not developing information which contributes fully to the EIS process. Much of the information that they could provide is omitted or lost in complex archives. That which is provided receives minimal use. In considering how to bring about improvement, many suggestions can be made and discussed that are both practical and easily implemented concerning archival, taxonomic resolution, design, and analysis. There is, however, a pervasive problem inherent in the mutual failure of the EIS process and benthic studies to identify those ecological processes which must be understood to predict impact and to determine the ecological value of soft-bottom environments. Development of such a process-oriented approach will be difficult.

In order to suggest improvements, the minimal purpose of preimpact surveys may be defined as providing managers with a systematically correct inventory of species sufficiently complete to predict (i) the spatial distribution, abundance and variance of dominant species and (ii) the extent to which the fauna contains rare forms. Once this minimum has been obtained, the optimal purpose may be sought. At this optimum, managers would be provided information about the sensitivity of the fauna, and the relative importance of different regions of the seafloor.

Adequacy of Archived Data Sets

Archive retrieval, data set recreation, and simple confirmation of previous results proved to be difficult due to three general types of problems. Troublesome, but surmountable was the first problem that the archives do not use a contemporary relational data base structure. They are fixed length ASCII files employing mixed formats of headers and data records that require conversion and restructuring in order to be useful. Second and increasingly critical, exact replication of published results is impractical because there is no cross referencing of reports and final data archives. There is little or no coding in the archive to indicate subproject and main project design, and analyses discussed in final reports cannot be duplicated due to inadequate detail. Third and most damaging, the archived data contained obvious errors in omissions and conflicting entries with no clues or warnings to the user.

A fundamental weakness of NODC (National Oceanographic Data Center) data sets is that they are card-image files severely limited by retention of the 80 characters per record fixed format (NODC 1991). The partially offsetting strength is that the prescribed format has been consistent through time. Each individual data set begins with a header giving pertinent information such as cruise, investigator, and research vessel. This header is followed in sequence by a header card for each sample giving position, date, equipment information, etc. Following each sample header card are a variable number of faunal data cards giving count and/or weight of each taxon in the sample. On the faunal data cards, taxa are identified by a taxonomic code of up to 12 characters, and samples by a code of up to 11 characters. Since data are often archived by cruise, a complete survey may contain more than one data set.

Errors in archived data became evident during creation of sample/taxa data arrays. This is the most common structure used to examine faunal data, and lends itself to scientific visualization of array data (Precision Visuals Workstation Analysis and Visualization Environment, PV-WAVE Precision Visuals 1992). The rows (samples) of these arrays were established by finding all unique sample headers in a data set and assigning a row to each. Columns (taxa) were established by reading all faunal data cards, extracting a list of the unique taxon codes and assigning a column to each. The values in the array were then read and mapped into the correct location according to the row and column subscript associated with each particular sample and taxon. The worst data set was that of the MAFL study in which approximately 10% of the 91,000 faunal counts were conspicuously in error. The most common errors were the actual absence of the count data on a record and multiple cards with different counts for the same taxon and sample. The rate of less conspicuous errors such as miscoding could not be determined. The error rate for the other three studies was far better with less than one suspect record out of every 10,000. Much stricter data management and quality assurance requirements now imposed by MMS should reduce this error rate.

Production of useful inventories with minimal statistical information for surveys containing over one thousand taxa is impossible in the peer-reviewed literature and of limited use if appended to final gray literature. Therefore, the public data archives must assume the role of providing access to inventory information. The current structure of NODC archives, even if accessible on-line, is ill-suited to this task since the user with little or no documentation is required to read a data set, extract unique taxon codes, build up statistical summaries by searching for those codes, and finally attach a name to the numerical code through reference to a master list of codes and names. Concurrent with an upgrading of public data bases, verification, documentation, and cross referencing of data sets with contract reports must be made a contractual obligation. Such information must be available on line with the data set.

Adequacy of the Species Inventory

The OCS studies reviewed here, and similar preimpact studies in other systems, are primarily intensive species inventories devoting far more effort to sorting and counting species than to ecological analysis. The apparent high endemism in the Gulf studies is very worrisome in that it suggests an absence of adequate taxonomic quality control. Unfortunately, the simple need to produce better quality inventories has been obscured by the overemphasis placed on dubious analyses and overly simple conclusions about diversity, abundance, and spatial pattern. Inventories can be greatly improved if guidelines are stipulated and quality assurance and quality control devised to correct the problems found in this review: (i) omission of species-specific information in a useful form, (ii) inconsistent taxonomic resolution, and (iii) possible taxonomic error.

Current quality control measures are inadequate. Neither the use of specialists nor of a curated voucher collection provide the level of assurance needed. Even if correct and consistent, identification of examples by a specialist is no assurance that all specimens are correctly sorted. Similarly, archived vouchers are a small fraction of the actual samples, and there are so few active taxonomists that corrections may be made many decades after the end of a project. Three steps should be taken now to greatly improve taxonomic quality. First, there needs to be formal study of the consequences of taxonomic error in impact prediction and assessment to aid in the establishment of priorities. Second, regional faunal guides must be produced which emphasize identification of high priority species and estimation of the likelihood of identification error. Third, selection of material for curation should be part of a formal quality assurance program designed to estimate the actual error rate of identifications.

Changing the Level of Systematic Resolution

Is it necessary to work at the species level? Certainly, the alternative to reduction of taxonomic error at the species level is to lower taxonomic resolution to an error-free level. Indeed, there are other arguments in favor of lower resolution. This review and other studies have produced results that support working well below the species level (class, phylum, and even kingdom). Herein it was found that total abundance of animals (kingdom level) produced statistically well behaved data. In some impact studies, pooling of species to reduce taxonomic resolution left conclusions unchanged and actually increased analytical power (Warwick 1988, Ferraro et al. 1989, Ferraro and Cole 1990). In addition to these statistical arguments, a biological basis has been suggested postulating that there are responses to environmental stress common to all species in certain higher taxa (Pearson and Rosenburg 1978).

Dropping the requirement that impact prediction and impact testing studies work at the species level is such a departure from traditional views of what constitutes a comprehensive and adequate study that it requires considerable investigation and criticism. Under scrutiny, the arguments in favor of lower resolution become less convincing, but cannot be discounted fully. Gross pooling at the kingdom level is statistically nice since it increases sample mean and lowers covariance. However, the positive covariances among species are more likely due to the zeros in the data than to ecological relationships. So, the reduction in variance that pooling produces (pooled variance equals the sum of variances less the sum of the covariance) reflects elimination of zeros and may have no ecological relevance. The advantages of pooling to the level of family, order, or class may be similarly illusory. Pooling many rare species in with a dominant species only hides the fact that the dominant species still contribute the most to variance. Trial pooling applied to the STBS data showed no appreciable change in the extent to which dominant taxa controlled total variance until the number of pooled groups reached approximately 25% of the original number of species in the data set.

These warnings that "other than species groups" may be nothing more than analysis of the most common species with the illusion of being more comprehensive is only intended as a call for caution, not abandonment. Pooling of taxa has definite advantages, but must be explored in much more detail before species-level surveying is abandoned. Foremost among the issues to be resolved is the question of most appropriate poolings. Pooled groups created just to provide specific statistical benefits, may lack the desired biological traits and not reflect important changes in the fauna (Underwood and Peterson 1987). This may be seen in a comparison of desirable biological and statistical traits. Biologically, a good pooled indicator would be one in which all component species responded to a stressing agent in a similar manner and similar degree. The guild approach to pooling polychaete worms (Fauchald and Jumars 1979) is an example of a biological indicator. Statistically, a good indicator is one with the greatest power to detect impact in the presence of natural variation. By this statistical definition with its emphasis upon power, ideal groups might be created by pooling to minimize zeros, minimize variances and maximize means. Such a statistical indicator must pool species which do not always co-occur (minimize zeros), whose population changes are in opposite directions where they do co-occur (minimize variance), and which include the dominant species (maximize means). In effect, pursuit of statistical power leads to the combining of species which show opposite responses.

Adequacy of Design

The primary difficulty in determining if preimpact studies employ adequate designs lies in the contrast between the complex questions the studies are intended to ask, and the scant use of those answers in the EIS process. The EIS process now makes very little use of the results. If coring and detailed macrofauna analysis is to be used simply to map habitats and confirm the extent of soft bottom, then very different designs should be employed. The information content of the studies, though poorly presented, is much more extensive. If a species inventory is to be produced, distributions delineated, and mean and variance estimates developed for use in future monitoring, then the design, but not the analyses, of current studies is adequate. Finally, if impacts are actually to be predicted, then current designs must be considered inadequate. However, improvement requires far more attention to development of a conceptual basis than to statistical and technical issues.

Current preimpact surveys do not actually seek to predict impacts, but are designed to show the resource manager what lives in the environment, where it is, and how it varies in time and space. It is then a matter of judgment as to whether an impact might occur and how it can be avoided, minimized, or mitigated. While adequate in the general sense of providing information to support the manager's judgment, the spatial scale of current designs is quite gross, and a better range of scales would improve the applicability of results. For instance, current survey practices take replicate cores a few meters apart, at stations tens of kilometers apart, over transects roughly 100 km long. The actual scale of interest around an oil and gas platform, approximately 1000 m, is not even directly considered. Since the resources needed for these studies are limited, increasing the range of scales must come at the cost of either replication or extent of geographic coverage. Since future impact monitoring needs the variance estimates produced by replication, restriction of geographic coverage is the preferred reduction.

Rather than simplifying the design of preimpact studies to the level of their current utilization, it would seem preferable to assure their better utilization. The failure of the EIS process to make better use of data is easily attributed to the burdensome complexity which now characterizes the EIS process. Environmental Impact Statements are not scientific documents, nor are they an effective means of presenting and considering scientific evidence. They have become a compendium of *ad hoc* concerns developed during a process of scoping in which all interested parties can express reservations about the proposed project's possible impacts. These concerns, which reflect a wide range of ecological and political sophistication, determine the implicit list of possible impacts to be predicted and avoided on the basis of survey data. MMS includes the identified information categories in a request for proposals and stipulates the manner of collection in a binding contract. In the 5 or more years before a final report is submitted by the contractor and the related EIS issued by MMS, the linkage between study components and original concerns may have become lost or moot.

Adequacy of Analyses

The failure of adequately designed and executed studies to produce the intended useful results can be attributed to the overuse of analyses intended to simplify results. Simplification would be useful and desirable, if based upon concise ecological concepts and conveying a well understood aspect of synthesis. However, the calculations of diversity and classifications (clusters, ordinations, etc.) that dominate these studies neither have the needed basis nor properties of synthesis.

The claim that diversity lacks an adequate conceptual basis is not new (Peters 1991), but diversity is so resilient in the face of criticism that the claim warrants repeating. Diversity is enigmatic. Kept vague, it is one of the fundamental descriptions of a biota or a fundamental attribute of an ecosystem; finding the unknown links between diversity and function is one of the highest priorities of ecological research (Lubchenco et al. 1991). Yet, given exact mathematical form, diversity's repeated calculation and presentation adds little to the description of the faunal patterns and apparently nothing to an understanding of the underlying systems. The original appealing idea that diversity might relate to stability has not been substantiated (Goodman 1975), and links between diversity and pollution stress have been found to be inconsistent (Schindler 1987).

Diversity lacks the conceptual basis needed to predict impacts, but as a variance-like term expressing how specimens are allocated to species, it can still be used as a basis of comparison between locations. The question then arises as to how it should be estimated. The Shannon-Weaver information theory diversity index, H', has become a contractually mandated, sometimes empirically useful (Ferraro et al. 1989), and statistically understood (Tong 1983) tool in impact studies; it serves as a simple summary statistic. Its adoption can be attributed to the advocacy of its additive properties (Pielou 1975) rather than tenuous ecological significance (Margalef 1958). In practice, the additivity principle has not been utilized, and has been shown to conflict with development of an unbiased estimate of diversity (Smith et al. 1979). Other than consistency, there are no strong arguments for retention of H'.

The major alternative to the calculation of any simple diversity index is to consider diversity to be the complete distribution of abundance for all species. When a parsimonious description of the diversity distribution is needed for hypothesis testing, the parameters of the associated distribution can be estimated, the probable error of that estimate determined and appropriate tests developed. This statistical approach was first advocated by Fisher et al. (1943), but usually rejected since prior assumptions about the actual distribution needed to be made. With the ubiquity of log-series, log-normal and similar species abundance distributions in large, multispecies data sets, this approach is now better justified (May 1975, 1981, 1984).

In practical terms, the presence of ubiquitous species abundance distributions indicates that the distribution of relative abundance of the most common to the rarest species can be reasonably well predicted given the number of species and total specimens. Therefore, species number and the abundance of the most abundant species in the samples are potentially more informative than any more complex index. When a more detailed comparison of diversity is needed, the most informative approach may be to look at departures from the expected distributions. This is effectively the "neutral model approach" of Caswell (1976), which has already had some application in marine benthic studies (Lambshead and Platt 1988) and warrants much further investigation.

The ease, speed, and ingenuity of data reduction and pattern finding algorithms are seductive in many unrelated fields where research must deal with multiple categories and complex sampling schemes. And, there can be no argument that large benthic surveys need objective means of classifying samples according to faunal composition, degree of impact, etc. The main problem is that the method selected to make such a classification may be ill-suited to the task and/or so deeply founded in nonintuitive mathematics as to defy ecological rationale. Cluster analysis was extensively used in the reports reviewed either in lieu of analysis of variance or to aid in the production of maps. Actual analysis of variance would be preferable, and ordination procedures better suited for mapping. Cluster analysis also poses the subjective questions of which similarity index and which clustering method should be used. Gradient analysis (Whittaker 1973), a special case of ordination, is an obvious choice of analysis to explore for ecological application. However, unlike terrestrial plant assemblages (Gosz 1992) or butterflies and plants (Kremen 1992) where gradient analysis has proved useful, there is so little basic natural history information on marine macrofauna that it is not yet clear what gradients exist and are ecologically important.

A primary use of estimates of means and variances derived from field surveys is determination of the statistical power to detect impacts in monitoring programs begun after the potentially impacting activity has commenced (Green 1979). Power analysis was not a component of the studies reviewed, but is a contractual specification in more recent MMS surveys (Carney, personal observation). Ideally, the design of future impact-testing studies should be specified at the time of the pre-impact survey, and the survey design structured to help refine the testing studies. For example, should the BACIPS design be employed (Stewart-Oaten, Chapter 7, Stewart-Oaten et al. 1986; Underwood 1991), then the survey must provide the usual estimates of within-site variation and must also provide information on selection of Control and Impact sites which will be coherent in the response of the indicator variable to natural environmental fluctuations (Osenberg et al., Chapter 6).

Toward Process-Oriented Prediction of Impact

Scant use of survey information in EIS, the general failure of surveys to present their most important results, and the lack of improvement in surveys all point to the absence of a strong conceptual basis for the work. In the soft-bottom environment, benthic ecology offers the manager few concepts that convert statistical summaries into predictions about sensitivity to impact. Lacking such predictive capability, a manager might prudently allocate more protection to regions that are of greater value. But, here again, few concepts convert survey data to conclusions about relative importance. Development of appropriate concepts about benthic processes is a necessity. Concepts of soft-bottom structure and function are so poorly developed relative to those for hard substrate systems (Sebens 1991) that this review cannot identify a single direction of obvious merit. However, a productive starting point could be testing of two simple hypotheses of direct management relevance.

Since all areas of continental shelf soft bottom are implicitly assumed to be of equal value, testing these assumptions could be a first step towards a process-oriented approach. Ecological value can be considered from the perspective of recruitment and productivity to pose two hypotheses. First, all areas of the soft-bottom continental shelf benthos are of equal importance in providing larvae for recruitment by settling and adults for recruitment by immigration into the populations of the shelf environment. Second, all areas of the soft bottom are equal with respect to secondary production. Testing the first hypothesis by tracing of recruits to their location of origin may not be feasible. However, strong circumstantial evidence could be gathered from fecundity studies. Testing the second hypothesis would require estimation of turnover rates rather than standing stock inventories. Such work might focus upon the contribution of specific areas to the production of commercial species of fish and crustaceans.

More elaborate development of process-oriented study might look at the possibility that oil and gas activity could disrupt a physical-biological link. Since the substrate of soft bottoms is subject to physical transport and the primary food source is detritus, successful identification of impact-sensitive links may be expected to be derived from some combination of transport studies (Nowell et al. 1987), detritus competition studies (Lopez et al. 1989), and fauna-bioturbation interaction studies (Rhoads 1974, Rhoads and Boyer 1984).

Conclusions

Environmental Impact Statements now make minimal use of preimpact benthic ecology data for two related reasons. First, the EIS process accepts description of a system in lieu of a true understanding of how the system functions and its components interact. Second, benthic ecology offers little alternative information. Thus, neither science nor policy drive the need for improved information, and preimpact surveys remain unchanged.

Lacking useful scientific information, the EIS process proceeds with assumptions rather than facts. The main assumption is that the vast area of soft bottom is homogenous faunistically and process-wise and so vast that the limited impacts of oil and gas operations could never significantly impact the whole. This may be a valid assumption, but the degree to which it is really supported by survey data is questionable. The analyses used have not tested the assumption and have probably been based only on less than 10% of the fauna.

Improvement in preimpact studies can be brought about by more useful archiving, taxonomic quality control, more complete reporting of results, some design modifications, and research into the appropriate taxonomic resolution for sampling and analysis. However, greater utilization requires that a process-oriented approach be adopted that will allow for prediction of impacts and recognition of ecological importance.

Summary Recommendations

1. Gray literature reports and poorly documented data archives are ill-suited for the complex EIS process and for benthic research. Enhancement of the archiving function may be the best solution. Agencies undertaking preimpact surveys should be obligated to archive data sets so that they are accurate, available for use, and contain detailed information on the design of the project and the use of the archives.

2. The extent of taxonomic error is unknown, but seriously compromises long-term and wide-area syntheses of results. A well-designed, taxonomic, quality-control program must first formally assess the consequences of error, then implement measures to minimize the most serious problems.

3. Species-level surveys should continue, but the advantages of pooled species groups in analyses must be explored. Rather than dependence upon creating pooled indicators which meet certain statistical criteria, ecologically relevant groups must be identified and then tested for statistical suitability.

4. The current sampling design of cross-shelf transects, stations, and replicates is adequate for a general survey, but does not provide information about variation on the scale of 1 km which is most appropriate to platform effects. Due to limited resources, inclusion of this scale of sampling may require a decrease in geographic coverage.

5. There is little benefit to be gained from addition of new statistical simplifications of survey data. Descriptive species-level statistics combined with easy access to data archives will be more useful in the EIS process and benthic studies than most pattern analysis. When sampling design allows a partitioning of variance, the appropriate analyses should be performed.

6. Species-level or lower taxonomic resolution surveys do not provide managers with a means of predicting impact or of determining relative ecological importance since there is no information about how the species inventoried participate in benthic processes. Bridging this statistics-process gap might be initiated by testing hypotheses about the contribution of local bottom areas to larval pools and foraging.

7. The EIS process must be reformed to assure that broad and varied public concerns do not distract from the pursuit of well-defined and critical scientific information. The success of such a reform is dependent upon successful adoption of a process-oriented approach to pre-impact studies.

Acknowledgments

The examination of NODC data archives was supported by the Minerals Management Service through the University Initiative Program. The staff of the New Orleans MMS office provided assistance in identifying available data archives and project reports. M. Hollinger of NODC provided assistance in the use of data archives. E. Evers undertook the task of archive verification and reformatting. L.A. Hayek provided advice and useful discussions on data analysis and statistics. The editors of this volume provided invaluable assistance in clearing clutter and focusing discussion.

References

Bedinger, C. A. ed. 1981. Ecological investigations of petroleum production platforms in the central gulf of Mexico. Report to BLM Contract AA551-CT8-17. Southwest Research Inst., San Antonio, Texas.

Carney, R. S. 1987. A review of study designs for the detection of long-term environmental effects of offshore petroleum activities. Pages 651–696 in D. F. Boesch and N. N. Rabalais, editors. Long-term environmental effects of offshore oil and gas development. Elsevier, New York, New York.

Caswell, H. 1976. Community structure: neutral mode analysis. Ecological Monographs 46:327–354.

Clarke, K. R., and R. H. Green. 1988. Statistical design and analysis for a 'biological effects' study. Marine Ecology Progress Series 46:213–226.

Dames and Moore. 1979. Final Report, Mississippi, Alabama, Florida Outer Continental Shelf baseline Environmental survey. Prepared for BLM under contract AA550-CT7-34.

Environmental Science and Engineering, LGL Ecological Research Associates, and Continental Shelf Associates. 1987. Southwest Florida shelf ecosystems study. Vol. II. Data Synthesis Report. Prepared for MMS contract 14-12-0001-30276.

Fauchald, K., and P. A. Jumars. 1979. The diet of worms: an analysis of polychaete feeding guilds. Oceanography and Marine Biology 17:193–284.

Ferraro, S. P., and F. A. Cole. 1990. Taxonomic level and sample size sufficient for assessing pollution impacts on the Southern California Bight macrobenthos. Marine Ecology Progress Series 67:251–262.

Ferraro, S. P., F. A. Cole, W. A. DeBen, and R. C. Swartz. 1989. Power-cost efficiency of eight macrobenthic sampling schemes in Puget Sound, Washington, USA. Canadian Journal of Fisheries and Aquatic Sciences 46:2157–2165.

Fisher, R. A., A. S. Corbet, and C. B. Williams. 1943. The relationship between the number of species and the number of individuals in a random sample of an animal population. Journal of Animal Ecology 12:42–58.

Flint, R. W., and N. Rabalais, editors. 1981. Environmental studies of a marine ecosystem: south Texas outer continental shelf, 1st Edition. University of Texas Press, Austin, Texas.

Goodman, D. 1975. The theory of diversity-stability relationships in ecology. Quarterly Review of Biolgy 50:237–266.

Gosz, J. R. 1992. Gradient analysis of ecological change in time and space: implications for forest management. Ecological Applications 2:248–261.

Green, R. H. 1979. Sampling design and statistical methods for environmental biologists. Wiley and Sons, New York, New York.

Kremen, C. 1992. Assessing the indicator properties of species assemblages for natural areas monitoring. Ecological Applications 2:203–217.

Lambshead, P. J. D., and H. M. Platt. 1988. Analyzing disturbance with the Ewens/Caswell neutral model: theoretical review and practical assessment. Marine Ecology Progress Series 43:31–41.

Lopez, G., G. Taghon, and J. Levinton. 1989. Ecology of marine deposit feeders. Vol 31, M.J. Bowman, R.T. Barber, C.N.K. Mooers and J. Raven (series editors). Lecture notes on coastal and estuarine studies. Springer-Verlag, New York. 322 p.

Lubchenco, J., A. M. Olson, L. B. Brubaker, S. R. Carpenter, M. M. Holland, S. P. Hubbell, S. A. Levin, J. A. MacMahon, P. A. Matson, J. M. Melillo, H. A. Mooney, C. H. Peterson, H. A. Pulliam, L. A. Real, P. J. Regal, and P. G. Risser. 1991. The sustainable biosphere initiative: an ecological agenda. Ecology 72:371–412.

Margalef, D. R. 1958. Information theory in ecology. General Systems 3:36–71.

May, R. M. 1975. Patterns of species abundance and diversity. Pages 81–120 in J. Diamond and M. Cody, editors. Ecology and evolution of communities. The Belknap Press of Harvard University Press, Cambridge, Massachusetts.

May, R. M. 1981. Patterns in multispecies communities. Pages 197–227 in R. May, editor. Theoretical ecology, 2nd Edition. Sinauer Associates, Sunderland, Massachusetts.

May, R. M. 1984. An overview: real and apparent patterns in community structure. Pages 3–18 in D. R. Strong, D. Simberloff, L. Abele and A. Thistle, editors. Ecological communities: conceptual issues and the evidence. Belknap Press, Cambridge, Massachusetts.

MMS (Minerals Management Service). 1983a. Environmental impact statement for the proposed 1984 North Atlantic outer continental shelf oil and gas lease sale. Department of Interior, Minerals Management Service.

MMS (Minerals Management Service). 1983b. Environmental impact statement for the proposed California lease offering April 1984. Department of Interior, Minerals Management Service.

MMS (Minerals Management Service). 1984a. Draft environmental impact statement for the proposed 1985 outer continental shelf oil and gas lease sale offshore the Mid-Atlantic states. Department of Interior, Minerals Management Service MMS-84-0039.

MMS (Minerals Management Service). 1984b. Draft environmental impact statement for the proposed St. George Basin Sale 89, Alaska OCS region. Department of Interior, Minerals Management Service MMS-84-0017.

MMS (Minerals Management Service). 1992. Gulf of Mexico sales 142 and 143: central and western planning areas. Final environmental impact statement. Department of Interior, Minerals Management Service MMS-92-0054.

NODC (National Oceanographic Data Center). 1991. NOCD users guide. U.S. Department of Commerce. National Oceanographic and Atmospheric Administration.

NRC (National Research Council). 1992. Assessment of the U.S. outer continental shelf environmental studies program. II. Ecology. National Academy Press, Washington D.C.

Nowell, A. R., P. A. Jumars, and J. H. Kravitz. 1987. Sediment transport events on shelves and slopes (STRESS) and biological effects of coastal oceans sediment transport (BECOST). EOS 68:722–724.

Pearson, T. H., and R. Rosenberg. 1978. Macrobenthic succession in relation to organic enrichment and pollution of the marine environment. Oceanography and Marine Biology Annual Review 16:229–311.

Peters, R. H. 1991. A critique for ecology. Cambridge University Press, Cambridge, UK.

Pielou, E. C. 1975. Ecological diversity. Wiley-Interscience, New York, New York.

Precision Visuals. 1992. PV-WAVE command language technical reference manual. Version 3.1. Precision Visuals, Inc., Boulder, Colorado.

Rhoads, D. C. 1974. Organism-sediment relations on the muddy sea floor. Oceanography and Marine Biology 12:263–300.

Rhoads, D. C., and L. F. Boyer. 1982. The effects of marine benthos on physical properties of sediments: a successful perspective. Pages 3–43 in P. L. McCall and M. J. S. Tevesz, editors.

Animal-sediment relations: the biogenic alteration of sediments. Plenum Press, New York, New York.

Schindler, D. W. 1987. Detecting ecosystem responses to anthropogenic stress. Canadian Journal of Fisheries and Aquatic Sciences **44**:6–25.

Sebens, K. 1991. Habitat structure and community dynamics in marine benthic systems. Pages 211–234 *in* S. Bell, E. McCoy and H. Mushinsky, editors. Habitat structure: the physical arrangement of objects in space. Chapman and Hill, London, England.

Shannon, C. E., and W. Weaver. 1949. The mathematical theory of communication. University of Illinois Press, Urbana, Illinois.

Smith, W., J. F. Grassle, and D. Kravitz. 1979. Measures of diversity with unbiased estimates. Pages 177–191 *in* J. F. Grassle, G. P. Patil, W. K. Smith and C. Taillie, editors. Ecological diversity in theroy and practice. Vol. 6. Satellite program in statistical ecology international statistical ecology program. International Cooperative Publishing House, Burtonsville, Maryland.

Stewart-Oaten, A., W. W. Murdoch, and K. R. Parker. 1986. Environmental impact assessment: "pseudoreplication" in time? Ecology **67**:929–940.

Taylor, L. R. 1961. Aggregations: variance and the mean. Nature **189**:732–735.

Tong, Y. L. 1983. Some distribution properties of the sample species diversity indices and their applications. Biometrika **39**:997–1008.

Underwood, A. J. 1991. Beyond BACI: experimental designs for detecting human environmental impacts on temporal variations in natural populations. Australian Journal of Marine and Freshwater Research **42**:569–587.

Underwood, A. J., and C. H. Peterson. 1988. Towards an ecological framework for investigating pollution. Marine Ecology Progress Series **46**:227–234.

Vezina, A. F. 1988. Sampling variance and the design of quantitative surveys of the marine benthos. Marine Biology **97**:151–156.

Warwick, R. M. 1988. Analysis of community attributes of the macrobenthos of Friersfjord/Langesundfjord at taxonomic levels higher than species. Marine Ecology Progress Series **46**:167–170.

Whitaker, R. H. 1973. Direct gradient analysis. Pages 1–52 *in* R. H. Whitaker, editor. Ordination and classification of communities. Dr. W. Junk, The Hague.

CHAPTER 16

ORGANIZATIONAL CONSTRAINTS ON ENVIRONMENTAL IMPACT ASSESSMENT RESEARCH

Frederick M. Piltz

Scientific uncertainty about environmental effects of human activities can impede scientifically defensible management of marine resources. In theory, there are research tools, such as the Before-After-Control-Impact (BACI) design (Stewart-Oaten et al. 1986, Stewart-Oaten, Chapter 7) that can be used to reduce this uncertainty, and thus lead to greater understanding of environmental effects. The most powerful of these tools requires information on environmental conditions before activity occurs and at several sites that vary in their exposure to the proposed activity (e.g., Control and Impact sites). Previous chapters have laid out scientific issues that underlie these approaches (e.g., Stewart-Oaten, Chapter 7, Underwood, Chapter 9), and other chapters have delineated statistical constraints (Thrush et al., Chapter 4, Mapstone, Chapter 5, Osenberg et al., Chapter 6) and biological considerations (Jones and Kaly, Chapter 3) that may hamper successful application of such studies. There are also many nonscientific constraints, which are often distinct from theoretical biological and statistical issues, that can also interfere with execution of an assessment study. These organizational constraints arise from project uncertainties about the timing, location, and nature of the activity, as well as institutional uncertainties about the ability of an agency to commit to a long-term research endeavor.

In this chapter, I analyze some nonscientific uncertainties that potentially limit the application of environmental impact assessment studies. I will draw examples from offshore oil production activities in California to illustrate how project and organizational constraints can hinder acquisition of scientific information even when appropriate scientific tools are available and can, in theory, be implemented by a governmental agency. The lessons I highlight are important to scientists involved with conducting assessment studies as well as agency personnel charged with administering such projects; scientists need to be aware of these

Detecting Ecological Impacts: Concepts and Applications in Coastal Habitats, edited by R. J. Schmitt and C. W. Osenberg

obstacles so that they can better incorporate appropriate contingencies into their research design. Policy makers and agency personnel need to recognize possible impediments so that they can better plan and facilitate productive research endeavors. I hope these lessons will aid scientists and governmental agencies in planning and implementing long-term research on environmental effects, and therefore help reduce the uncertainty we face regarding human impacts on natural ecosystems.

The Opportunity

The Minerals Management Service (MMS), an agency of the United States Department of the Interior, is charged with managing mineral resources on the Outer Continental Shelf (OCS) of the United States (MMS 1988). One program of MMS, the Environmental Studies Program, is responsible for collecting information on environmental and social effects arising from activities related to extraction of OCS minerals (primarily oil and natural gas). This information is used in the decision-making process for the sale (and later the exploration and development) of offshore leases (see Carney, Chapter 16 and Lester, Chapter 17).

In the late 1970s, MMS recognized the need for better information on the environmental effects of production platforms in OCS waters. Information existing at that time was deemed insufficient for assessment of multiwell production platform impacts (NRC 1983). First, prior research primarily focused on effects of exploratory drilling rigs, not production platforms. It was not prudent to extrapolate from the measured effects of a small, single-well exploratory rig, which remains at a site for only a few months, to those arising from a huge, multiwell (up to 60+ wells) platform that remains in place for several decades. Second, the only existing study of a production platform, in the Buccaneer gas field off Texas (Middleditch 1981), was compromised due to the lack of Before data (e.g., see Underwood 1991, Osenberg and Schmitt, Chapter 1) and the occurrence of several sources of large, natural disturbances near the Buccaneer field (e.g., discharge from the Mississippi River), which confounded or overwhelmed all but the most obvious near-field effects of the platforms (Middleditch 1981, Carney 1987).

Therefore, in 1979, MMS began planning for a long-term study of the effects of multiwell production platforms and associated discharges (NOAA 1981). An opportunity to initiate a major research program on this issue arose the following year when Chevron Oil Company and its partners bid $350 million for a lease in the Pt. Arguello oil field off Pt. Conception, California. Because this sale indicated a strong potential that production activities might arise several years in the future, MMS proceeded with detailed planning aimed at collecting Before and After data. Although the lease sale did not guarantee that production activities would occur, the agency accepted the risks associated with implementing a multiyear research program (see below) because of the recognized need for adequate Before impact data.

MMS devised a research program, based on a BACI-type assessment design, to explore the effects of production platforms in the Pt. Arguello field. As conceived, the California Monitoring Program (CAMP) was to consist of three sequential phases: a broad spatial survey/reconnaissance of habitats and organisms in the region (Phase I), which was to help guide the design of a BACI-type study (Piltz 1986); a time series of Before impact surveys at several expected Impact and Control (or reference) locations, culminating in a series of After surveys at these same locations to examine "acute" impacts arising from the release of muds and cuttings from the drilling of wells (Phase II); and a further series of After impact surveys at these sites following cessation of drilling discharges to explore any longer-term, chronic effects arising from production operations (Phase III). Three platforms were proposed by oil companies in areas of rocky substrates, and one was proposed to occur in a soft sediment area; the CAMP research effort originally was divided equally between these two habitat types. The approximate schedule for leasing through commencement of platform operations, together with that for the various CAMP research activities, is shown in Figure 16.1.

As events unfolded, execution of CAMP research did not follow the idealized scheme. Setting aside scientific issues that arose (and are dealt with generically in other chapters), the CAMP experience provides an excellent example of how nonscientific constraints can hinder the conceptually simple goal of estimating environmental effects of a development project using a BACI-type approach. Some of these constraints are discussed below, and are illustrated by the CAMP experience. Details regarding the design and results of the California Monitoring Program are presented elsewhere (e.g., Lissner and Shoakes 1986, Brewer et al. 1987, 1991, Hyland and Neff 1987, Steinhauer and Imamura 1990).

Project Uncertainties

For a proposed development project, it often is not possible to anticipate precisely where, when, and even if the development will occur. Further, details of a project typically change during the permitting process from those originally proposed, which can alter the nature of potential effects on the environment. Thus, although a BACI-type study can, in theory, provide excellent information on environmental impacts, these "project-specific" uncertainties can constrain the ultimate success of such a venture.

Uncertainty in the Occurrence of a Proposed Project

Perhaps the largest uncertainty is whether a proposed project is likely to occur, and therefore whether (expensive) assessment research is warranted. Proposed projects often are abandoned or canceled well before implementation, even when considerable financial investments have been made by the applicant.

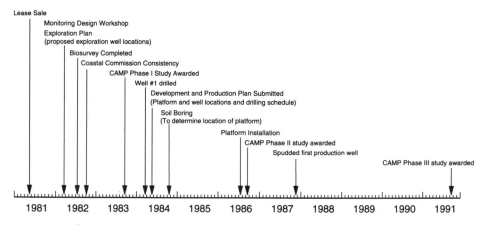

Figure 16.1. Time line showing major events in the planning and implementation of the California Monitoring Program (CAMP).

The offshore oil and gas industry provides notable illustrations. Decisions to explore and develop offshore mineral reserves are influenced by global economic factors, especially the price of the commodities. In the 1980s, oil companies invested heavily in leasing OCS tracts in the United States based largely on economic projections of the global market. When these projections were not realized, companies abandoned lease tracts for which they had paid millions of dollars because it was not economically feasible to develop them. Viewed in this light, the decision of MMS to commit to a platform effects study (i.e., the California Monitoring Program) in the Pt. Arguello field was risky indeed. The mere sale of the lease tracts for $350 million did not insure that economics would warrant extraction of any oil reserves found.

Proposed development projects, especially those impinging on coastal and marine environments, can be highly controversial (Lester, Chapter 17). Thus, social and political considerations also are prime elements affecting whether a proposed project will be built. For example, there is a great deal of public discomfort with offshore oil and gas production in the United States, much of which reflects the perceptions of the public about risks to the natural environment. With respect to assessment research by an agency, these concerns can be a double-edged sword. It can translate into broad-based support for more research on unresolved issues. However, public outcry can result in the cancellation of a development project to which an agency has devoted considerable time and resources to study. This is precisely what occurred in the California Monitoring Program. Between 1986 and 1990, about 50% of the CAMP effort was devoted to the collection of Before impact data for the proposed platform (Julius) on soft sediments. Although this platform was permitted by all federal, state, and local agencies, it was never installed because a ballot referendum prohibited the construction of onshore facilities to receive its oil. The agency discontinued Before impact sampling of soft sediments in the area, despite the crucial need for such information if the platform were to be installed in the future. Thus, the agency accepted the possible "cost of lost opportunity" in return for benefits obtained by diverting limited resources towards more pressing information needs that had greater probabilities of return. In either case, the extensive resources that were used to design and conduct the initial studies may never be used in the context of the original goal: assessment of effects of Platform Julius on the marine environment.

Uncertainty in Project Timing

The period of time available for Before impact sampling is crucial to the success of a BACI-type assessment program: the greater the number of temporal replicates, the greater the probability of distinguishing putative impacts from natural variation (see Osenberg et al., Chapter 6). The timing of when a project comes on line relative to the initiation of Before sampling is therefore of paramount importance, but this consideration can be affected severely by uncertainties inherent to the project. For example, if a project proceeds more quickly than anticipated, or planning of the research proceeds more slowly, then effort expended to initiate a study and collect preimpact samples might be in vain, particularly if the realized duration of the Before period is too short to yield high statistical power.

The California Monitoring Program illustrates this point. Although oil companies expressed interest in developing the Pt. Arguello field in 1980, final detailed development plans were not submitted until 1983. By the time the agency finished planning the assessment design (based on the development plans), completed the necessary initial field work (i.e., Phase I), and obtained funding commitments for the next phase, a limited amount of time was available to conduct the preimpact sampling program. As a result, only three preimpact surveys were completed before the primary platform under study (Hidalgo) began discharging drilling muds and cuttings in late 1987 (Figure 16.1).

Social concerns, often transmitted through the permitting process, can also modify the timing of project initiation. Applicants must satisfactorily address concerns of permitting agencies and the public, which can lead to delays of varying lengths in gaining approval to proceed. Delays in the projected start of a project can have both positive and negative effects on assessment research. Clearly, assessment research can benefit if the delay provides additional time for Before impact sampling. As an example, a new oil processing facility at Gaviota, California was scheduled to commence discharging produced water (an aqueous waste product of oil processing) in late 1988. Because virtually no preimpact data had ever been collected to assess the effects of produced water on coastal environments, Before sampling had been initiated in February 1988 at two Impact and one Control sites (Osenberg et al., Chapter 6). However, because of a series of permitting delays, commencement of produced water discharge was delayed until at least 1995. As a consequence, the Before impact sampling period of this assessment study was increased from a few months to over 5 years. Similarly, a delay of about 2 years in the start of the San Onofre Nuclear Generating Station resulted in a much longer and better set of pre-impact data to evaluate effects of that power plant (Murdoch et al. 1989, Bence et al., Chapter 8, Ambrose et al., Chapter 18).

The advantage gained by a project's delay can be tempered by factors that diminish institutional support for the related assessment program. Interest can wane, research priorities can shift, and budget conditions can degrade. An agency can find it increasingly hard to justify continued support of assessment research for a delayed project, where there is little immediate payoff for making decisions. A case in point is the assessment study of Gaviota produced water mentioned above. This promises to be one of the most powerful assessments of the effects of produced water on the coastal marine environment, yet by the very nature of most funding agencies, there can be no guarantee that the study will be supported to fruition. Most of the factors that come into play in this context can be classified as institutional uncertainties, which are considered in more detail in a later section.

Uncertainty in the Location of Potential Project Impacts

Assessment studies using BACI and other designs require that environmental conditions are estimated at locations that will be affected by the project (i.e., Impact sites) and also at similar areas that will remain unaffected (i.e., Control or reference sites). Obviously, it is crucial to know precisely where impacts are likely to occur, and where they will not. Sometimes this basic information is unavailable for a project, or the exact location of the source of impact is moved from that originally planned.

Again the CAMP assessment program provides a nice illustration of the nature of this problem. Each individual OCS lease tract is roughly a 3 mile by 3 mile square, and the final location of the platform within that square will be driven in part by the distribution of the resource (oil and/or gas) and the technology available to extract it. Occasionally, the location is moved from that originally proposed by a company to alleviate concerns raised in the permitting process. Thus, it is possible that the principal effects of a project occur in an area not subjected to preimpact sampling. Consideration of potential effects on rocky substrates in the Pt. Arguello field raise an additional concern in this context. The locations of rocky reefs are fixed and limited in number; platform location could have been such that all hard bottom in the area would have been affected, effectively eliminating all Control sites.

Often it is desirable to estimate the spatial extent of effects from point sources of disturbances such as an oil platform (Raimondi and Reed, Chapter 10). This typically involves sampling along a gradient of distance away from such a structure, and again the most rigorous assessment requires preimpact sampling. Clearly, knowledge of the exact location where that structure will be situated is crucial for the correct placement of the sampling grid(s). In the face of these uncertainties, a sampling program might yield data that have little or no application to the documentation of platform impacts.

Uncertainty in the Nature of Impacts

A research plan should include consideration of the specific activities that will occur, and the ways in which these activities might mediate environmental impacts. Some activities might suggest measurement of particular suites of parameters or sampling designs focused at particular spatial scales. The nature of these anticipated impacts can change during the permitting and development processes, and therefore undermine an otherwise well-conceived research plan. For example, the amount of wastewater discharged through a diffuser might change from that originally proposed and therefore change the predicted spatial scale over which impacts should occur. Sites originally sampled in anticipation of one discharge condition might not be appropriate for another. Similarly, technological innovations can lead to reductions in the types or concentrations of particular contaminants or the way in which those contaminants are discharged into the environment.

Another uncertainty can arise if the research is designed to study impacts associated with different phases of development. For example, part of the goal of CAMP was to separately study impacts associated with initial platform development activities (due to discharge of drilling muds and cuttings associated with drilling wells) and impacts associated with production (e.g., the long-term discharge of operational wastes such as produced water). Understanding the effects of different activities has obvious implications for improving our ability to minimize environmental effects. As it has turned out, drilling activity for the platform under examination was not concentrated in time as proposed, but has been much more drawn out and intermittent. Further, only a fraction of the planned wells have been drilled to date. Additionally, some production operations have occurred simultaneously with drilling activities, rather than being sequential events as proposed. Therefore, Before and After periods were not clearly differentiated, and the idealized conceptualization of the BACI design was never realized. Complications like these will always arise, and they pose an important challenge to the design and application of BACI-type assessment studies.

Institutional Uncertainties

Even when details of a particular project are completely specified, success of assessment research can be hindered by uncertainties that arise institutionally, and which limit the ability of an agency to conduct or oversee environmental research. For example, resource agencies need to balance two conflicting components of their environmental research programs: (i) commitment to investigations that require long-term study; and (ii) responsiveness to new issues that arise. These two components compete for limited resources, and commitment to long-term studies can wane as a result of turnover in agency personnel, shifts in perceived information needs, or changes in budgetary conditions. For example, reviews of an agency's research agenda, such as the National Research Council's reviews of MMS's Environmental Studies Program (NRC 1983, 1990, 1992a, 1992b, 1992c), can create pressures for sudden shifts in research priorities within the agency. While critical review of an agency's programs are absolutely vital, care must be exercised to avoid overreacting to suggested shifts in emphasis; a successful environmental research program must include long-term studies that encompass sufficient time to obtain adequate time series of data Before and After planned interventions occur. Indeed, one of the greatest shortcomings of many previous studies is the lack of adequate Before data. Thus, if appropriate long-term studies are not safeguarded against institutional uncertainties, we rarely will be able to assess rigorously environmental impacts. To do so requires an appropriate balance between commitment to ongoing projects and responsiveness to new needs. Below, I briefly consider several sources of institutional uncertainty, which are independent of the particular intervention being examined, that can affect the success of assessment research.

Agency and Funding Stability

Perhaps the most obvious source of institutional uncertainty (and therefore a constraint to long-term study) stems from the fact that agencies tend to be dynamic: personnel turnover and priorities change through time. If the life span of a particular agency's research agenda is shorter than the required life span of an assessment study, the successful completion of that research is threatened. For example, the tenure of many agency personnel and decision makers (particularly in the Executive Branch of the U.S. government) is often shorter than the time needed to complete scientific programs that they influenced. A research program embraced by one administration may not be embraced by a subsequent administration, or it may be given lower priority and lower funds in subsequent years. This suggests that projects extending over more than four years, for example, might be particularly vulnerable. Unfortunately, collecting adequate environmental data and developing appropriate models for any particular intervention often will require more than four years (Osenberg et al., Chapter 6).

This imbalance between institutional and research time scales is further evidenced by the budget process in many governmental agencies. For example, almost all U.S. federal agencies operate on annual appropriations from Congress, which are subject to changes from year to year, and can be modified further by shifts in priorities within an agency. Financial commitment to programs extending more than 2 or 3 years often is uncertain, which can hamper the quality of long-term research programs.

Mandated Research Scope and Agency Authority

All agencies have legislated or delegated mandates and language which define the scope of their authority and the nature of research they can conduct or support. For example, in the United States, several federal agencies have authority to regulate various oil and gas activities. These agencies may also conduct environmental research on the effects of offshore oil and gas production (e.g., MMS in the Department of the Interior, Department of Energy, National Marine Fisheries Service, Environmental Protection Agency). In some cases, there are clear definitions that delineate the roles of these different agencies; the Environmental Projection Agency, for example, issues permits for drilling discharges from oil and gas platforms. The EPA also occasionally funds environmental research into the effects of such discharges. In some situations, the delineation of research roles is not clear and this can obstruct natural extensions of research conducted by a particular agency. Unless these "competing" agencies are part of a well-integrated network, including state and local agencies, then valuable research proposals may fall between the cracks due to lack of institutional authority.

To some extent, these limitations on authority necessitate more broadly defined research scopes in other areas. For example, the Environmental Studies Program of MMS is required to investigate physical, chemical, biological and social impacts related to oil and gas activities on the outer continental shelf. Thus, while the agency has a geographically narrow scope, its mandate requires broad expertise in a variety of natural and social science disciplines. There are two possible costs that arise from the breadth of this mandated research scope: (i) it creates a potential for sudden and extreme shifts in research priorities, arising, for example, in response to highly visible reviews of the program (NRC 1983, 1990, 1992a, 1992b, 1992c), and (ii) resources and expertise can be spread too thin, resulting in information that is not sufficiently focused to yield especially valuable understanding within a particular discipline.

Comfort Level Associated with Scientific Uncertainty

The research priorities of an agency are determined largely by its evaluation of existing scientific information, and its determination of the extent to which existing information can be applied to new issues. The more environmentally important an issue, and the more uncertain the scientific information, the higher new research on that issue will be ranked. Differences in rankings can arise via legitimate disagreement about the adequacy of existing scientific information and/or "importance" of the issue. In some cases, uncertainty can also arise from lack of familiarity with the body of scientific information available on a particular subject. This can apply equally to staff within the agency or to groups empowered to influence the research directions of the agency. For example, in designing the California Monitoring Program, MMS reviewed existing literature and felt there was little reason to conduct a broad survey of metals (excluding barium) in drilling muds and in sediments around platforms; MMS believed that sufficient information already existed on this subject. However, state agencies and public interest groups urged MMS to analyze for metals specifically listed in the California Ocean Plan (legislation that places specific restrictions on discharge of metals and other pollutants). Based on this insistence, MMS measured these metals and found results consistent with those reported in previous studies. Thus, resources were diverted from other activities to measure environmental parameters that did not enhance the agency's capability to explain and predict environmental impacts of oil exploration and production; the information has, however, increased the comfort level of managers and the public that unexpected, catastrophic events are not taking place. Nonetheless, the more we aim to detect events that previous scientific information suggest are of little concern, the more we hamper the study of other, less resolved issues.

Conclusions and Lessons

The greatest obstacles to the practice of good assessment research are not scientific in nature, but rather involve nonscientific factors that constrain the application of available tools. As a consequence, the state of assessment research on planned interventions is not substantially different from that of chance episodes such as accidental spills, to which substantial research effort (and funds) of an agency often are diverted. This is in some sense ironic since we have a very limited ability to estimate impacts from unplanned events (e.g., lack of preimpact data), and such chance events are far less common, and contribute much less to environmental degradation, than planned interventions.

If we are going to improve the assessment of environmental impacts arising from planned events, then resource agencies and scientists must recognize how institutional and project uncertainties constrain the quality of science that gets done. To reduce institutional constraints: (i) agencies must first appreciate that the resolution of certain fundamental issues will require long-term research.

Beyond recognition of this need, institutional uncertainties must be reduced. Appropriate actions include: (ii) developing means to dampen unproductive shifts in research priorities. This can be achieved by better communication of research priorities and their rationale by an agency. (iii) Further, ways to reduce funding instability for long-term research programs must be developed. This can be achieved by better justification of long-term goals, which would reduce reallocation of limiting funds (see point (ii)). It also requires greater long-term commitment by an agency, as well as increased willingness to co-sponsor research with other agencies with mutual information needs.

Agencies, industry and environmental scientists need to work in concert to reduce project uncertainties. Appropriate actions include: (iv) better anticipation of project design and timing. This requires enhanced communication among the applicant, agency and scientist. (v) Sampling programs that are better able to deal with project uncertainties, and which can be implemented rapidly when an opportunity arises need to be designed (i.e., to maximize the amount and kinds of Before impact information). This might require an agency to have a general research plan on hand that can be molded quickly to any particular project. The plan should emphasize an extensive sampling program (many sites and/or parameters) for the Before impact period, which can be streamlined as more project-specific information is obtained.

Finally, even when project-specific and institutional uncertainties are reduced, all parties must recognize that regulatory decisions still will be made with limited information (Ruckelshaus 1983). Our knowledge cannot be perfect, and extrapolations must be made from less than ideal information. The lack of perfect information provides a basis for conflict among different interest groups (Lester, Chapter 17). To the extent that conflict arises from scientific uncertainty, better information will facilitate, but not guarantee, a greater consensus among interest groups. Thus, rigorous scientific information is a necessary, but not sufficient, condition to reduce conflict and ensure that decisions appropriately balance environmental and competing risks.

Acknowledgments

My appreciation is extended to Dr. Russell Schmitt, Dr. Craig Osenberg, Dr. Mark Pierson, and Dr. Ken Weber who reviewed the manuscript for this chapter and suggested improvements. The observations, conclusions, and recommendations are my own based upon my experience in planning, implementing, and managing the CAMP research program. I also wish to thank Dr. Gary Brewer who managed the CAMP Phase II research program. This chapter does not represent any formal policies of the Department of the Interior or the Minerals Management Service.

References

Brewer, G. D., F. M. Piltz, and J. Hyland. 1987. Monitoring changes in benthic communities adjacent to OCS oil production platforms off California. Proceedings Ocean 87 **4**:1593–1597.

Brewer, G. D., J. Hyland, and D. D. Hardin. 1991. Effects of oil drilling on deep-water reefs offshore California. American Fisheries Society Symposium **11**:26–38.

Carney, R. S. 1987. A review of study designs for the detection of long-term environmental effects of offshore petroleum activities. Pages 651–696 *in* D. F. Boesch and N. N. Rabalais, editors. Long-term environmental effects of offshore oil and gas development. Elsevier, New York, New York.

Hyland, J., and J. Neff. 1987. California OCS Phase II Monitoring Program, Year One Annual Report, Vol. I. Prepared by Battelle Ocean Sciences.

Lissner, A., and R. Shoakes. 1986. Assessment of Long-Term Changes in Biological Communities in the Santa Maria Basin and Santa Barbara Channel. Phase I. Vol. I and Vol. II; OCS Study MMS 86–0012. National Technical Information Service No. PB86240363 and PB86240371.

Middleditch, B. S. 1981. Environmental effects of offshore oil production: The Buccaneer Gas and Oil Field Study. Plenum Press, New York, New York.

MMS (Minerals Management Service). 1988. MMS in perspective 1982–1988. U.S. Department of Interior. 33 pages.

MMS (Minerals Management Service). 1991. California OCS phase II monitoring program. Final report. OCS Study No. MMS 91-0083. U.S. Department of Interior. 306 pages.

Murdoch, W. W., R. C. Fay, and B. J. Mechalas. 1989. Final report of the Marine Review Committee to the California Coastal Commission. Marine Review Committee, Inc.

NOAA (National Oceanic and Atmospheric Administration). 1981. National Marine Pollution Program Plan, Federal Plan 1981–1985. U.S. Department of Commerce. 185 pages.

NRC (National Research Council). 1983. Drilling discharges in the marine environment. National Academy Press, Washington, D.C.

NRC (National Research Council). 1990. Managing troubled waters: the role of marine environmental monitoring. National Academy Press, Washington, D.C.

NRC (National Research Council). 1992a. Assessment of the U.S. outer continental shelf environmental studies program: I. physical oceanography. National Academy Press, Washington, D.C.

NRC (National Research Council). 1992b. Assessment of the U.S. outer continental shelf environmental studies program. II. ecology. National Academy Press, Washington D.C.

NRC (National Research Council). 1992c. Assessment of the U.S. outer continental shelf environmental studies program. III. social and economic studies. National Academy Press, Washington, D.C.

Piltz, F. M. 1986. Monitoring long-term changes in biological communities near oil and gas production platforms. Proceedings Ocean 86 **3**:856–861.

Ruckelshaus, W. D. 1983. Science, risk, and public policy. Science **221**:1026–1028.

Steinhauer, W., and E. Imamura. 1990. California OCS Phase II Monitoring Program, Year Three Annual Report, Vol. I. Prepared Battelle Sciences.

Stewart-Oaten, A., W. W. Murdoch, and K. R. Parker. 1986. Environmental impact assessment: "pseudoreplication" in time? Ecology **67**:929–940.

Underwood, A. J. 1991. Beyond BACI: experimental designs for detecting human environmental impacts on temporal variations in natural populations. Australian Journal of Marine and Freshwater Research **42**:569–587.

CHAPTER 17

ADMINISTRATIVE, LEGAL, AND PUBLIC POLICY CONSTRAINTS ON ENVIRONMENTAL IMPACT ASSESSMENT

Charles Lester

A primary purpose of the National Environmental Policy Act (NEPA) and similar state laws is to reduce uncertainty about the environmental impacts of major governmental actions. The prescribed mechanism for achieving this purpose is the requirement that government agencies comprehensively assess the environmental impacts of their actions under full purview of the judiciary and the public. Unfortunately, the administrative, legal, and public policy contexts established by the NEPA assessment mechanism can impose significant constraints on the environmental impact assessment process. Thus, while the participation of government agencies, the judiciary, and the general public may enhance the democratic character of environmental assessment, the interaction of multiple and diverse interests in the assessment process can also work at cross-purposes to the ultimate goal of reducing environmental uncertainty.

To illustrate this problem, this chapter draws on a case study of a highly contentious environmental policy issue—the leasing and development of U.S. Outer Continental Shelf oil resources. It should be noted, though, that the extent to which the lessons of this case study are generalizable to other public policy contexts may be a function of the degree of public conflict surrounding a policy problem. This is because many of the constraints on environmental assessment discussed in this chapter are partly a function of the degree of public concern and thus participation in the question of Outer Continental Shelf oil development. To be sure, this public policy issue would fall at the high end on a spectrum measuring the "contentiousness" of public policy issues, along with such issues as nuclear power and the siting of hazardous waste facilities. Nonetheless, most if not all significant environmental assessment problems do take place in the administrative, legal, and public policy contexts established by NEPA-type environmental assessment laws. Thus, the lessons of this chapter may hold true *to some degree* for every public environmental assessment process.

Resolving Conflict through Comprehensive Environmental Assessment: The U.S. Offshore Oil Leasing Program

The case of U.S. Outer Continental Shelf (OCS) oil development presents an excellent opportunity to evaluate the theory and practice of public environmental impact assessment. First, because offshore oil development has been politically controversial for over twenty years, its environmental impacts are relatively well studied. Between 1973 and 1990, the federal government spent about 500 million dollars to identify and evaluate both the environmental and the socio-economic impacts of offshore oil development on the OCS (MMS 1991a), which is generally defined as the zone 3–200 miles offshore. Further, a multitude of Environmental Impact Statements (EIS) have been prepared by the federal government under NEPA that assess the potential impacts of proposed OCS development on the marine environment.

This substantial history of OCS environmental study, though, points to a second and far more significant reason that OCS development is a good case study for evaluating environmental assessment: the modern OCS program was explicitly designed to take advantage of comprehensive environmental assessment to help calm political conflict surrounding offshore oil development. This conflict began in 1969, when an offshore oil spill in the Santa Barbara Channel inundated local California beaches with thousands of barrels of crude oil (Nash et al. 1972). Almost immediately environmentalists and coastal states appealed to Congress for stricter laws governing OCS developers, particularly requirements for more comprehensive environmental assessment. In fact, the Santa Barbara spill is often noted by scholars as being partially responsible for the passage of the 1969 National Environmental Policy Act legislation that requires the preparation of EISs for all major governmental actions (Kallman 1984).

The event that finally motivated Congress to mandate more OCS environmental assessment was the oil embargo of 1973 that increased the importance of OCS oil to the energy security of the United States. In response to the embargo, the federal government attempted to accelerate the pace of OCS development to provide more domestic oil supplies. In the meantime, environmental groups such as the Natural Resources Defense Council (NRDC) began to vigorously challenge the accelerated federal OCS leasing program in court arguing, in part, that more comprehensive environmental assessment was needed before OCS development could move forward in a rational manner (for example, *Natural Resources Defense Council, Inc. v. Morton*, 458 F.2d 827 [D.C. Cir. 1972]). In conjunction with political lobbying from coastal states, legal challenges of the environmental community delayed the OCS development program, and thus undermined a quick response to the energy crisis of the mid-1970s. It was this delay of the OCS program that focused the attention of Congress.

In 1978, Congress amended the federal Outer Continental Shelf Lands Act (OCSLA) to require the preparation of Environmental Impact Statements under

NEPA at the overall planning, individual lease sale, and post-exploration proposal stages of OCS development (43 U.S.C. 1330 et seq.). In addition, even though the Department of the Interior (DOI) already had instituted a program to conduct basic research on the OCS environment, the amendments officially established an Environmental Studies Program (ESP) to garner more complete information about the OCS environment, and to assist in making decisions regarding OCS leasing, exploration, development and production (43 U.S.C. 1346). Overall, the new statutory goal of the federal OCS program was to facilitate "expeditious" yet "environmentally-sound" OCS development.

The general theory underlying the Congress's 1978 amendments to the OCS Lands Act was that "top-down" comprehensive planning, based on the assessment of geological, environmental, economic, and other relevant information, would help resolve differences among the various interests that were in conflict over the development of OCS oil resources. In particular, Congress felt that, through the EIS process, comprehensive assessment of public concerns, geological resource estimates, environmental impacts, and economic benefits of various OCS leasing proposals would produce proposals that were not only well considered and administratively feasible, but also were agreeable to all parties and therefore not subject to litigation delays. In short, Congress thought that the OCSLA's "revised procedures [would] limit frivolous lawsuits, and expedite all court actions." This, in turn, would lead to "[c]ooperation, and thus more certainty" of OCS development (U.S. Congress 1977).

Unfortunately, implementation of the OCS Lands Act Amendments did not unfold as Congress expected. The OCS oil development program continued to be mired in the courts, and between 1978 and 1990 there were no less than 19 separate lawsuits challenging OCS development lease sales (Lester 1991). The California Coastal Commission, for example, challenged four of the five lease sales proposed by the federal government for the Pacific OCS region. The state of Massachusetts, meanwhile, took every lease sale proposed for the north Atlantic to court, and other states such as Florida, Alaska, and North Carolina either threatened or took legal action as well. More broadly, each of three comprehensive 5-year OCS development plans prepared by the federal government between 1978 and 1990 was litigated. Congress also began to prohibit OCS leasing activities for increasingly larger areas of the OCS through the 1980s. By 1990, President Bush himself was forced to place a moratorium over most OCS leasing activities due to the continuing and extreme political, legal, and scientific disagreements about offshore oil development (Lester 1991). In short, the institution of new comprehensive assessment and decision making processes had not ameliorated the OCS conflict in the least.

Many reasons for the continuing OCS conflict have been identified, including extreme differences in political values (Heintz 1988), the local concentration of environmental costs versus the national distribution of development benefits (Cicin-Sain 1986), and the highly political implementation of the OCS program throughout the 1980s (e.g., Short 1989). Part of the explanation, though, also may lie in the inability of the comprehensive OCS environmental assessment process to reduce *scientific uncertainty*, by which is meant any of the range of factors, such as experimental design and measurement errors, data gaps, and extrapolation problems, that may make the answer to a scientific question somewhat indeterminate. Thus, one might hypothesize that continuing OCS political and legal conflict is in part a function of continuing disagreement or uncertainty about the environmental impacts of OCS development. That is, all things being equal, continuing environmental uncertainty provides a reason or focus for continued conflict among various interests.

One method for testing this hypothesis is to evaluate the relationship between the amount of scientific and environmental assessment devoted to a policy question (which in theory reduces uncertainty) and the degree of political/legal conflict for that question. In the case of OCS development, one measure of the amount of environmental assessment is the size of the administrative record for individual OCS lease sales. The administrative record for a lease sale typically contains the various scientific, environmental, geological, and economic documents (or summaries) generated to support a lease sale decision. It also contains the public comments and input from other governmental agencies which itself is a measure of both the degree of interest in a decision, and the extent of agreement on the various aspects of an environmental assessment. In short, a larger administrative record for a lease sale suggests, in part, that more extensive environmental assessment was conducted prior to a leasing decision. As for the degree of political/legal conflict, a fairly good indicator is whether or not an OCS lease sale decision is subsequently litigated.

Figure 17.1 tests the relationship between the degree of assessment and degree of conflict by plotting the average lease sale administrative record size against the lease sale litigation rate for each of the four OCS regions: the Gulf of Mexico, Alaska, the North Atlantic, and the Pacific. According to the logic of the 1978 OCSLA amendments, one might suppose that the lease sales with more comprehensive environmental assessment supporting them would be less controversial and therefore be litigated less often, due to decreased factual uncertainty. Figure 17.1, however, might suggest the opposite conclusion. Granted, a correlation between the two variables alone cannot reveal a direct causal link between the "extent of assessment" and litigation rates. Indeed, the increasing litigation rate also mirrors the degree of basic value conflict or "contentiousness" found in each OCS region, with the Gulf of Mexico and its "oil production" culture exhibiting the least amount of litigation and the "environmental" Pacific region (i.e., California) exhibiting the highest (Lester 1991).

Still, the fact that lease sale litigation rates increase proportionally with the extent of environmental impact assessment indicates that comprehensive environmental assessment has contributed little, if any, to the resolution of OCS lease sale conflict. At the very least, Figure 17.1 suggests that any contribution of

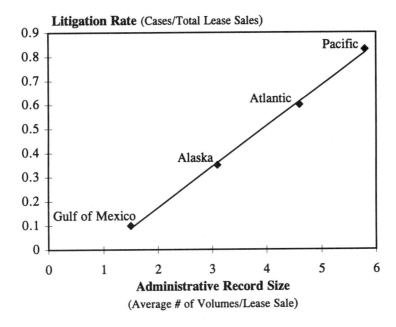

Figure 17.1. Relationship between litigation rate (cases / lease sale) and administrative record size (an index of the "thoroughness" of the environmental impact report / statement) using data from each of the four OCS regions.

environmental assessment to such conflict resolution has been overwhelmed by other legal and political factors that may be in play in the OCS leasing decision process. The remainder of this chapter discusses some of the reasons that may underlie the relationship in Figure 17.1.

The Limitations of Public Environmental Assessment

The debate about whether NEPA-type environmental review processes have promoted improved environmental decision making is an ongoing one (Sax 1973, Wenner 1989). Many scholars believe that requiring the preparation of environmental impact statements for major governmental actions has, in fact, improved the public policy process. To be sure, governmental decisions that require extensive environmental review and assessment do not invariably end up in litigation; one survey found that not more than 10% of decisions made under NEPA between 1970 and 1981 ended up in court (Taylor 1984). Thus, as mentioned at the beginning of this chapter, the extent to which the limitations of public environmental assessment illustrated by the OCS case study or other controversial issues are generalizable may be a function of the degree of contentiousness of a given public policy issue. Again, Figure 17.1 suggests just such a continuum of relevance.

Nonetheless, there are certain limitations inherent in the problem of environmental uncertainty and the response of NEPA-type public environmental assessment that can actually increase the prospects for political and legal conflict where the parties to a dispute already are so inclined. Such is the case with the question of OCS development. Generally, these limitations can be summarized under three headings: scientific constraints, administrative and legal constraints, and public policy constraints.

Scientific Constraints

The first limitation on the environmental assessment process for policy problems like OCS development is rooted in the nature of environmental scientific assessment itself, some aspects of which are discussed in more detail in other chapters of this book. Notwithstanding the tremendous resources and scientific effort devoted to learning about the OCS environment, scientific uncertainty still is pervasive. There are at least three reasons for this.

First, the OCS environmental assessment research agenda is broad, and it takes considerable time and resources to produce "answers" for even a few of the multitude of impact questions that may be raised by an OCS lease sale or development plan. Many variables must be taken into account, and complicated causal interactions must be identified, in order to adequately assess potential environmental impacts (Aurand 1988, NRC 1990). Moreover, the scientific study of a problem often raises more questions than it answers, further broadening the research agenda. For example, in 1975, after reviewing available information about the impacts of petroleum in the marine environment, the National Research Council (NRC) concluded that more data were needed on many issues. In 1980, the NRC concluded that many of the issues raised in 1975 needed reevaluation. In 1985 it published a 700-page tome of information on OCS oil impacts that concluded that while considerable progress had been made, many issues remained unresolved (NRC 1975, 1985, Aurand 1988). Overall, the sheer immensity of the OCS research agenda means that the amount and quality of scientific information varies by question and thus, information gaps are inevitable.

The second reason for continuing scientific uncertainty is simply that it is not possible to eliminate uncertainty regarding either the existence and size of an impact, or the underlying cause(s) of an impact. Indeed, most "identified" impacts are really no more than generally agreed upon probabilistic estimates. Further, scientists may have legitimate and quite specific disagreements among themselves concerning appropriate definitions, analytic models, or research methodologies that leave a given estimate even more indeterminate. For example, a 1991 OCS "Information Transfer Meeting," revealed numerous and conflicting definitions of the basic term "environmental sensitivity," each based on a different disciplinary perspective: ecological, biological, sociological, and environmental (MMS 1991b). At a 1981 Senate hearing concerning OCS development, a biologist challenged the surface water and wind movement parameters of an oil spill assessment model for a proposed California lease sale (U.S. Senate 1981). Still others have questioned the use of laboratory studies rather than *in situ* testing to evaluate the potential impacts of OCS oil on the marine environment–an extrapolation problem (Jenkins 1985, NRC 1985).

Finally, even if everyone agrees on the existence and/or size of an impact, there still remains the problem of whether the impact constitutes an "ecologically substantial" change; that is, does it matter? This problem has been raised in all areas of environmental policy making and perhaps is most prominent in the recent discussions of "risk perception" and "risk management" by the U.S. Environmental Protection Agency (e.g., U.S. EPA 1993). In a now infamous report, the EPA found major differences between what the public and scientific experts perceived as the most significant environmental risks (U.S. EPA 1990). In the case of OCS development, the problem of drilling muds provides a good example. Whereas the scientific community has tended towards the conclusion that the impacts of muds are localized and short lived, and therefore probably not of serious concern (Aurand 1988), the California Coastal Commission's regulation of drilling muds for example, grew increasingly strict, due to a risk averse response to the range of uncertainty about the significance of the impacts of drilling muds (Lester 1991). Unfortunately, there are few "objective" or scientific criteria for making such evaluations of significance and thus, a certain amount of subjective interpretation is injected into the environmental assessment process.

Each of the three basic reasons for continuing scientific uncertainty about the OCS environment have played a role in perpetuating conflict. Thus, notwithstanding the extent of environmental assessment that has taken place, many scientific questions about the OCS have remained indeterminate. This, in turn, has provided an opportunity for those already inclined to disagree to identify further points of disagreement rather than resolve their differences. Still, the scientific constraints just discussed merely provide the context for disagreement. The next two sections discuss the administrative, legal, and public policy aspects of the environmental assessment process that allow this disagreement to come to fruition.

Administrative and Legal Constraints

The responsibility of preparing an environmental impact statement or report lies primarily with the governmental agencies undertaking a major action or decision. The environmental assessment process, therefore, may be significantly shaped by the organizational imperatives of an agency which, in turn, may or may not be related to an interest in producing a high quality environmental assessment. For example, during the 1980s the U.S. Department of the Interior—

the agency responsible for preparing EISs for OCS lease sales—often was criticized for letting political and bureaucratic imperatives, rather than scientific imperatives, dictate the quality of its environmental assessments for OCS lease sales. Most notably, the controversial Secretary of the Interior James Watt promoted an expanded and "streamlined" OCS leasing program in an effort to increase the pace of OCS development. Unfortunately, many coastal states felt that this approach either precluded the inclusion of the "most up-to-date and accurate resource information ... to produce a meaningful analysis of the impacts of [an OCS] sale" (State of Alaska 1982), or downplayed environmental risks by averaging potential impacts over larger areas of the OCS (for example, State of Massachusetts 1983, State of Louisiana 1981).

Bureaucracies also have a way of embracing rigid rules and procedures in the day-to-day implementation of their programs and often an administrative task such as the preparation of an EIS, or the adoption of a general implementation schedule, may be driven by its own bureaucratic momentum (Lester 1991). For example, the "Lease Evaluation" staff for the OCS program in the Gulf of Mexico region has used a proscribed 60 to 80 step sequence of decision points and deadlines to guide its production and use of the EIS in leasing decisions. These steps included such items as the schedule for disseminating the document through the agency, required consultations with other federal agencies like the U.S. Fish and Wildlife Service, dates for official interaction with the public, and even times to send the documents to the printer. Although such rigid schedules are no doubt necessary to help an agency meet statutorily-prescribed deadlines, they can also limit the effectiveness of environmental assessment. Bureaucratic imperatives, then, can defeat the purpose of what may otherwise be high quality scientific assessment, as the comments of a representative of the state of Massachusetts illustrate:

> In reviewing the research conducted in the North Atlantic, I believe that the program has been excellent—the objectives of the research program have been well focused and the principal investigators are among the top in their fields. We have been troubled by the leasing schedule, not the Environmental Studies Program, because [the federal government] has been leasing in areas where the research had yet to be completed, or progressed to a stage where some preliminary conclusions are available. A good example of this is the proposed leasing of the submarine canyons in Sale 52 where the draft EIS had been printed before preliminary data was available on the sediment transport mechanisms of [the undersea] canyons (State of Massachusetts 1985).

The most significant bureaucratic limitation on the environmental assessment process, however, is that which has evolved out of the interaction between courts and administrative agencies, and their respective responses to the scientific uncertainty discussed earlier. As mentioned at the beginning of the chapter, the judicial system plays a primary role in the oversight of government agency decision making. Thus, when a policy question is controversial, as with OCS development, judges often become the "final arbiters" of whether or not an environmental assessment conducted under NEPA is adequate. The legal system, though, is not necessarily well-suited to judge complex scientific questions because judges are, generally speaking, legal experts, not scientific experts (e.g., O'Brien 1987).

Because of this, a judge will often defer to the judgment of agency experts when he or she is reviewing an agency's decision making because of their acknowledged expertise in the question at hand. This judicial doctrine of deference to the administrator is rooted in a long tradition of administrative law in the United States (Shapiro 1988). What this doctrine means, however, is that a judge will not overturn an agency decision, such as a scientific finding based on an EIS analysis, unless the agency has failed to "reasonably" support its analysis. Thus, judges tend to look for reasonable explanations or documentation of decisions, not whether or not a particular decision is scientifically correct.

For example, in the judicial review of a federal OCS 5-year leasing plan in the 1980s, a court was "left uncertain" by the Department of the Interior's calculation of the net economic value of various lease sales. Accordingly, the court mandated that DOI "present a coherent explanation of how net economic value [was] being determined." Once this explanation was provided, however, the court found that DOI's analysis was reasonable (*State of California by and through Brown v. Watt*, 668 F.2d 1290, 1318 [1981]).

In the case of complex scientific and technical policy areas, the doctrine of deference to the administrator has evolved even further to what is termed the "frontiers of science" doctrine. Thus, when scientific facts or methodologies are so uncertain that even scientists may disagree, as is the case with many OCS issues, a judge will grant even more deference to the expertise of the administrator. So, for example, in the case of the DOI's use of cost-benefit analysis, a court has stated:

> [t]he facts used by the Secretary [of the Interior] in performing the [cost-benefit] analysis are largely predictive in nature, and the methodology utilized was necessarily novel because this type of analysis has not been performed extensively in the past. Thus, ... great deference is afforded to the Secretary in these areas (State of California v. Watt, 712 F.2d 584 [1983]).

Because of the complex questions raised by this OCS case, the court went on to defer to DOI's analysis so long as it was "not unreasonable." In short, under the "frontiers of science" rationale, the court would not presume to resolve differences among the experts so long as the decisions of DOI were reasonable and supported by the record.

In time, most government agencies will learn that in order to pass safely through judicial review they simply need to provide well-reasoned and supported scientific analyses in the administrative records that they present to courts in litigation. Similarly, however, those that wish to successfully challenge a government agency's decision or environmental assessment in court will learn that they need to become increasingly sophisticated. For example, in the case of OCS 5-year leasing plans, litigants have challenged not only the use of decision techniques such as cost-benefit analysis, but also the methodological assumptions and parameters underlying a cost-benefit analysis. In response, the federal government now supplies analyses of the "sensitivity" of cost-benefit outcomes to changing assumptions in the analysis. A memorandum in the administrative record of an OCS five year plan plainly acknowledges this response to judicial review, stating:

> [b]ased on [an] Appeals Court ruling, the ... analysis [of OCS leasing options] was substantially expanded. More elaborate estimates of net economic value, external costs, environmental sensitivity, and marine productivity were developed (DOI 1984).

This explicit concern for judicial review and potential legal claims helps to explain the correlation between administrative record size and lease sale litigation rates presented in Figure 17.1. Over time, the specter of judicial review under the standards of "reasonableness" and the "frontiers of science," shapes the environmental assessment process into a hyper-rationalized exercise in justification. In the past, DOI has even established a working group for the 5-year OCS planning process whose sole purpose was to ensure the proper documentation for the administrative record (DOI 1984). In 1985 one DOI staff member expressed frustration with "overanalyzing and presenting everything conceivable in excessive detail." To him, it seemed that the object of producing a 5-year plan was to "write something so long that no judge [could] possibly read it" (DOI 1985).

The most troubling aspect of this administrative defensiveness, though, is what it says about the OCS lease sale planning and environmental assessment process. Rather than interacting substantively concerning the various policy issues raised in the environmental assessment process, DOI, as well as non-agency participants, engage in administrative record building, each making sure that their positions are adequately supported in the documents that the courts will ultimately examine (Farrow 1990). The most telling evidence is the conclusion of DOI's own task force on improving the 5-year planning process for OCS development:

> ... DOI has progressively found it necessary to produce data designed to defend court action along with that which is needed for program decisions (MMS 1988).

This process, of course, is the antithesis of the ideal model of comprehensive environmental assessment embraced by the Congress in the 1978 OCSLA amendments, whereby well-considered analysis would help to *resolve* differences and *decrease* conflict. Instead, the dynamic established between agencies and courts encourages everyone to look for, and accentuate, scientific differences and problems, in order to "win" in the judicial forum. Thus, even though the administrative-judicial dynamic leads to the creation of more information, it is not information that decreases uncertainty or that resolves conflict. Rather, conflict merely escalates, further limiting the effectiveness of environmental assessment. The next section discusses certain general constraints of the public policy process that may aggravate further the limitations caused by bureaucratic and legal imperatives.

Public Policy Constraints

A final set of limitations on environmental assessment inhere in the fact that most scientific and environmental assessments, particularly those concerning sensitive environmental contexts such as the OCS, are tied to a public policy process. Because of this, the imperatives and goals of interest groups and advocacy organizations, other federal, state, and local agencies, as well as the general public, may also determine the effectiveness of the assessment process. Although this may seem to be an obvious point, the OCS environmental assessment process has suffered for failure to sufficiently acknowledge it. Thus, since almost the beginning of the OCS conflict, coastal and marine scientific research has been inadequately integrated with the public policy process at all levels of OCS decision making. As a consequence, OCS policy decisions have not necessarily been well supported by scientific information and this has enhanced political and legal conflict.

For example, the Department of the Interior's early attempt to establish an effective Environmental Studies Program was criticized as being too unfocused to contribute to actual OCS decision making; the U.S. General Accounting Office concluded that "research results were not adequately integrated and geared to decision points ..." and thus, that the Department of Interior had to "base decisions on less information than could be available" (U.S. GAO 1978). Similarly, the National Research Council concluded that the early studies program did not "effectively contribute to leasing decisions or to the accrual of sound information adequate for OCS management" (NRC 1978, Burroughs 1981). In retrospect, even the administrator of DOI's Studies Program in the 1980s described the early assessment program as "naive," inasmuch as it did not develop scientific information useful for actual public policy decisions (Aurand 1988).

Notwithstanding these early criticisms, the General Accounting Office made similar conclusions about the new Environmental Studies Program, revamped after the 1978 OCSLA amendments. It concluded that while the studies program did meet most user needs, its research was still poorly integrated with decisions, at least from the perspective of coastal states and environmentalists (U.S. GAO 1988). Much of this criticism was due to the highly political and bureaucratic implementation of the OCS leasing program under Secretary Watt that was mentioned earlier. Some of this continuing criticism was also due, however, to the fact that the various parties involved in the OCS dispute, particularly the geologically-trained decision makers with DOI and those concerned about the OCS environment, had always maintained differing perceptions both about the types of information needed for effective policy decisions at the OCS leasing stage, and about the respective functions of the Environmental Studies Program and the public EIS process.

For example, federal OCS decision makers and the oil industry generally believed that the leasing stage of OCS development was too early in the process to assess specific environmental impacts of potential development projects, because of the uncertainty of both the location of leases that may be acquired by industry in the lease bidding process, and whether there may be any oil at these locations (generally only one in ten exploration wells actually yield a significant amount of producible oil; e.g., see discussion in Piltz, Chapter 16). Rather, they felt that such specific studies should take place once development projects were more tangible (Lester 1991). In contrast, coastal states and environmentalists believed that once an OCS lease was granted to a company, it would be difficult if not impossible to stop future OCS development in the event that OCS oil was found because of the large capital investment of the oil company. They felt, therefore, that more complete information about the OCS environment was needed prior to the leasing of specific OCS acreage (NRC 1989).

Moreover, whereas the Department of the Interior envisioned the Environmental Studies Program to be the appropriate programmatic context for such original, basic environmental research, coastal states wanted original research tied to the EIS assessment processes that accompanied the public OCS leasing decision processes. That is, it was important to coastal states and environmentalists that OCS environmental assessment be integrated directly with tangible information needs identified through public review and comment on proposed OCS lease sales. Unfortunately, most government agencies, including the DOI, treat the NEPA EIS process primarily as an opportunity to make findings about a project's potential environmental impacts based on a review of *existing* information, not as a chance to conduct new, original research.

These conflicting perspectives on the appropriate context and time for certain kinds of environmental research have been aggravated still further by the nature of the public environmental assessment process under NEPA. As discussed earlier, the 1978 OCSLA amendments appealed to a traditional assessment model under which public input is sought, various alternatives and the associated costs and benefits are developed, public comment on these alternatives and scientific study are synthesized, and a final, well-planned decision is promulgated by the federal government, relatively free from conflict (see MMS 1987). In contrast, however, the actual assessment process might be described as a perpetual, uneven, and increasingly adversarial, process of public learning, in which the issues to be addressed not only keep changing but keep expanding as well.

First, the NEPA assessment process is a 2-year period. Over this time, experts and interested parties outside of the federal government develop an increasingly sophisticated understanding of environmental problems. In part, this is a symptom of the basic scientific uncertainty described earlier—studies expand and new studies spin off as researchers learn. It is also a function of many people simply learning about the issue. At the beginning of the process people are unaware of basic facts, like what a "jack-up" drilling rig is, and how much space it takes up on the ocean. As time passes, however, they begin to get into the analysis of details. What species of fish may be affected by the proposal? What is the causal link underlying such effects? How might these effects be reflected in costs to us? What discount rate should be used for these costs?

Second, other state and federal agencies comprised of their own scientific experts naturally contribute increasingly sophisticated comments—the Alaska Department of Fish and Game for example was instrumental in extending a conflict over a proposed OCS lease sale for the Bristol Bay, simply by identifying increasingly specific information needs, right down to knowledge about the effects of oil on particular species. Indeed, at the beginning of the assessment process for Bristol Bay, the state had identified eight "information gaps" that it thought needed to be addressed before leasing could go forward. Before too long, however, this short list had expanded to 38 specific information needs, including original studies, that it felt should be completed prior to the lease sale (Lester 1991).

Finally, the EIS public policy process is, by definition, a "critical" one, which can also limit the ability of assessment process to decrease uncertainty. Thus, the dynamic of information assessment under NEPA is one where state experts and other interested parties *comment* on the assessment made by the federal government's experts. First, a draft EIS is produced and comments are requested on its weaknesses. Because the DOI's framework of analysis is already in place by the time an EIS is under way, the process inevitably becomes one of critique and of identifying information gaps—the Alaska Department of Fish and Game, for example, has little choice but to point out the inadequacies of the draft assessment as it sees them. Moreover, as suggested earlier, many public advocacy groups are not interested so much in resolving factual uncertainty but in using this uncertainty to their political advantage; thus, they are actively looking for inadequacies in the EIS document.

While the DOI responds to the first round of "comments," parties external to DOI continue to focus on the "problems" of the federal government's assessment. When the final EIS comes out, still more specific weaknesses are identified. Of course, like all experts, federal experts also take pride in their analysis; and thus a dynamic is established in which both sides of a controversy feel that their expertise is being challenged. Rather than narrowing differences, differences escalate to the point where each detail is argued over. Ultimately, the parties are left with differences concerning costs and benefits and assumed discount rates—essentially value choices—with only the courts left to resolve differences. And, as already discussed, the courts are not well equipped to address such subjective information differences that may be identified through the NEPA public environmental assessment process.

Conclusion

This chapter has briefly discussed some of the constraints on environmental assessment generated by the administrative, legal, and public policy contexts of NEPA and similar state laws. While these constraints have been drawn out of a case study of a highly contentious public policy issue and thus, are probably most apparent in contentious cases, nearly all significant environmental assessment problems take place in contexts similar to those established by NEPA-type environmental assessment laws. The lessons of this chapter may hold true *to some degree,* therefore, for every public environmental assessment process.

With this caveat in mind, the case of OCS development suggests that public environmental assessment processes are necessarily limited by the participation of many diverse interests, each responding in piecemeal fashion to the problem of scientific uncertainty. Bureaucracies, courts, scientists, public advocates, and politicians each have their own agendas, which may lead to increased conflict and uncertainty. This suggests further that scientists and public policy actors need to concentrate more on the integration of their respective disciplines and goals if environmental assessment is to serve its intended purpose (see also Piltz, Chapter 16). Scientific environmental assessment cannot take place in a vacuum if it is to be useful, and public policy cannot maintain its fragmented, adversarial character if science is to be used successfully in policy decisions.

Both scientists and public policy makers are gradually acknowledging this need for integration, particularly in the OCS policy context. Ad hoc task forces and scientific committees that consolidate the various perspectives at play in the OCS debate are being increasingly used by the federal government in the consideration of OCS environmental problems (Lester 1991). More emphasis is being placed on applied research, tied to specific policy questions raised by the public. Perhaps most important, the federal government is paying more attention to effectively communicating the results of OCS scientific studies to the public (MMS 1990). There is hope, then, that the original goal of laws such as NEPA, namely, that our environment will be better understood by scientists, government decision makers, and the public, can be achieved.

References

Aurand, D. 1988. The future of the interior OCS studies program. Pages 161–165 *in* Oceans '88: a partnership of marine interests, proceedings. Marine Technology Society, Baltimore, Maryland.

Burroughs, R. H. 1981. OCS Oil and Gas: Relationships between Resource Management and Environmental Research. Coastal Zone Management Journal **9**.

Cicin-Sain, B. 1986. Offshore Oil Development in California: Challenges to Governments and to the Public Interest. Public Affairs Report 27. Berkeley, CA: Institute of Governmental Studies.

Farrow, S. 1990. Managing the Outer Continental Shelf lands: oceans of controversy. Taylor and Francis, New York, New York.

Heintz, T. 1988. Advocacy coalitions and the OCS leasing debate: a case study in policy evolution. Policy Sciences **21**:213–238.

Jenkins, K. 1985. Muds and cuttings. *In* W. N. J. Tiffney, editor. Proceedings of the 1985 California offshore petroleum conference: resources, technology, the environment, and regulation. Pallister Resource Management, Ltd., Malibu, California.

Kallman, R. E. 1984. Coastal crude in a sea of conflict. Blake Printery and Publishing Co., San Luis Obispo, California.

Lester, C. 1991. The Search for Dialogue in the Administrative State: The Politics, Policy, and Law of Offshore Oil Development. Ph.D. Dissertation. University California, Berkeley. (Unpublished)

MMS (Minerals Management Service). 1987. Leasing energy resources on the Outer Continental Shelf. U.S. Department of Interior, Washington, D.C.

MMS (Minerals Management Service). 1988. Improving the process for developing the 5-year OCS oil & gas leasing program. Report of the Task Group OCS Policy Committee. Washington, D.C.

MMS (Minerals Management Service). 1990. The offshore environmental studies program (1973–1989): a summary of minerals management service research conducted on the U.S. Outer Continental Shelf. U.S. Department of Interior, Herndon, Virginia.

MMS (Minerals Management Service). 1991a. Pacific OCS Region. Sixth information transfer meeting. U.S. Department of Interior, Los Angeles, California.

MMS (Minerals Management Service). 1991b. OCS national compendium: Outer Continental Shelf oil & gas information through October 1990. U.S. Department of Interior, Herndon, Virginia.

Nash, A. E. K., D. Mann, and P. G. Olsen. 1972. Oil pollution and the public interest: A study of the Santa Barbara Oil Spill. Institute of Governmental Studies, University of California, Berkeley, California.

NRC (National Research Council). 1975. Petroleum in the marine environment. National Academy Press, Washington, D.C.

NRC (National Research Council). 1978. OCS oil and gas: an assessment of the department of interior environmental studies program. National Academy Press, Washington, D.C.

NRC (National Research Council). 1985. Oil in the sea: inputs, fates, and effects. National Academy Press, Washington, D.C.

NRC (National Research Council). 1989. The adequacy of environmental information for Outer Continental Shelf oil and gas decisions: Florida and California. National Academy Press, Washington, D.C.

NRC (National Research Council). 1990. Managing troubled waters: the role of marine environmental monitoring. National Academy Press, Washington, D.C.

O'Brien, D. 1987. What process is due? Courts and science-policy disputes. Russell Sage Foundation, New York, New York.

Sax, J. 1973. The (unhappy) truth about NEPA. Oklahoma Law Review **26**:239.

Shapiro, M. 1988. Who guards the guardians? Judicial control of administration. University of Georgia Press, Athens, Georgia.

Short, C. B. 1989. Ronald Reagan and the public lands. Texas A & M University Press, College Station, Texas.

State of Alaska. 1982. Position on OCS 5-year Leasing Program, to Senator Lowell Weicker. September 28.

State of Louisiana. 1981. Correspondence. Louisiana Geological Survey to Department of Interior.

State of Massachusetts. 1983. Comments on Sale 82. DEIS, 5 August.

State of Massachusetts. 1985. Correspondence. Massachusetts Coastal Zone Management Office to MMS Environmental Studies Program Director, June 4.

Taylor, S. 1984. Making bureaucracies think: the evironmental impact statement strategy of administrative reform. Stanford University Press, Stanford, California.

U.S. Congress. 1977. Report of the Ad Hoc Select Committee on the Outer Continental Shelf. H.Rep. No. 95-590. 95th Congress, 1st session.

U.S. DOI. 1984. Adminstrative Record. Index 1, V.I, II March.

U.S. DOI. 1985. Administrative Record. V.12, September.

U.S. EPA. 1990. Reducing Risk: Setting Priorities and Strategies for Environmental Protection. Washington, D.C., SAB-EC-90-012.

U.S. EPA. 1993. EPA Journal. 19.

U.S. GAO. 1978. Benefits Derived from the Outer Continental Shelf Environmental Studies Program are Questionable. Washington, D.C., CED-78–93.

U.S. GAO. 1988. Environmental Studies Program Meets Most User Needs but Changes Needed. Washington, D.C., R-CED-88-104.

U.S. Senate. 1981. Statement of Dr. Ronald Ritschard. Senate Hearings, Subcommittee on Environment, Energy, and Natural Resources, 1 Apr.

Wenner, L. 1989. The courts and environmental policy. Pages 238–260 *in* J. P. Lester, editor. Environmental politics and policy: theories and evidence. Duke University Press, Durham, North Carolina.

CHAPTER 18

PREDICTED AND OBSERVED ENVIRONMENTAL IMPACTS

Can We Foretell Ecological Change?

Richard F. Ambrose,[1] Russell J. Schmitt,[2] and Craig W. Osenberg

Environmental Impact Reports and Statements (EIR/S) provide predictions about the likely effects of a particular project. These predictions are often used to modify the project's design, impose a variety of permit conditions (e.g., limits on the timing, duration, or quantity of discharged material), or require specific mitigation to compensate for impacts (e.g., construction of artificial reefs or wetlands). Thus, the predictions of the EIR/S can have serious legal, economic, and environmental consequences. If the predictions of the EIR/S are accurate, the information can be used effectively to safeguard the environment, while simultaneously permitting environmentally sound projects to move forward. However, if the predictions are inaccurate, considerable resources may be lost on unnecessary project modifications and mitigation for impacts that never occur. Inaccurate predictions, which include both predictions that differ from the actual impacts and the absence of predictions about impacts that do occur, will not only fail to protect the environment, but they will also hamper industry. Despite these serious consequences, relatively few attempts have been made to compare actual impacts with those predicted during the EIR/S process.

Recently, however, there has been greater interest in "predictive techniques audits" (sensu Tomlinson and Atkinson 1987a) that attempt to compare predicted impacts with actual impacts (Beanlands and Duinker 1984, Culhane 1987, Tomlinson and Atkinson 1987b, Luecht et al. 1989, Bailey and Hobbs 1990, Buckley 1991a, 1991b). These audits have found that many or most forecasts in EIR/Ss are imprecise and thus difficult to audit. This is particularly true for

---

[1] Richard Ambrose served as technical advisor to the Marine Review Committee and developed recommendations for mitigating the impacts of the San Onofre Nuclear Generating Station.

[2] Russell Schmitt served as consultant and Scientific Advisor to the Chairman of the Marine Review Committee. The opinions expressed in this chapter do not necessarily reflect those of the Marine Review Committee.

ecological parameters, which are often not quantitative. As a result, existing audits have focused on the more quantitative predictions, often derived from engineering considerations, regarding air emissions, traffic, water quality, and so forth. An additional problem is that the actual ecological impacts of a project are rarely determined. Postproject assessments are rarely conducted (Larkin 1984) or they are based on inadequately designed monitoring programs that fail to separate impacts from natural variability (Gore et al. 1977, Carney 1987, Osenberg and Schmitt, Chapter 1).

Auditing predictions made in EIR/Ss can contribute to decision making and environmental sciences in at least two ways. First, the process will help evaluate how much stock to put in such predictions and clarify conditions under which predictions might be more or less accurate. Second, it will help refine predictive techniques, essentially providing tests of the models used to generate predictions. In this way, predictive techniques audits can further environmental sciences by discriminating among approaches that vary in their predictive ability. This is particularly important at the present time because approaches that aim to predict population and community consequences are becoming increasingly sophisticated (e.g., Suter 1993). Indeed, "ecological risk assessments" (e.g., which predict field effects based on laboratory-derived toxicological data and transport models) run the risk of being able to use complex models to precisely specify predictions that may not bear any resemblance to actual impacts that arise under field conditions. It is critical that predictions and tests of those predictions be compared to facilitate the development of appropriate approaches and models.

There are few existing studies of ecological impacts that allow explicit comparison of predictions with well documented effects. In this chapter, we focus on the predicted and measured ecological impacts from one well-studied project, the San Onofre Nuclear Generating Station. We summarize predictions generated in three different ways: predictions made in the Final Environmental Statement (i.e., the EIR/S), predictions made by scientists testifying before the permitting agency, and predictions made by an independent group of scientists charged with conducting a comprehensive study of the power plant. We then compare these predictions with actual effects that were documented as the result of a 15-year study, and discuss the reasons for variation in the accuracy of these predictions.

The San Onofre Nuclear Generating Station

The San Onofre Nuclear Generating Station (SONGS), located on the California coast between San Diego and Los Angeles, is owned by Southern California Edison Company (SCE), San Diego Gas and Electric, and two Southern California cities (Anaheim and Riverside). SONGS consists of three units (Figure 18.1); each uses seawater for cooling. Unit 1 began operating in 1968. At full operation Unit 1 generates 436 MW of power and takes in about 1.1 million liters of seawater per minute through its once-through cooling system.

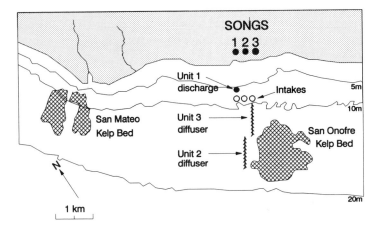

Figure 18.1. San Onofre region showing the intakes and discharges for SONGS Units 1, 2, and 3 and the two nearby kelp beds (San Mateo Kelp and San Onofre Kelp).

Unit 1's single intake structure is located in 9 m of water about 980 m from shore; its discharge is a single vertical pipe in 8 m of water about 800 m from shore.

In the early 1970s, Southern California Edison and the other owners of SONGS proposed building two new units, Units 2 and 3, with outputs of 1100 MW each. This proposal went through various environmental reviews, as discussed below, and the new units began commercial operation in 1983–1984. Like Unit 1, Units 2 and 3 use a once-through cooling system, with each taking in about 3.1 million liters per minute. The intakes for Units 2 and 3 are near the Unit 1 intake (Figure 18.1). Units 2 and 3 discharge water through diffusers that are designed to allow the units to meet the thermal standards mandated by the California Thermal Plan. Each diffuser is 762 m long, with 63 diffuser ports spaced along its length. The diffuser ports discharge water in a powerful jet upwards towards the surface and in an offshore direction. Unit 2's diffuser ends 2500 m from shore in 15 m of water, whereas Unit 3's diffuser ends 1800 m from shore in 12 m of water. The diffusers limit the rise in temperature around the discharge by mixing the heated effluent with large volumes of ambient seawater, about ten times the volume of the discharge. The volume of water passing through all three units in the course of a week is equivalent to all the water out to the 11 m depth contour (about 1500 m from shore) along a 9.7 km stretch of coast.

Impacts from SONGS could be expected to occur through two basic mechanisms. First, organisms could be killed by being drawn into the plant along with the water used for cooling. Two terms are used to describe this type of mortality. "Entrainment" refers to organisms being drawn into the cooling system along with the cooling water; small fish, fish larvae, and plankton, which are drawn passively along with water masses, are entrained. "Impingement" refers to organisms being trapped ("impinged") on screens that are employed to prevent large objects from entering the condensers; larger fish are impinged. Second, organisms could be affected by the discharge, through increased water temperature or turbidity in the plume from Units 2 and 3, increased deposition of organic material, discharge of radionuclides and metals, or entrainment and offshore transport in the plume (i.e., "translocation").

These potential impacts were addressed in two main environmental reviews associated with Southern California Edison's application for federal and state permits to construct and operate Units 2 and 3 of the plant. These were conducted by the U.S. Atomic Energy Commission (now the Nuclear Regulatory Commission) and the California Coastal Zone Conservation Commission (now the California Coastal Commission).

Federal permits were required from the U.S. Atomic Energy Commission (AEC). The National Environmental Policy Act of 1969 established procedures for assessing the potential environmental impacts of federal projects, including the need for an Environmental Impact Statement. For SONGS, the AEC was the lead agency responsible for preparing and certifying the EIS. The AEC prepared a draft Environmental Statement and held a number of public hearings. In the Final Environmental Statement on the Construction of SONGS Units 2 and 3, produced in March 1973 (and hereafter referred to as the "1973 FES"), the Atomic Energy Commission recommended that the new units be built (AEC 1973). When construction of the new units was nearly completed, the AEC, which by then had become the Nuclear Regulatory Commission (NRC), produced a Final Environmental Statement on the Operation of Units 2 and 3 (the "1981 FES")(NRC 1981). The 1981 FES updated the 1973 report to consider some design changes to the discharge system and to incorporate additional information, but the basic conclusions from the earlier document were unchanged.

The California Coastal Zone Conservation Commission (now the California Coastal Commission; hereafter we refer to both as the "Coastal Commission" or "Commission") also reviewed Southern California Edison's application to build Units 2 and 3 under the newly enacted California Coastal Zone Conservation Act (Proposition 20). The proposal to build new nuclear power plants in California caused considerable controversy and the Coastal Commission held a number of public hearings on the matter. Somewhat predictably, at the public hearings utility scientists testified that the new units would have minimal impacts on the marine environment while environmental groups testified that the operation of the new units would cause an ecological disaster. Academic scientists for the most part testified that there would be tremendous ecological consequences for the marine environment, and so were aligned more closely with the environmentalists. The environmental groups and academic scientists who testified against the power plant are hereafter called intervenors. The Coastal Commission felt it could not determine the likely effects of the new units given available information, largely because of the lack of consensus about the effects. The Commission ultimately issued a permit for construction of Units 2 and 3, but with the conditions that (i) an independent review committee be established to study the impacts of Units 2 and 3 on the marine environment, and (ii) the Commission could require the utilities to change the plant to reduce or eliminate impacts if this committee determined that substantial adverse effects did result. The Marine Review Committee (MRC) was thus established in 1974 as a condition of the Coastal Commission permit for construction and operation of Units 2 and 3. To insure a balanced examination of the impacts, the Marine Review Committee was composed of three scientists, one representative of the intervenors, one representative of the utilities, and one representative of the Commission.

Predicted Impacts

Predictions about the effects of SONGS Units 2 and 3 on the marine environment are available from three sources (Table 18.1). First, the federal assessment of environmental impacts resulted in two Environmental Statements (AEC 1973; NRC 1981). The 1981 FES updated the 1973 FES to consider design changes and to incorporate additional information, including the initial results of the MRC and its predictions about the effects of Units 2 and 3. The predictions in the 1973 and 1981 versions are not substantively different. Because the 1973 version is most similar to the level of analysis included in most EIR/Ss, we have used its predictions for this analysis. Second, there was extensive testimony from utility scientists and intervenors during the Coastal Commission public hearings. Finally, in 1980, after conducting preliminary studies of the effects of Unit 1, the Marine Review Committee predicted the likely effects of Units 2 and 3 (Murdoch et al. 1980). Unlike the other predictions, the MRC's 1980 predictions were based on site-specific data from studies of a similar, albeit much smaller, power plant (Unit 1).

Final Environmental Statement

The Final Environmental Statement reported a wide range of possible effects. Much of the focus of the FES evaluation was on thermal effects, although other mechanisms such as entrainment, impingement, turbidity, induced circulation, and release of chemicals were also evaluated. We have summarized the FES conclusions under five taxonomic headings: plankton, kelp, benthos (invertebrates), fish and other effects.

Plankton. It was predicted that plankton could be affected, although a great deal of uncertainty was noted. For phytoplankton, the main effects were expected to be a change in species composition and perhaps a stimulation of red tides arising from increased water temperature, experienced during entrainment in the intakes and discharge plume. The FES concluded that these effects would be localized and that the phytoplankton population as a whole would not be affected adversely (AEC 1973, p. 5–31).

Table 18.1. Predicted Impacts of San Onofre Nuclear Generating Station Units 2 and 3 Extracted from Four Sources[a]

		Predictions		
Ecological characteristic	Final Environmental Statement	Intervenors (CCC testimony)	Utility scientists (CCC testimony)	Marine Review Committee
Phytoplankton	Possible change in species composition; stimulation of red tides	Massive impact on plankton, with devastating ecosystem consequences	No adverse effect	Increased production of 84,000 tons/yr
Zooplankton	240–950 tons/yr entrained; local depression possible, but no effect on larger aquatic community	Massive impact on plankton, with devastating ecosystem consequences	No adverse effect	1200 tons/yr killed in intakes; unknown (perhaps large) amount killed in discharge; decrease in local density and change in species composition
Mysids	No prediction	Massive impact on plankton, with devastating ecosystem consequences	No adverse effect	50–60 tons/yr killed in intakes; probably more lost in discharge plume; substantial decline in density
Kelp	No adverse effect	No prediction	No adverse effect	Adverse effect on San Onofre Kelp Bed
Benthos				
Intertidal	No adverse effect	(Serious effects implied to arise from impacts on plankton)	No adverse effect	No prediction
Subtidal: soft-bottom	No adverse effect		No adverse effect	Increased density
Subtidal: hard-bottom	No adverse effect		No adverse effect	Possible change in species composition (not likely)
Fish				
Amount impinged	18–39 tons/yr	No prediction	No prediction	50–84 tons/yr
Sport and commercial fish	No major impact	(Serious effects implied to arise from impacts on plankton)	No adverse effect	Loss of 27–60 tons of productivity/yr; not a significant effect on populations
Fodder fish	No major impact		No adverse effect	Loss of ~320 tons of production/yr
Kelp bed fish	No major impact		No adverse effect	Loss of additional 0–3 tons of production/yr
Regional fish stocks	No major impact		No adverse effect	Probable declines in So. California Bight
Other effects	No adverse effect	No prediction	No adverse effect	No prediction
Proposed mechanisms	Thermal effects	Thermal effects, loss of recruitment and trophic consequences	Not applicable	Intake losses of larvae; discharge effects on turbidity and organic matter

[a]The Final Environmental Statement for SONGS Units 2 and 3 (AEC 1973), testimony of intervenors and utility scientists as recorded in the minutes of various Coastal Commission hearings on SONGS Units 2 and 3 (including meeting of October 18, 1973), and the report of the Marine Review Committee (Murdoch et al. 1980). "No prediction" indicates that no specific prediction was made for that taxon, and that a general prediction could not be clearly inferred from other predictions. Predictions from the FES and MRC for entrainment and impingement losses include effects of Unit 1, which contributes less than a third of the total losses.

For zooplankton, the main impact was also expected to stem from entrainment, resulting in thermal shock and mechanical damage. All three units were expected to entrain about 240–950 metric tons of zooplankton per year and kill about 30% of these organisms. Again, substantial uncertainty was cited, but the FES concluded that the local density of zooplankton would decline, but that "because of recruitment from the open ocean, the larger aquatic community will not be affected significantly" (AEC 1973, p. 5-32).

Kelp. Giant kelp (*Macrocystis pyrifera*) is a key species in southern California. Not only is giant kelp itself a valuable resource, but it provides food and shelter for many other species. As a result, kelp receives considerable attention in most EIR/Ss concerned with potential impacts on nearshore environments in southern California.

At the time of the 1973 analysis, the San Onofre Kelp Bed, adjacent to the Units 2 and 3 diffusers, was not present. Thus, the 1973 FES evaluation focused on the next nearest kelp beds: San Mateo Kelp, 5 km from SONGS, and Barn Kelp, 10 km from SONGS. The FES recognized that the discharge from Units 2 and 3 could increase turbidity, and estimated that turbidity from all three units could cover an area of 2 km^2 (AEC 1973, p. 5-16). However, increased turbidity was not discussed as a possible mechanism affecting giant kelp. This contrasts with the predictions of the Marine Review Committee (see below).

Despite being overlooked in the 1973 report, the San Onofre Kelp Bed had historically been an important bed in the area, and in fact it became reestablished in the mid-1970s. The San Onofre Kelp Bed was a conspicuous element of the local marine environment when the 1981 FES was being prepared, so the 1981 FES revised the analysis of effects on giant kelp. However, the analysis remained focused on thermal effects, although turbidity was mentioned parenthetically. The 1981 FES concluded that the heated water discharged from the diffusers might result in the destruction of a portion of San Onofre Kelp Bed during the summer months, but that under average conditions the impact would not be detectable or would result only in a noticeably earlier seasonal decline in the canopy (NRC 1981, p. 5-18). Although the FES discussed other more severe possible impacts, it concluded that they were unlikely to occur.

Benthos. In considering possible impacts of Units 2 and 3 on intertidal and subtidal organisms, the FES again focused on sensitivities to temperature. Because intertidal organisms are adapted to temperature changes and because the intertidal environment exhibits substantial change naturally, the FES concluded that Units 2 and 3 would have no significant effect on the intertidal biota (AEC 1973, p. 5-33).

The FES considered that increased temperature, turbidity, entrainment and induced currents could affect the subtidal biota. Because elevated temperatures were predicted not to reach the ocean floor, no adverse effect from temperature was expected. Mortality was expected from entrainment, but "[b]ecause of the reproductive capabilities of most of the benthic organisms, no widespread significant effect from entrainment of eggs and gametes is expected" (AEC 1973, p. 5-34). The FES stated that turbidity might cause some change, although no supporting evidence was given. The FES also noted that some changes occurred around the Unit 1 outfall and suggested that similar changes might be expected around Units 2 and 3. The FES concluded that "although some changes will occur in the benthic biota in the vicinity of the outfall and discharge structure, no major adverse effect is expected on the sublittoral benthic biota" (AEC 1973, p. 5-34).

Fish. The FES considered whether entrainment of fish eggs and larvae would have important consequences for fish populations. Although the analysis indicated that "some mortality and damage would be expected for entrained fish and fish eggs," the FES concluded that "no major impact on the fish population should result from entrainment of juvenile fish and fish eggs" (AEC 1973, p. 5-12).

The FES considered mortality of adult fish to be more important than mortality of younger life stages. During normal operation, Units 2 and 3 were expected to entrain 32 metric tons of adult fish per year, but some of those fish were expected to survive due to operation of the Fish Return System installed by Southern California Edison. The FES calculated that entrainment would result in the death of 13–25 tons of fish each year (AEC 1973, p. 8-2). In addition to the loss during normal operations, fish were expected to be killed during heat treatment of the cooling system, which is required to remove fouling organisms from the intake and discharge conduits. Heat treatment losses were predicted to range from 5–14 metric tons/yr. Therefore, total losses of adult fish were estimated to be 18–39 metric tons/yr for Units 2 and 3, which would increase to 37–46 tons/yr if the Fish Return System was completely ineffective (AEC 1973, p. 13-4).

The FES considered the overall impact of the operation of the new units on 15 species of fish: sardine (*Sardinops caerulea*), northern anchovy (*Engraulis mordax*), jack mackerel (*Trachurus symmetricus*), yellowtail (*Seriola dorsalis*), California barracuda (*Sphyraena argentea*), kelp bass (*Paralabrax clathratus*), white seabass (*Cynoscion nobilis*), opaleye (*Girella nigricans*), walleye surfperch (*Hyperprosopon argenteum*), white surfperch (*Phanerodon furcatus*), black surfperch (*Embiotoca jacksoni*), California sargo (*Anisotremus davidsonii*), blacksmith (*Chromis punctipinnis*), white croaker (*Genyonemus lineatus*), and queenfish (*Seriphus politus*). The specific impacts varied by species, but for most species it was judged that the main impact would be through the entrainment of adults (or juveniles) (AEC 1973, pp. 5-34 to 5-42). Relatively large numbers of adults, and sometimes large numbers of juveniles, of queenfish and all the kelp bed species (kelp bass, opaleye, black perch, California sargo and blacksmith) were expected to be entrained, but even these losses were not expected to have a major impact on the populations (AEC 1973, pp. 5-34 to

5-42). Indirect effects on kelp bed fishes due to reductions in kelp cover were not considered.

Other Effects. The FES also considered the possibility that there would be radiological effects, but concluded that "no detectable effect is expected on the aquatic biota or waterfowl as a result of the quantity of radionuclides to be released in the liquid effluents of Units 1, 2 and 3 of the San Onofre Nuclear Generating Station" (AEC 1973, p. 5-45).

Summary. In summary, the FES predicted that the principal effect of Units 2 and 3 would be the loss of fish and other organisms that would be entrained in the cooling-water system (AEC 1973, p. 8-2). Losses of adult and juvenile fish were expected to be 18–39 metric tons/year. About 30% of the other marine organisms (i.e., plankton and larvae) passing through the cooling system were expected to die (due largely to thermal effects), but the FES judged that the effect of this mortality on the populations would be small (AEC 1973, p. 8-2). Effects on kelp were predicted to be minimal.

Testimony to the Coastal Commission

The testimony presented to the California Coastal Commission in 1973–1974 focused on a narrower scope of possible impacts than the Final Environmental Statement. Virtually all of the expert testimony regarding SONGS' likely effects concerned the possible effects on zooplankton. We distinguish testimony from two groups of scientists: intervenors (primarily academic scientists from the Scripps Institution of Oceanography) and industry scientists representing Southern California Edison.

Intervenors. Vast numbers of zooplankton were predicted to be killed by being taken into the plant (Table 18.1). Even larger numbers were predicted to be transported far offshore by the discharge, where they would die because of inhospitable conditions. The consequences would be "impoverishment of the nearshore fauna, including clams, lobsters, and mussels [whose larvae are in the plankton], and everything that depends on the plankton" (Minutes of California Coastal Zone Conservation Commission hearing, October 18, 1973, p. 13). The tremendous effect of the new units on the zooplankton would have dramatic and far-reaching consequences for the entire ecosystem, devastating species whose larvae occur in the plankton as well as species who depend on the plankton for food; the ultimate consequence would be the creation of a nearshore "marine desert" of several square kilometers.

Utility Scientists. The Commission also heard testimony from the utilities. At the October 18, 1973 hearing, Southern California Edison stated that "[t]here will be an impact, of course, of a facility such as this, but that impact is not severe nor adverse." Edison argued that less environmentally damaging alternatives such as cooling towers were not necessary because the adverse environmental effects of the proposed units would be minimal. Edison also testified that the marine biological monitoring program required by its permit to construct and operate Unit 1 had revealed no substantial adverse effects on the marine environment (Minutes of December 5, 1973 hearing) (see Osenberg and Schmitt (Chapter 1), for general discussions of the inadequacies of these monitoring programs).

Predictions of the Marine Review Committee

After several years of studying the effects of Unit 1, the Marine Review Committee reported its predictions of the effects of Units 2 and 3 to the Coastal Commission (Murdoch et al. 1980). The MRC based its predictions on the results of its initial studies, predictions of the movement and turbidity of the water discharged from Units 2 and 3, consideration of the likely mechanisms of impact, projections of losses due to the intakes and diffusers, and analysis of the population and community consequences of these losses.

Plankton. The MRC predicted that 50–60 tons of mysids (small shrimp-like organisms) would be killed in SONGS' intakes each year, plus an unknown amount (but projected to be up to several hundred tons) killed by translocation in the diffuser plume. This loss was expected to cause a reduction in mysid density of about 50% within several kilometers of SONGS, with smaller depressions in density out to 10 km. The loss of these mysids as well as similar species was tentatively placed at 350 tons. The MRC reasoned that a loss of this magnitude could result in the loss of as much as 30 tons of fodder fish production per year. However, the MRC expected that the actual loss would be much less, in part because much of the mysid biomass killed and moved offshore would be eaten by the same fish species as occur near the diffusers. The MRC predicted there would *not* be a significant effect on sport and commercial species due to SONGS' impacts on mysids.

SONGS' intakes were predicted to kill about 1200 tons of zooplankton per year (excluding mysids). It was expected that some additional zooplankton would be killed by being moved offshore to unfavorable habitats by the diffusers; the MRC could not estimate the loss precisely, but suggested it might be on the order of 4000 tons. The MRC predicted that SONGS would probably reduce the local density of zooplankton species that are restricted close to shore, and that there would be changes in the relative abundances of species in this assemblage. Because the nature of these changes would depend on mixing rates, the ability of the populations to compensate and interactions between species, the MRC could not be certain of their magnitude and extent, but expected them to be less extensive than the predicted effects on mysids.

In addition to effects on zooplankton, the MRC anticipated effects on phytoplankton. The MRC expected that the diffuser discharges would bring nutrient-rich water to the surface offshore of the diffusers, and that this would result in the production of an extra 84,000 tons of phytoplankton each year. The MRC expected this to lead to the production of an extra 460 tons of anchovy per year, but only a negligible effect on sport and commercial fish production.

Kelp.   SONGS was predicted to reduce the density of kelp plants in the San Onofre Kelp Bed due to increased turbidity from the diffusers. The increased turbidity was predicted to reduce light on the bottom by about 40%, which was predicted to reduce the recruitment of young kelp plants and reduce kelp growth. Other mechanisms were discussed, including effects of fouling organisms, sea urchins, sedimentation, temperature, nutrients, competitors and toxic substances. SONGS was predicted to lengthen the periods during which the kelp bed is absent or very sparse following unfavorable oceanographic conditions (such as El Niño events). The MRC also predicted that mysids living in the canopy of the kelp would be reduced in abundance, and that this change might alter the diets of fish in the bed. Other impacts on kelp bed fishes were also expected (see below), based on the predicted loss of kelp habitat.

Benthos.   For soft-bottom habitats, the MRC predicted that SONGS would alter the bottom sediments, making them richer in organic content and resulting in an increase in the abundance, number of species and annual production of invertebrates in the enriched area. The enrichment was not expected to influence the production of sport and commercial fish. Some negative effects were anticipated, but they were not expected to have a significant effect on the overall production of the community. For hard-bottom habitats, the MRC noted that increased turbidity in the San Onofre Kelp Bed could lead to a change in the species composition of kelp bed invertebrates, and that this shift could slow the recruitment of kelp by intensifying competition for space. However, the MRC noted that there was no strong evidence to suggest that those changes would occur.

Fish.   Much of the MRC's report focused on fish. The MRC predicted that 50–84 tons of juvenile and adult fish would be killed each year by being taken into the plant with the cooling water and becoming impinged on screens that "sieve" out large particles prior to delivery to the cooling system. In addition to impacts from the loss of these older life stages, the MRC considered the population consequences of fish larvae being killed by SONGS and estimated the indirect effect of habitat loss due to effects on the San Onofre Kelp Bed. In contrast to the Final Environmental Statement, the entrainment and loss of eggs, larvae and immature stages in the intakes (expected to equal 392 tons/yr) were predicted to be more important than impingement of older life stages. Large losses due to translocation (movement of larvae farther offshore due to entrainment in the discharge plume) were also expected but could not be quantified. In their analyses, the MRC distinguished two primary categories of fish: nearshore sport and commercial fish, and fodder fish.

Nearshore sport and commercial fish were projected to experience a total loss of 27–63 tons of production per year, with the lower figure being more probable (which would be equivalent to ~1% of the total production in the Southern California Bight, the area between Cabo Colnett in Baja California and Point Conception). Halibut was predicted to be the most affected. Overall though, the MRC judged that this level of impact would not have a significant effect on sport and commercial fish populations. However, they did caution about the risks of cumulative effects of many "small" impacts.

Fodder fish form a major portion of the prey of the sport and commercial fish species; the most common fodder species are northern anchovy, queenfish and white croaker. The MRC predicted that SONGS would cause the loss of over 300 tons of production/year of nearshore fodder fish, mostly queenfish and white croaker. The MRC expected these losses to be spread throughout the Bight, and projected that they might equal ~ 7% of the annual production of these fish in the Bight. The MRC cautioned that they did not know enough about compensation to be able to estimate how the populations would be affected, but stated that "the accumulation of effects of this order would be expected eventually to cause declines in these stocks." Among the local stocks of fodder fish, the MRC expected changes (mainly in queenfish) out to a distance of several kilometers from SONGS. These predictions about local stocks were based on the MRC's study of Unit 1, which showed changes in the age structure and sex ratio of fodder fish near the power plant. Although large numbers of anchovy larvae would be killed by SONGS, the population is so large that the MRC predicted that SONGS would have no effect on anchovies.

Comparison of Predictions

The four sets of predictions vary considerably, particularly with regard to plankton. The intervenors testifying before the Coastal Commission expected a massive impact on the plankton, whereas industry scientists predicted no adverse effects. Both the Environmental Statement and the MRC expected some detectable changes in local plankton populations, but no large-scale adverse effects. In addition, the intervenors inferred from the fact that plankton form an important part of marine food chains that destruction of the local plankton community would result in devastation of all nearshore organisms. This was not predicted by the other three groups.

These differences stem from the focus of each of the analyses behind the predictions. No quantitative analytical basis is presented in the testimony to the Coastal Commission, but the intervenors' focus was clearly on the large volume of water that would pass through the new units, an assumption that the plankton entrained into the units would be killed, and an assumption that the discharge would create a "river" transporting plankton into inhospitable waters where they would die. In contrast, the FES was concerned largely with temperature effects and did not consider the possibility of offshore transport. The MRC focused more on the mechanisms proposed by the intervenors, such as entrainment, than those examined in the FES because they believed that the diffuser design proposed for Units 2 and 3 made it unlikely that elevated water temperature would be a problem. Unlike the intervenors, the MRC concluded that local depressions in plankton density would be relatively small (with no far reaching implications) due to the short generation time of plankton and the large dilution that occurs when the effluent is discharged from the diffusers.

The MRC predicted that the new units would adversely affect the San Onofre Kelp Bed, whereas no adverse effect was predicted in the Final Environmental Statement, and kelp was not even mentioned in the testimony to the Coastal Commission. One reason the FES did not predict an effect on the kelp bed was that it was not present when the 1973 FES was completed, although historical records were available to document its presence in the past. However, even in 1981, when the kelp bed was present, the revised FES predicted no significant impact. The difference between the 1981 FES and the MRC predictions stem from the expected mechanisms of impact: the FES focused on thermal effects while the MRC also considered the effects of turbidity. The FES recognized that SONGS would increase turbidity in the region and the NRC staff was aware of the MRC's predictions about the effects of turbidity on kelp, but the FES nonetheless did not consider turbidity to be likely to affect the kelp bed.

The FES and MRC also differed in their predicted effects on fish. The FES judged the main impact for most species to be due to the impingement of adults and minimized the potential loss of larvae. In contrast, the MRC considered larval losses to be more important, particularly for fodder fish such as queenfish and white croaker. In addition, the FES considered the significance of impacts in terms of the overall population for each species throughout its range. Therefore, substantial reductions in the local abundance of a species would not be significant if the species was widespread and abundant at other locations. In contrast, the MRC explicitly considered how SONGS might affect local populations of fish, especially fodder fish and kelp bed fish, as well as regional stocks.

Finally, neither the AEC nor the scientists testifying before the Coastal Commission expected impacts on mysids. However, based on studies conducted at Unit 1, the MRC concluded that Units 2 and 3 were likely to cause substantial decreases in mysids.

The Impacts That Were Detected

To check the accuracy of impact predictions, we need to know what the impacts of a project actually are. As pointed out in other sections of this book, this is not a trivial task, and there remains considerable discussion of the relative merits of different approaches. There have been two programs designed to detect the effects of SONGS. First, as a requirement of the National Pollution Discharge Elimination System (NPDES) permits for SONGS, Southern California Edison has conducted compliance monitoring. The design of most compliance monitoring limits its ability to detect impacts (Osenberg and Schmitt, Chapter 1, Mapstone, Chapter 5). Second, as a condition of its Coastal Commission permit allowing the construction of Units 2 and 3, Southern California Edison was required to fund the independent study of SONGS conducted by the Marine Review Committee. Because it was an *independent* study and was much more extensive and rigorous than Edison's NPDES monitoring, we have based our discussion of the effects of SONGS Units 2 and 3 on the Marine Review Committee's findings (Murdoch et al. 1989).

The Marine Review Committee's findings were based on a 15-year study that cost about $45 million. Most of the MRC's studies used the Before-After-Control-Impact Paired Series design (BACIPS: Stewart-Oaten et al. 1986, Stewart-Oaten, Chapter 7), although the MRC also relied on other approaches when BACIPS could not be applied (e.g., to extrapolate intake losses of young fish into impacts on regional fish stocks: Nisbet et al., Chapter 13). The MRC's study detected a number of substantial adverse effects of SONGS (Schroeter et al. 1993, Bence et al., Chapter 8, Nisbet et al., Chapter 13). Table 18.2 summarizes the impacts for each of the groups for which predictions were reported in Table 18.1.

Plankton. There were few, if any, adverse effects of SONGS on plankton. For example, although detailed measurements of species composition were not made, there was no suggestion that the total density of phytoplankton changed appreciably due to the operation of SONGS. Further, although the Marine Review Committee estimated that 1400 tons (dry weight) of zooplankton were killed per year due to entrainment in the intakes, the abundance of zooplankton near SONGS was largely unaffected by the power plant. The MRC concluded that these losses were rapidly dispersed by the rapid mixing of discharged water with ambient seawater. The MRC concluded that this loss did not constitute a substantial adverse effect.

Mysids, along with soft-bottom invertebrates (such as worms and clams) and benthic fish, actually increased in abundance due to the operation of SONGS. These increases were probably due to an increase in organic material, which originated from organisms killed in the cooling system and discharged in the plume.

Kelp. The most significant effects of SONGS' discharge were on the San Onofre Kelp Bed community. The turbid plume created by the discharge from Units 2 and 3 caused a significant reduction in the amount of light that reached the ocean floor and also increased the flow of sediments in the kelp bed. As a result of these changes in the physical environment, fewer young giant kelp

Table 18.2.   Results of Marine Review Committee Study on the Ecological Effects of SONGS Units 2 and 3

Ecological characteristic	Impact
Phytoplankton	No increase in abundance
Zooplankton	1200 tons (dry mass) entrained/yr; no evidence of local reduction
Mysids	13 tons killed/yr; abundances generally increased
Kelp	**60% reduction in San Onofre Kelp Bed**
Benthos	
Intertidal	No evidence of adverse effects, but only one species investigated
Subtidal soft-bottom	Invertebrates increased in density
Subtidal hard-bottom	**30–90% reduction in kelp bed invertebrates**
Fish	19–51 tons/yr killed from impingement
	3–5 billion fish larvae killed per year, leading to substantial adverse effects on regional fish stocks
	70% reduction in kelp bed fish
	30–70% reduction in local midwater fish
	Increase in local benthic fish
Other effects	Low-level discharges of metals and radionuclides, but no detectable ecological effects
Mechanisms	Entrainment in intakes; discharge effects on turbidity and organic material

Note: MRC results are from Murdoch et al. (1989), except for phytoplankton, which is from Kastendiek and Parker (1988). Substantial adverse effects are shown in boldface type. Entrainment and impingement losses include effects of Unit 1, which contributes less than a third of the total losses.

plants survived and grew into adults, and there were fewer adults in the kelp bed than there would have been if SONGS had not been operating. The MRC estimated that the San Onofre Kelp Bed had been reduced on average by 60%, or about 80 ha (200 acres) (see Bence et al., Chapter 8). These declines in kelp cover contributed to significant declines in the abundances of invertebrates and fish living on or near the bottom of the kelp bed.

Benthos.   As mentioned above, soft-bottom invertebrates (such as worms and clams) increased in abundance, probably due to increases in organic material caused by mortality of organisms in SONGS' intakes. In contrast, hard-bottom invertebrates (such as gastropods and echinoderms) declined in abundance, probably due to the loss of kelp and increased organic material on hard substrates. Effects on the intertidal community were not investigated.

Fish. Another important effect of SONGS occurs because the plant kills large numbers of organisms when it takes in water for cooling. The most important effect of this entrainment concerns fish. Three to five billion fish larvae are killed by Units 2 and 3 each year. Since some of these fish would have survived to become adults, the short-term effect of this impact is to reduce the recruitment of fish into the adult populations. For the most vulnerable fish species, the MRC estimated the amount by which SONGS increases the death rate of immature stages in the Southern California Bight. SONGS increases the death rate of immature queenfish and white croaker, two of the most common nearshore species, by 13 and 6%, respectively; the reduction is around 5% for six other species.

The long-term effects on the number of adults in the population (standing stock) are more difficult to determine. It is possible that the surviving fish "compensate" for the short-term losses by growing faster, surviving better, or reproducing more. Unfortunately, it is not possible to measure this potential impact directly, in part because fish larvae disperse widely so that any impact would be spread over a large area and hence be "diluted." Instead of direct measurements of reduced abundances, the Marine Review Committee relied on simple models, which incorporated likely mechanisms of compensation, to project the long-term effect of SONGS on fish populations (see Nisbet et al., Chapter 13). The amount of the decline could not be estimated with much certainty. However, reasonable assumptions indicate that the long-term reduction in average abundance would be equal to the additional mortality caused by SONGS. Thus, the standing stocks of the common species queenfish and white croaker would decline by about 13 and 6%, respectively. Taking into account the uncertainties associated with the various estimates and models used in their analyses, the Marine Review Committee concluded that SONGS would cause a decline of several hundred tons in the standing stock of fish in the Southern California Bight.

Besides entraining large numbers of larval and juvenile fish, SONGS kills 21 to 56 tons of larger fish each year when they are impinged on the screens in the intakes of Units 2 and 3. The impingement loss itself was not considered by the MRC to be a substantial effect. However, this loss contributes to the significant decline in local midwater fish abundances detected by the MRC, although other factors (such as the local increase in turbidity, which might cause fish to leave the area) also appear to be involved. The MRC considered the reductions in local fish populations to be a substantial, but local, impact.

Predicted Effects versus "Reality": Do We Get It Right?

No set of predictions was entirely accurate (Table 18.3). Indeed, accuracy was very difficult to measure owing to discrepancies between the types of impacts that were predicted and the types of impacts that were investigated, the qualitative nature of many of the predictions, and errors in the proposed mechanisms of impact when, in fact, the general outcome was correct. Further, the predictions made in the FES and by the MRC were often complex and multifaceted and do not lend themselves to a simple summary of accuracy. It is also hard to determine the weight that should be given to each of the predictions. For example, from the perspective of protecting environmental quality, predicting no impact when in fact there is an impact would be a more serious mistake than predicting an impact when there is none. It also could be argued that assigning several "correct" ratings to the utility scientists is unfair because it gives credit for an accurate prediction about particular taxa when in fact the utility simply made a blanket assertion, without clear scientific evidence or specific rationale, that there would be no substantial adverse effects on *any* taxon. Similarly, it is unfair to credit intervenors with "correct" predictions when the MRC detected adverse effects, because the environment did not turn into a "marine desert" as predicted. The assignment of "correct" predictions to the FES is also somewhat inflated, in this case by accurate predictions made for the wrong reasons (e.g., they predicted no adverse effects on soft-bottom benthos due to thermal stratification, rather than an increase in organic matter). Finally, this analysis does not include the predictions *missed*; for example, the AEC did not make a prediction about mysids, so effects on mysids do not contribute to its accuracy rating. Nonetheless, this crude analysis indicates substantial differences in the accuracy of the different predictions.

Although not easily quantified, the overall tone of the different predictions also varied in their accuracy. The intervenors and the utility scientists were clearly wrong, predicting that SONGS would be either a total and complete disaster or completely benign, when in fact it is neither. The FES was closer to the impacts actually reported by the MRC, predicting adverse effects in some groups but not others. However, the FES provided an overall picture of a relatively benign project, with essentially no far-reaching effects and relatively minor local effects. The FES minimized the importance of local effects, particularly in the discussion of impacts to fish. In these respects, the tone of the FES differed considerably from that of the MRC's 1980 predictions. Like the FES, the MRC predicted adverse impacts in some groups but not in others. However, the MRC did not dismiss the importance of local impacts and it did recognize that there could be some regional effects. Although the specific predictions about particular taxa were not all correct, the overall tone of the MRC's predictions matches very well the impacts eventually reported.

Why was there such a great range in the accuracy of these predictions? Previous comparisons of predicted and actual impacts have also found a range of

Table 18.3.   Accuracy of Predictions Made by Four Different Groups about the Ecological Impacts of the San Onofre Nuclear Generating Station

	Accuracy of predictions			
Environmental characteristic	Final Environmental Statement	Intervenors (CCC testimony)	Utility scientists (CCC testimony)	Marine Review Committee
Mysids				
Amount killed	No Prediction	Incorrect	No Prediction	**Correct-Qual.**
Effect on local population	No Prediction	Incorrect	No Prediction	Incorrect
Effect on fishes	No Prediction	Incorrect	No Prediction	Incorrect
Other Zooplankton				
Amount killed	Correct-Qual.	Incorrect	No Prediction	**Correct-Quant.**
Effect on local population	Incorrect	Incorrect	**Correct**	Incorrect
Species composition	**Correct-Qual.**	Incorrect	**Correct**	Incorrect
Phytoplankton				
Abundance, production or composition	Incorrect	Incorrect	**Correct**	Incorrect
Prevalence of red tides	Incorrect	Incorrect	**Correct**	Incorrect
Kelp (abundance or coverage)	Incorrect	Incorrect	Incorrect	**Correct-Qual.**
Intertidal Benthos	Insuff. Data	Insuff. Data	Insuff. Data	Insuff. Data
Subtidal Benthos				
Soft bottom	Correct-Qual.	Incorrect	Incorrect	**Correct-Qual.**
hard bottom	Incorrect	Incorrect	Incorrect	Incorrect
Fish				
Amount impinged	**Correct-Quant.**	No Prediction	No Prediction	Correct-Quant.
Sport and commercial fishes	Incorrect	Incorrect	Incorrect	**Correct-Qual.**
Fodder fish	Incorrect	Correct-Qual.	Incorrect	**Correct-Qual.**
Kelp bed fish	Incorrect	Correct-Qual.	Incorrect	**Correct-Qual.**
Midwater fish	Incorrect	Correct-Qual.	Incorrect	**Correct-Qual.**
Benthic fish	**Correct**	No Prediction	Correct	Incorrect
Regional fish stocks	Incorrect	Correct-Qual.	Incorrect	**Correct-Qual.**
Other Effects	**Correct**	No Prediction	Correct	No Prediction
Underlying Mechanisms	Incorrect	Incorrect	Incorrect	**Correct**

Note: Actual impacts are based on the results of the Marine Review Committee study (see Table 18.2). For each ecological characteristic, the most accurate prediction is indicated in bold. When adverse effects were correctly predicted, they were distinguished as qualitative or quantitative. Correct predictions that were originally given in quantitative terms and were within the observed range of effects (or within 30% of the final estimate) are denoted by "Correct-Quant.". Quantitative predictions that were not quantitatively accurate, but were in the correct direction, are indicated by "Correct-Qual.", as are correct predictions that were only qualitative. Correct predictions of "no adverse effect" are simply indicated by "Correct". When there were insufficient data to evaluate an effect, "Insuff. Data" is given. Also note that testimony to the California Coastal Commission (CCC) by intervenors and utility scientists was often less specific than the categories used below; however, most of this testimony consisted of blanket statements (e.g., "no adverse effects", or "devastating effects on the entire ecosystem"), which allowed us to extract these more specific predictions. Accuracy was based on the final prediction, and not necessarily the underlying mechanisms, which were generally incorrect, except for those proposed by the MRC (also see text).

accuracies (Culhane 1987, Tomlinson and Atkinson 1987b, Buckley 1991a, 1991b), but these studies have not included different sets of predictions for the same project. Furthermore, these studies have provided few clues about *why* predictions often go awry. We can make some tentative suggestions from our observations about SONGS.

The testimony before the Coastal Commission was based on "professional judgment", with little explicit scientific analysis. Almost all of this testimony was wrong. A number of academic scientists predicted a major impact on the plankton of the region; the MRC detected no such effect. These scientists also predicted catastrophic consequences for the entire ecosystem, and it is clear that no such catastrophe occurred, and the area is not, even after nine years of operation of Units 2 and 3, a marine "desert". On the other hand, Southern California Edison testified that there would be no ecological impact of consequence. This also was not correct. The MRC detected a number of impacts that, while not catastrophic, were judged to be ecologically important (Table 18.2).

An obvious problem with the testimony before the Coastal Commission is that it lacked appropriate scientific rigor and oversight. Interested parties were testifying on behalf of a particular position. Utility scientists had the interest of their employer to consider. They also relied heavily on conclusions "supported" by the monitoring that Southern California Edison (or its contractors) had conducted of Unit 1. However, self-monitoring programs not only often suffer from many design flaws (Osenberg and Schmitt, Chapter 1), but can also lack statistical power (Mapstone, Chapter 5). These problems arise, in part, by requiring dischargers to monitor the effects of their own activities.

Scientists who testified against the utility may also have been responding to their own interests, but in addition they were volunteering their time and may not have had the opportunity to consider all aspects of the problem, nor access to all of the available information. A consideration of the dilution and rapid dissipation of the heated effluent might have changed their conclusions; in addition, it is now known that the discharge plume does not move as far offshore as they assumed. A more thorough, independent, and objective analysis of the likely power plant impacts might have altered their testimony to the Commission.

The predictions in the Final Environmental Statement (AEC 1973) were produced with a procedure that is similar to that used for the majority of environmental impact predictions in the United States. An extensive review of the relevant literature was conducted, including reference to impacts detected at other power plants. There is no standardized, universally applicable technique for determining ecological impacts, in part because of the complexity of ecological systems. However, the AEC considered the major types of impacts that were likely and made judgments about the level and severity of each, with brief explanations of the reasons behind the decisions. The accuracy of this procedure was mixed, but generally not very high. Where the predictions were based on simple, straightforward calculations, such as the biomass of plankton entrained, they were reasonably accurate. However, few potential impacts are this straightforward, and even in the case of plankton the amount of biomass entrained is just a midpoint to the ultimate judgment about the impact of entrainment on population or ecosystem processes. Where judgment is involved without a solid theoretical or empirical underpinning, bias is always a possibility, and it is possible that the AEC was predisposed to minimizing possible impacts. Even if this were not the case, the lack of information and well-accepted models upon which to base judgments about impacts inevitably led to uncertainty about the likely impacts. This uncertainty was frequently stated explicitly in the Final Environmental Statement. Time has shown that the judgments made in the face of uncertainty were often wrong. Finally, many of the inaccuracies were due to a misunderstanding of the underlying mechanism(s) of impact. For example, the importance of thermal effects was greatly overstated, leading to inaccurate predictions about phytoplankton and kelp.

In at least one case, the analysis performed by the AEC seems inadequate. The discussion of the likely impacts to different fish populations is quite superficial, being based largely on generalizations and qualitative statements. No quantitative analysis is evident in the FES. Instead, the discussion typically concluded that due to a species' "wide range, impact to the ... population is not expected to be significant" (e.g., p. 5-38). The FES did not consider the loss of larval fish to be significant, seemingly dismissing it out-of-hand without careful consideration. This judgment appears to be quite subjective and without an adequate supporting argument. Although the issue of the consequences of impingement of large numbers of fish larvae by power plants is controversial (Saila et al. 1987, Barnthouse et al. 1988) and very different conclusions can be supported by different analyses, the FES is deficient because it lacks *any* careful analysis of the population consequences of larval losses.

The inaccuracies in the predictions of the Final Environmental Statement are particularly troublesome. The Atomic Energy Commission followed established standards and procedures for predicting the impacts—the same basic standards and practices that are currently used. Despite access to a great deal of information about the local biology, impacts of other coastal power plants, and changes to the physical environment that were likely to occur, the AEC frequently cited the uncertainty about possible effects of the power plant. The information base and procedures used by the AEC were not sufficient to lead to accurate predictions. Current EIR/Ss frequently make judgments about likely impacts in the face of similar uncertainty.

Finally, the Marine Review Committee's predictions were the most accurate. Three factors contributed to the improved accuracy of the MRC's predictions. First, the MRC was able to draw on more detailed baseline information about the area. The AEC staff had access to a great deal of information in various reports and publications, but little or no first-hand knowledge of the San Onofre marine environment. MRC scientists had conducted field studies of biological (fish, kelp and invertebrates) and oceanographic features around San Onofre before the MRC made its predictions. In addition, the MRC had completed its analysis of the impacts of Unit 1, which undoubtedly gave it greater insight into the likely effects of Units 2 and 3. (Information alone cannot explain the differences between the FES and MRC predictions, however, since the 1981 FES considered all of the information available to the MRC, including the MRC's predictions, yet the predictions were not modified substantively.)[3] Second, the MRC paid particular attention to the potential mechanisms of impact. Finally, the MRC was an independent body, so self-interest was not involved in its predictions. The MRC was charged with measuring impacts, but its composition was carefully balanced to avoid bias.

Ironically, the additional information the MRC had available, which generally led to more accurate predictions, also led the MRC to some incorrect predictions. The MRC was wrong about its predicted effect on mysids, which was one of the predictions the MRC was able to make because of its more detailed, site-specific studies. In this case, it turns out that the MRC based its predictions on an incorrect model of nearshore oceanography and plume dynamics as well as overestimates of the rate of entrainment of mysids. The MRC also did not anticipate the adverse effect of Units 2 and 3 on the invertebrates living in the San Onofre Kelp Bed, apparently because it did not understand the mechanisms by which the power plant would affect these species, although it did expect adverse effects to giant kelp and kelp bed fish.

Despite its successes, the inaccuracies of the MRC's predictions point to limitations in predicting ecological impacts. For example, considerable resources were spent by the MRC in developing its predictions of the likely impacts of Units 2 and 3. Resources were spent conducting site-specific studies of the San Onofre environment, investigating ongoing effects of Unit 1, synthesizing literature from other power plant studies, and extracting general predictions from basic ecological studies that could be applied to this particular problem. This effort involved several years of work by five teams of scientists. For other proposed projects, it is likely that far fewer resources would be available, and the prospects for accurate predictions in those situations probably are not high. For SONGS, the amount of effort spent in developing predictions was a good predictor of the accuracy of the predictions: the MRC, with the highest accuracy, spent the greatest effort; the FES had the second-highest accuracy and the second-highest level of effort; and the testimony to the Coastal Commission was the least accurate, but was based on the least effort. To facilitate accurate predictions we must change the way in which EIR/Ss are conducted and increase the basic understanding of the processes that drive ecological change.

[3]Perhaps some insight into the mindset of the NRC can be gained from the staff's evaluation of the MRC's 1980 report, including its predictions. The 1981 FES states: "The conclusions of the MRC are essentially consistent with those ... described in Section 5 of this Statement. Although noting uncertainties, the MRC ... concluded that it does not predict at this time that substantial adverse effects on the ... environment are likely to occur from the operation of the SONGS cooling system" (p. 6–7). As noted in Table 18.2, although there are similarities, there are also substantial differences between the FES and MRC predictions, including predicted impacts that the MRC concluded were substantial adverse effects.

Conclusions and Recommendations

The analyses presented here and in other predictive audits (Culhane 1987, Tomlinson and Atkinson 1987b, Buckley 1991a, 1991b) demonstrate that currently we cannot rely on most predictions of ecological impacts to be accurate. This is especially true for Environmental Impact Reports/Statements, which often cover an immense range of possible impacts and draw on limited information. To counteract the failures of prediction, we need to conduct more tests of impact predictions. Testing predictions will be the fastest, most cost-effective way to improve the models and approaches that we use to generate predictions.

An essential step for testing predictions is to make sure that the predictions are stated in explicit, clear terms. Past predictive audits (Tomlinson and Atkinson 1987a) found that most ecological predictions could not even be included in the audit because they were stated imprecisely or were not documented well enough. We need to make sure that predictions are scientifically grounded, that the bases of predictions are explicitly stated, and that the predictions themselves are explicit, clear, and as quantitative as possible. Most EIR/Ss fail these requirements. We need to demand more for EIR/Ss. Forcing analysts to lay out explicitly the assumptions and logic that led to their stated predictions would be a useful beginning.

Similarly, predictions should not be phrased as subjective interpretations (such as "no adverse effect" or "significant effect"), without explicit documentation of the predicted magnitude and nature of the impacts. Although categorization of impacts in subjective categories (e.g., "adverse, but mitigatable" versus "not adverse") might be useful for the EIR/S process, they obscure the predictions and make testing difficult. For example, in the context of SONGS, it is unclear if the utility scientists were actually incorrect about the effects of SONGS on fishes, or if they simply judged the size of these impacts to be "not adverse" or "insignificant".

Our predictions are not poor just because we have not stated them clearly; there is also a fundamental lack of knowledge. Improving our ability to predict the environmental impacts of a project depends on improving our understanding of how ecosystems function and how different projects change the physical and biological environments and thus affect the ecological properties of interest. Yet in most cases, we simply do not know enough about the important processes. In discussing why some of the effects predicted by the MRC in 1980 did not occur, the MRC concluded that:

> The failures of prediction can be explained largely by a lack of quantitative information about crucial processes, especially the rate of oceanic mixing and the amount of offshore transport by the plume, and about the quantitative details of complex ecological processes. Our oceanographic studies eventually gave us much of the needed quantitative physical information. But the quantitative details of interactions in the biotic community remain in practical terms unknowable for most marine situations. [Murdoch et al. 1989, p. 46]

To facilitate better understanding of relevant processes, we need to gain greater understanding of the impacts of existing projects. This can lead to general insight on likely impacts in new situations. The MRC conducted one of the most intensive and comprehensive studies of a particular area anywhere on the west coast of the U.S. The understanding of physical and biological processes that has come out of the MRC's study can be applied to other potential impacts. In fact, the MRC results have already been used to evaluate potential impacts and mitigation for additional coastal power plant development in Southern California (Ambrose 1990). Recall, also, that the greater accuracy of the MRC's predictions stemmed, in part, from its detailed study of the ongoing effects of Unit 1.

Explicit predictions do no good if they are never checked, and an inadequate test of predictions may be worse than no check at all. Although a tremendous volume of compliance monitoring reports exists, compliance monitoring generally is not designed to test specific impact predictions. Further, as illustrated by Southern California Edison's monitoring reports (Swarbrick and Douros 1990; see also Osenberg and Schmitt, Chapter 1), these studies probably are not sufficient to indicate what impacts have actually occurred. Well-designed environmental assessments, specifically geared toward testing predicted impacts, are needed. Application of better tools, such as BACIPS (Chapters 6–8) and more mechanistic studies would be a considerable improvement. However, regulatory agencies appear reluctant to require rigorous postconstruction monitoring and developers are understandably opposed to funding it. Cost is clearly a major concern. However, it is a false economy, not to mention dangerously misleading, to require inadequate studies.

Because resource management and planning decisions are being based on *predicted* impacts rather than measured impacts, it is essential that we work towards improving predictive abilities. But we must also recognize that, right now, our predictive abilities are not very good. The impacts that are predicted in environmental assessments are *not* the actual impacts of the proposed projects, although they become *de facto* actual impacts because there are so few follow-up studies to determine the true impacts. Until we have more confidence in our ability to make accurate predictions, we need to take steps to integrate this uncertainty into the decision-making and permitting processes. For example, more permits could carry conditions requiring impact monitoring—not mindless compliance monitoring, but focused studies to test particular predictions (which was the motivation behind the formation of the MRC). In these cases, the permit could contain a "re-opener" clause that allowed the project to be modified in specific ways if the actual impacts were found to differ from those predicted. Developers who objected to the open-endedness of this approach would be free to support independent, unbiased studies to collect the background information needed for more accurate predictions (such as the MRC did before making its 1980 predictions). This approach would be more protective of our natural resources and, at the same time, help to develop our ability to make more accurate predictions in the future.

Acknowledgments

We thank S.J. Holbrook and C. St. Mary for helpful discussions, Bobette Nelson for assistance compiling the public record of the Coastal Commission hearings, and the UC Coastal Toxicology Program and the Minerals Management Service (U.S. Department of Interior under MMS Agreement No. 14-35-001-3071) for assistance in preparation of this chapter. The views and conclusions in this chapter are those of the authors and should not be interpreted as necessarily representing the official policies, either expressed or implied, of the U.S. Government.

References

Ambrose, R. F. 1990. Potential techniques for mitigating biological impacts of proposed combined-cycle coastal power plants in San Diego County. Report to California Energy Commission, prepared for City of Chula Vista and City of Carlsbad. 46 pp.

Atomic Energy Commission (AEC). 1973. Final Environmental Statement related to the proposed San Onofre Nuclear Generating Station Units 2 and 3. Southern California Edison Company and the San Diego Gas and Electric Company. Docket Nos. 50–361 and 50–362. U.S. Atomic Energy Commission Directorate of Licensing.

Bailey, J., and V. Hobbs. 1990. A proposed framework and database for EIA auditing. Journal of Environmental Management 31:163–172.

Barnthouse, L. W., R. J. Klauda, D. S. Vaughan, and R. L. Kendall, editors. 1988. Science, law, and Hudson River power plants: a case study in environmental impact assessment. American Fisheries Society, Bethesda, Maryland.

Beanlands, G. E., and P. N. Duinker. 1984. An ecological framework for environmental impact assessment. Journal of Environmental Management 18:267–277.

Buckley, R. 1991a. Auditing the precision and accuracy of environmental impact predictions in Australia. Environmental Monitoring and Assessment 18:1–23.

Buckley, R. 1991b. How accurate are environmental impact predictions? Ambio 20:161–162.

Carney, R. S. 1987. A review of study designs for the detection of long-term environmental effects of offshore petroleum activities. Pages 651–696 in D. F. Boesch and N. N. Rabalais, editors. Long-term environmental effects of offshore oil and gas development. Elsevier, New York, New York.

Culhane, P. J. 1987. The precision and accuracy of U.S. Environmental Impact Statements. Environmental Monitoring and Assessment 8:217–238.

Gore, K. L., J. M. Thomas, and D. G. Watson. 1977. Evaluation of nuclear power plant environmental impact prediction, based on monitoring programs. Prepared for the Nuclear Regulatory Commission by Battelle Pacific Northwest Laboratories. BNWL-2153/NRC-1. 32pp.

Kastendiek, J. and K. R. Parker. 1988. Interim technical report to the California Coastal Commission. 4. Plankton. Marine Review Committee, Inc.

Larkin, P. A. 1984. A commentary on environmental impact assessment for large projects affecting lakes and streams. Canadian Journal of Fisheries and Aquatic Sciences 41:1121–1127.

Luecht, D., L. Adams-Walden, R. Bair, and D. Siebert. 1989. Statistical evaluation of predicted and actual impacts of construction grants projects in three river basins of U.S. EPA Region 5. The Environmental Professional 11:160–170.

Murdoch, W. W., R. C. Fay, and B. J. Mechalas. 1989. Final report of the Marine Review Committee to the California Coastal Commission. Marine Review Committee, Inc.

Murdoch, W. W., B. J. Mechalas, and R. C. Fay. 1980. Report of the Marine Review Committee to the California Coastal Commission: Predictions of the effects of San Onofre Nuclear Generating Station, and recommendations. Part I: Recommendations, predictions, and rationale. Marine Review Committee, Inc.

NRC (Nuclear Regulatory Commission). 1981. Final Environmental Statement related to the operation of San Onofre Nuclear Generating Station Units 2 and 3. Southern California Edison Company, the San Diego Gas and Electric Company, The City of Riverside, and The City of Anaheim. U.S. Nuclear Regulatory Commission Office of Nuclear Reactor Regulation.

Saila, S. B., X. Chen, K. Erzini, and B. Martin. 1987. Compensatory mechanisms in fish populations: Literature Reviews. Volume 1: Critical evaluation of case histories of fish populations experiencing chronic exploitation or impact. Report prepared for Electric Power Research Institute, EA-5200, Volume 1. Research Project 1633-6.

Schroeter, S. C., J. D. Dixon, J. Kastendiek, R. O. Smith, and J. R. Bence. 1993. Detecting the ecological effects of environmental impacts: a case study of kelp forest invertebrates. Ecological Applications 3:331–350.

Stewart-Oaten, A., W. W. Murdoch, and K. R. Parker. 1986. Environmental impact assessment: "pseudoreplication" in time? Ecology 67:929–940.

Suter, G. W., II, editor. 1993. Ecological risk assessment. Lewis Publishers, Boca Raton, Florida.

Swarbrick, S. L. and W. Douros. 1990. Technical report to the California Coastal Commission. O. Water quality compliance. Marine Review Committee, Inc.

Tomlinson, P., and S. F. Atkinson. 1987a. Environmental audits: proposed terminology. Environmental Monitoring and Assessment 8:187–198.

Tomlinson, P., and S. F. Atkinson. 1987b. Environmental audits: a literature review. Environmental Monitoring and Assessment 8:239–261.

Glossary of Acronyms, Assessment Designs, and Organizations

AEC Atomic Energy Commission, a now defunct agency of the U.S. Department of Energy (DOE), was responsible for promoting and regulating nonmilitary uses of nuclear fission. The AEC has been replaced by the Nuclear Regulatory Commission (NRC).

ARMA Auto-Regressive Moving Average is a technique for analyzing time series data in which observations at a given time are estimated by a regression formula based on values at previous times, and errors of these estimates are based, in part, on a decaying sum of previous errors. The 'order' of such a model refers to the number of prior times that are considered.

Before-After Before-After monitoring design is sometimes used to detect localized environmental impacts from a point-source disturbance by sampling and comparing environmental conditions at an Impact site Before and After the intervention has occurred. This design confounds impacts with ubiquitous natural fluctuations in time. See Control-Impact, BACI, beyond-BACI, and BACIPS.

BACI Before-After-Control-Impact sampling design for detecting localized environmental effects of an unreplicated point source disturbance. One BACI design (Green 1979) requires estimation of parameter values at a Control and Impact site at a single point in time Before and After a disturbance; this design is likely to confound an impact with natural fluctuations that occur only at one location. In other cases, the term BACI is used as a generic reference to all designs that involve sampling at Control and Impact locations, on more than one date Before and After a perturbation. See BACIPS and beyond-BACI.

BACIP or BACIPS Before-After-Control-Impact Paired Series sampling design is the most powerful of the BACI-type assessment designs. Samples from Control and Impact sites are taken simultaneously ("Paired" samples) on several occasions spread through time (to provide a time "Series") both

Before and After the intervention has occurred. This design is able to isolate most sources of natural spatial and temporal variation from the impact. See Before-After, Control-Impact, BACI, and beyond-BACI.

beyond-BACI   beyond-Before-After-Control-Impact sampling design is a BACI design that includes multiple Control sites as well as multiple sampling dates within the Before and After periods. It is intended to enable detection of a greater variety of impacts than either the BACI or BACIPS designs, such as impacts on variances as well as means. Unlike the BACIPS design it does not use paired sampling and may be more prone to problems involving serial correlation.

BLM   Bureau of Land Management is an agency of the U.S. Department of Interior (DOI), involved in regulating terrestrial land use issues on Federal lands, and formerly involved in regulating offshore mineral extraction activities. The latter are now regulated by the DOI's Minerals Management Service (MMS), whose research needs are supported in part by DOI's National Biological Survey (NBS).

CCC   California Coastal Commission is a regulatory agency of California charged with ensuring that proposed activities in the coastal zone conform with statutory requirements of State and, where applicable, Federal environmental laws. Formerly the California Coastal Zone Conservation Commission (CCZCC).

CCZCA   California Coastal Zone Conservation Act, a state proposition that stipulates environmental objectives and public access rights associated with the coastal zone of California, and provides the statutory authority for the California Coastal Commission (CCC).

CCZCC   California Coastal Zone Conservation Commission, the former name of the California Coastal Commission (CCC).

CEQA   California Environmental Quality Act sets out and requires the administrative process for Environmental Impact Assessment (EIA) for planned development activities in the State of California. Also see NEPA.

CAMP   California Monitoring Program is a research study initiated by the Environmental Studies Program (ESP) of the Minerals Management Service (MMS) to estimate environmental effects from multiple offshore oil and gas production platforms. CAMP currently is being conducted under the aegis of the National Biological Survey (NBS).

COP   California Ocean Plan sets environmental objectives and stipulates discharge standards for wastewater discharged into aquatic environments of California, which are mandated by the Federal Clean Water Act and are used in the National Pollution Discharge Elimination System (see NPDES) permitting and compliance process.

CTP California Thermal Plan sets thermal discharge standards for receiving waters in California under requirements of the Clean Water Act; thermal standards now are part of the California Ocean Plan.

CGP Central Gulf Platform Study was a scientific research project, conducted during 1978 and 1979 and funded by the Bureau of Land Management (BLM) and the Minerals Management Service (MMS), designed to reveal environmental effects of offshore platforms.

Control-Impact Control-Impact monitoring design, a commonly used approach to detect localized environmental impacts from a point-source disturbance by sampling and comparing environmental conditions at Control and Impact sites After an intervention has occurred. This design confounds impacts with ubiquitous natural spatial differences between sites. Also referred to as an "After-only" design. See Before-After, BACI, and BACIPS.

CSIRO Commonwealth Scientific and Industrial Research Organisation is a branch of the Australian Federal government that is charged with conducting research with practical implications.

DOE Department of Energy is a branch of the U.S. Federal government charged with regulating and promoting energy developments. See AEC and NRC.

DOI Department of the Interior is a branch of the U.S. Federal government charged with regulating and promoting use of Federal lands and natural resources. See BLM, MMS, ESP, NBS, NMFS, NPS, and USFWS.

EIA Environmental Impact Assessment is a term for the administrative process conducted publicly to reveal likely consequences of proposed developments as a tool for decision making by regulatory agencies. In the U.S., the process is required and defined by the National Environmental Policy Act of 1970 (NEPA) for Federal projects, and by similar legislation for state projects (e.g., California Environmental Quality Act, CEQA). Other national governments also require a similar process. Throughout the book, EIA is also used as a generic reference to the entire arena of impact assessment (i.e., including the adminstrative EIA, EIR, EIS, as well as field assessments and their design). See NEPA, CEQA, EIR, EIS, BACIPS (and related designs).

EIM Environmental Impact Monitoring is a generic term for field studies to assess possible environmental impacts of a given activity. Some type of EIM often is required as a permit condition (e.g., compliance monitoring) for certain types of developments (e.g., wastewater outfalls).

EIR Environmental Impact Report is the information record prepared as part of the administrative Environmental Impact Assessment (EIA) of a proposed development under the purview of a state government.

EIR/S Environmental Impact Report/Statement is a generic reference to EIR and EIS.

EIS Environmental Impact Statement is the information record prepared under the National Environmental Policy Act (see NEPA) as part of the administrative Environmental Impact Assessment (EIA) of a proposed development under the purview of the U.S. Federal government.

EPA Environmental Protection Agency is a regulatory agency of the U.S. Federal government concerned with setting environmental standards, ensuring compliance and cleaning up polluted sites.

ESP Environmental Studies Program is a division of the Minerals Management Service (MMS), Department of the Interior (DOI), charged with gathering environmental and socio-economic information to support decisions on the sale of leases for mineral extraction (including oil and gas reserves) in the Outer Continental Shelf (OCS) of the United States.

FEIS or FES Final Environmental Impact Statement (also FIE Report) is an EIR/S document that has been certified by the agency in charge of an EIA for a specific proposed development as complete under the guidelines of NEPA (or equivalent state statute).

GAO General Accounting Office is a branch of the U.S. government concerned with compliance of Federal agencies with fiscal policies.

ICP-AES Inductively Coupled Plasma–Atomic Emission Spectrometry is an analytical chemistry technique to identify and quantify trace metals and other elements in an environmental sample.

LOF Linear Oceanographic Feature is a generic term for several localized physical oceanographic phenomena that result in more or less linear discontinuities in physical or chemical properties of the ocean. Surface "slicks" caused by internal waves are an example.

MRC Marine Review Committee was an independent, non-profit entity set up as a permit condition by the California Coastal Commission to design, oversee and interpret scientific research to estimate impacts on the marine environment of new seawater-cooled units of the San Onofre Nuclear Generating Station (SONGS).

MMS Minerals Management Service is a regulatory and research agency of the U.S. Department of the Interior (DOI) charged with stewardship and extraction of mineral resources (including oil and gas reserves) in the Outer Continental Shelf (OCS) regions offshore of the U.S. coasts. See also ESP.

MAFL Mississippi, Alabama, Florida Study was a study funded by the Environmental Studies Program (ESP) of the Minerals Management Service (MMS) designed to provide basic baseline information on the Gulf of Mexico to aid in predicting impacts associated with oil and gas development.

MWP Mussel Watch Program is an ongoing monitoring approach of the California Department of Fish and Game that uses a standard bioindicator (marine mussels) to assess and identify coastal regions that are polluted. A similar approach at the Federal level is taken in the Status and Trends Program of the National Oceanic and Atmospheric Administration (NOAA).

NBS National Biological Survey is a new research agency of the U.S. Department of the Interior (DOI).

NEPA National Environmental Policy Act sets out and requires the process of Environmental Impact Assessment (EIA) for planned development activities in the United States.

NMFS National Marine Fisheries Service is an agency of the Department of the Interior (DOI) responsible for conducting fisheries research on exploitable species of marine fin- and shellfish in the United States.

NPDES National Pollution Discharge Elimination System are regulatory requirements of the Federal Clean Water Act that stipulate compliance of wastewater discharges with set standards (such as by the California Ocean Plan).

NRC National Research Council is the part of the U.S. National Academy of Sciences, a non-profit society of distinguished scholars, that advises the Federal government on scientific and technical matters.

NRDC Natural Resources Defense Council is a non-profit environmental group with expertise in legal recourse.

NRC Nuclear Regulatory Commission is an agency of the Department of Energy charged with regulating the nuclear power industry of the United States. The NRC replaced the earlier Atomic Energy Commission (see AEC).

OCS Outer Continental Shelf is defined as the zone from 3 to 200 nautical miles offshore of the coastline of the United States, and development activities in this zone come under the auspices of the U.S. Federal government. Activities from shore to 3 nautical miles offshore generally are regulated by the State government.

OCSLA Outer Continental Shelf Lands Act, as amended in 1978, is a Federal law that requires the Minerals Management Service to manage oil and gas leasing programs in light of economic, social and environmental value of the renewable and non-renewable resources in the U.S. Outer Continental Shelf (OCS).

SONGS San Onofre Nuclear Generating Station, operated by Southern California Edison, is located near Oceanside CA, and consists of 1 older and 2 new, larger nuclear power generating units that are cooled by flow-through seawater systems. Environmental effects of the 2 new units were estimated by the Marine Review Committee (see MRC).

SCE   Southern California Edison, a public utility in California, is the primary owner and operator of the San Onofre Nuclear Generating Station (SONGS). Also see MRC.

STBS   South Texas Baseline Study was a research project funded by the Environmental Studies Program (ESP) of the Minerals Management Service (MMS) from 1975 to 1978, which was designed to provide baseline information on the environmental status of the Gulf of Mexico and to help predict impacts of gas and oil development.

SWFL   Southwest Florida Study was a research project funded by the Environmental Studies Program (ESP) of the Minerals Management Service (MMS) from 1980 to 1982, which was designed to provide baseline information and to help predict impacts of oil and gas development in this region of the Gulf of Mexico.

SAAP   Spatial Autocorrelation Analysis Program. Analytical software designed to analyze and display spatial data (Wartenberg 1989).

USFWS   U.S. Fish and Wildlife Service, an agency of the U.S. Department of the Interior (DOI), is responsible among other things for carrying out statutory requirements for rare, threatened, and endangered species and marine mammals. Many research functions conducted by USFWS are now done by the National Biological Survey (NBS).

Contributor Biographies

Khalil E. Abu-Saba conducted his Master's Thesis research with the Institute of Marine Sciences at the University of California, Santa Cruz. He is currently a PhD student in the Department of Chemistry and Biochemistry at UCSC.

Richard F. Ambrose is an Associate Professor in the Environmental Science and Engineering Program at the University of California, Los Angeles. His research interests range from basic ecological problems to the ecological and policy aspects of environmental problems. His current research on environmental issues centers on techniques for mitigating resource losses in coastal environments and ecological monitoring and assessment. His recommendations for mitigation requirements for the San Onofre Nuclear Generating Station were recently adopted by the California Coastal Commission, and he is Chair of the Scientific Advisory Panel convened by the Coastal Commission to oversee monitoring of this project.

James R. Bence is a quantitative fishery ecologist with primary interests in the dynamics, assessment, and management of harvested or perturbed populations. He recently has taken a position of Assistant Professor at Michigan State University as part of their new Partnership for Ecosystem Research and Management in the Department of Fisheries and Wildlife and will be directing his research efforts toward the fisheries of the Great Lakes. He has research interests in modeling population and ecosystem dynamics and in the analysis and interpretation of spatially or temporally correlated environmental and ecological data. He works actively with management agencies to apply research results to improve the management of natural resources

Kerry P. Black is is a Research Professor of Applied Numerical Modelling and Coastal Oceanography at Waikato University in New Zealand. His diverse studies of the physics of coastal processes, and the relationship of these processes to the earth and life sciences, have led to a wide variety of research interests. His present major programs encompass numerical simulations and studies of beach wave and sediment processes in mixed waved and current environments and the transport of pollutants, larvae, salt, or heat in estuaries and on open coasts.

Robert S. Carney is currently Director of the Coastal Ecology Institute at Louisiana State University and an Associate Professor in the Department of Oceanography and Coastal Sciences. He is a biological oceanographer whose original research interests focused on deep-sea benthic ecology but have increasingly been forced to deal with issues of science and policy linkage for continental shelf and slope management. Current field research involves submersible work in the Gulf of Mexico chemosynthetic communities. Dry-land research time is spent pondering the quirks of benthic data.

Jean Chesson is currently a Senior Research Scientist in the Fisheries Resources Branch of the Australian Bureau of Resource Sciences where she assesses the status of commercial fish stocks. At the time this work was carried out, Jean was Director of Research at the Resource Assessment Commission, an organization established by the Australian Government to examine controversial resource use issues from environmental, economic, social, and industry perspectives. Her research has involved the application of statistical and mathematical techniques to a wide variety of ecological and environmental problems.

William J. Douros, a Deputy Director for Santa Barbara County's Planning and Development Department, is responsible for managing the Energy Division, the regulatory agency for the County which oversees offshore and onshore oil and gas development. He earned a master's degree in ecology from the University of California, Santa Barbara, and his earlier research focused on intraspecific competition in abalone. As manager of the Energy Division, he has been responsible for environmental review, permitting, and compliance of multibillion-dollar oil and gas development projects. Mr. Douros has directed and been responsible for the successful implementation of various population and habitat restoration programs.

Russell Flegal is a Professor of Earth Science and Chair of the Environmental Toxicology Program at the University of California, Santa Cruz. He is a geochemist.

Charles A. Gray is a Fisheries Biologist at the Fisheries Research Institute of New South Wales. His primary research interests include the influence of sewage plumes on fish larvae and other plankters, the ecology of prawns, and fisheries ecology.

Judi E. Hewitt is a researcher at the National Institute of Water and Atmospheric Research, in Hamilton, New Zealand. Her primary research interest is the application of statistical techniques to further our understanding of the ecology of soft-bottom communities. Her research has been applied to identify and solve various issues of concern to managers of marine resources.

Sally J. Holbrook is a Professor of Biology in the Department of Biological Sciences, University of California, Santa Barbara. She is the Faculty Manager of the UC Santa Cruz Island Reserve and has served on the editorial board of

Ecology and Ecological Monographs. Her research explores factors that influence the abundance and dynamics of populations of temperate and tropical reef fishes. She has a particular interest in the role of biotic interactions and resource constraints, and in the links between individual and population-level responses to perturbations.

Geoffrey P. Jones is a Lecturer in Marine Biology at James Cook University. His primary interests include the population and community ecology of reef fishes and their interactions with other reef organisms, marine conservation biology and impact assessment. He has worked extensively on human impacts in reef environments, both in tropical and temperate regions of Australia and the South Pacific.

Ursula L. Kaly is a Lecturer in Marine Biology at James Cook University. Her interests include community responses of organisms to habitat degradation and restoration, and environmental impact assessment. She has worked in mangrove, rocky shore, and coral reef habitats, with a variety of taxa including algae, mangroves, polychaetes, mollusks, fish and corals in Australia, New Zealand, and the South Pacific.

Michael J. Keough is a Reader in the Zoology Department at the University of Melbourne. His primary research explores the influence of larval stages on the population ecology of marine invertebrates. His other research on the effects of natural disturbances on shallow-water marine assemblages has been expanded to examine the effects of a range of human activities. He is also Marine Science Coordinator at the University of Melbourne.

Michael J. Kingsford is a Senior Lecturer in the School of Biological Sciences at the University of Sydney, where he teaches Marine Ecology and Marine Zoology. His primary research interests include the influence of oceanography on the early life history stages of marine organisms, the population dynamics of fish, the influence of pollutants on marine assemblages, and fisheries ecology. He works in temperate and tropical waters.

Charles Lester is an Assistant Professor of Political Science at the University of Colorado, Boulder. He received his law degree from Boalt Hall of the University of California, Berkeley. His general area of expertise is Environmental Politics, Policy, and Law, and his primary research to date has focused on offshore oil development and coastal zone management. More recently, he has been studying the problem of cumulative environmental impacts in the coastal zone and has been working in San Francisco with the California Coastal Commission.

Bruce D. Mapstone was coordinator of research into effects of fishing in the Great Barrier Reef region when he wrote his chapter. He obtained his doctorate in marine ecology from the University of Sydney and subsequently took a National Research Fellowship at James Cook University of North Queensland, researching the use of statistical inference in ecology. He has been a consultant in environmental impact assessment, experimental design, and biostatistics to various government agencies. His research interests are increasingly statistical in direction, with emphasis on the use of statistics in ecological research and environmental issues. (Bruce is now a Senior Research Fellow at the CRC: Reef Research Centre, James Cook University.)

William W. Murdoch is a Professor of Biology at the University of California, Santa Barbara. He works on the development and experimental testing of theory for population dynamics and species interactions. He is especially interested in the mechanisms leading to the regulation of interactions between predators and prey. In recent years he has focused on individual-based models and the dynamical effects of spatial heterogeneity, and has exploited insects and freshwater plankton as experimental systems. He has long been interested in applying ecological approaches to the solution of environmental problems.

Roger M. Nisbet is a Professor of Biology at the University of California, Santa Barbara. He has a research program in theoretical population ecology, with particular focus on individual-based models which relate population dynamics to individual behavior and physiology.

Craig W. Osenberg is an Assistant Professor at the University of Florida. During much of this book's ontogeny, Craig was on faculty at the University of California, Berkeley or on research faculty at UC Santa Barbara. His principle interests include the application of size- and stage-structured models of predation and competition to population and community dynamics (exemplified by his ongoing research on sunfish–invertebrate interactions in lakes), and the design and implementation of whole-system experiments and impact assessment studies (of which this book is an outgrowth). His work strives to understand population and community dynamics by incorporating mechanisms that operate at the individual level.

Frederick M. Piltz is Chief of Environmental Studies for the Minerals Management Service Pacific Outer Continental Shelf Region. His responsibilities include planning and managing a diverse research program in marine and coastal science disciplines. His interests in invertebrate taxonomy and benthic ecology, in which he earned a PhD from the University of Southern California, have led to the design and implementation of a research program investigating the long term effects of offshore oil and gas production platforms in California. He has also been involved in the role of science in decision making within government. He has been involved in symposia, government task forces, and various interagency coordinating committees dealing with science and governmental policy.

Richard D. Pridmore is Research Director for the National Institute of Water and Atmospheric Research, New Zealand. Rick has varied research interests in

aquatic ecology ranging from productivity in lake ecosystems to biological interactions in marine benthic communities. An important component of his research
has involved providing information and solutions to environmental managers.

Peter T. Raimondi is an Assistant Research Biologist in the Marine Science
Institute at the University of California, Santa Barbara. His primary research is
directed at understanding the role of recruitment and recruitment processes in
determining population and community structure. Pete will join the faculty of the
University of California, Santa Cruz (Department of Biology) in 1996.

Daniel C. Reed is an Assistant Research Biologist in the Marine Science
Institute at the University of California, Santa Barbara. His primary research
interests lie in understanding processes of recovery in coastal marine habitats
following local and widespread disturbance. Much of his work has been done in
subtidal kelp communities where he has investigated dispersal and recruitment of
marine plant populations.

Russell J. Schmitt is Professor of Marine Ecology in the Department of
Biological Sciences at the University of California, Santa Barbara. He also is
Director of the Coastal Research Center of UCSB's Marine Science Institute, and
serves as Program Director of the UC Coastal Toxicology Teaching and Research
Program and two UC cooperative research programs with the Minerals
Management Service, U.S. Department of the Interior. His primary research
programs address the abundance and dynamics of benthic marine animals.
His interests in the application of basic ecology has led to his research on the
implementation of environmental impact assessment studies.

Stephen C. Schroeter is an Adjunct Professor in the Biology Department at
San Diego State University, and a co-partner of Ecometrics, an environmental
consulting company. He is an ecologist with more than 15 years experience
examining the ecological effects of human perturbation on the marine environment and the long-term consequences of episodic events on intertidal and
subtidal communities. He has experience in demographics and the experimental
study of interactions among animals and plants in a variety of habitats.

Allan Stewart-Oaten is Professor of Mathematical Biology at the University
of California, Santa Barbara. His main research is in applications of statistics,
probability and other mathematical methods to problems in biology, especially
impact assessment, optimality problems, and the dynamics of interacting
populations.

Simon F. Thrush is a researcher at the National Institute of Water and
Atmospheric Research, in Hamilton, New Zealand. His major research interests
center around the ecology of marine soft-bottom ecosystems. Research on patch
dynamics and recovery processes in soft-bottom communities has been applied
to a variety of environmental issues including the implementation, design, and
interpretation of ecological monitoring programs and assessments of the

ecological effects of urban runoff, pesticides and habitat disturbance by commercial fishing.

A. J. Underwood is the Professor of Experimental Ecology and Director of the Institute of Marine Ecology at the University of Sydney. His research is on experimental analyses of patterns and processes in shallow-water coastal habitats including rocky shores, kelp-beds, mangrove forests, and estuarine sediments. He is involved in numerous projects in fisheries, aquaculture, coastal management, and environmental assessment. He also teaches international workshops on the design and analysis of biological experiments.

Author Index[1]

[1]Junior authors on cited papers with more than two authors are not indexed. An entry gives the page number on which a citation occurs; an italicized entry gives the location of the full citation; a boldface entry gives the pages for chapters contributed by the author.

Subject Index